2015 Silicon Nanoelectronics Workshop (SNW 2015)

Kyoto, Japan
14-15 June 2015

IEEE Catalog Number: CFP15SNW-POD
ISBN: 978-1-4673-7604-4

Copyright © 2015, The Japan Society of Applied Physics (JSAP)
All Rights Reserved

This publication is a representation of what appears in the IEEE Digital Libraries. Some format issues inherent in the e-media version may also appear in this print version.

IEEE Catalog Number: CFP15SNW-POD
ISBN 13:   978-1-4673-7604-4
ISSN: 2161-4636

Additional Copies of This Publication Are Available From:

Curran Associates, Inc
57 Morehouse Lane
Red Hook, NY 12571 USA
Phone: (845) 758-0400
Fax: (845) 758-2633
E-mail: curran@proceedings.com
Web: www.proceedings.com

2015 Silicon Nanoelectronics Workshop

June 14-15, 2015

Rihga Royal Hotel Kyoto, Kyoto, Japan

Workshop Abstracts

2015 Silicon Nanoelectronics Workshop

On behalf of the Workshop Committee, I would like to welcome you to the 2015 Silicon Nanoelectronics Workshop, a Satellite Workshop of the VLSI Symposia. This is the twentieth anniversary out of a series of workshops, which have been alternating between Hawaii and Kyoto. Thanks to the support of the Japan Society of Applied Physics and the IEEE Electron Devices Society, the Silicon Nanoelectronics Workshop has become a major international workshop in the area of VLSI-related nanoelectronics and is providing us with an opportunity to discuss the emerging technologies challenging to replace traditional transistors and memories. The two-day program this year includes one plenary talk, four invited talks, 27 oral presentations, and 28 poster presentations, covering a number of emerging Si-based nanotechnologies. In addition, a panel discussion is held as a ceremony for the twentieth anniversary of this workshop.

The program Chair, Kazuhiko Endo, and I would like to express our special thanks to all the attendees, authors, and the Technical Program Committee members for holding this year's workshop. We hope that you enjoy the twentieth-anniversary workshop and have many fruitful discussions on nanoelectronics.

Yasuo Takahashi
General Chair

Committee Members for the 2015 Silicon Nanoelectronics Workshop

General Chair
Yasuo Takahashi
Hokkaido University

Program Chair
Kazuhiko Endo
AIST

Program Committee
Steve Chung, National Chiao Tung University
Kristin De Meyer, IMEC
Stephen Goodnick, Arizona State University
Toshiro Hiramoto, University of Tokyo
Ru Huang, Peking University
Adrian Ionescu, EPFL
Toshifumi Irisawa, AIST
Raj Jammy, Intermolecular
Kuniyuki Kakushima, Tokyo Institute of Technology
Dong-Won Kim, Samsung
Tejas Krishnamohan, Intel
Max C. Lemme, University of Siegen
Pei-Wen Li, National Central University
Bunji Mizuno, Panasonic
Hiroshi Mizuta, JAIST
Yukinori Ono, Toyama University
Byung-Gook Park, Seoul National University
Wolfgang Porod, University of Notre Dame
Takahiro Shinada, Tohoku University
Michiharu Tabe, Shizuoka University
Mitsuru Takenaka, University of Tokyo
Ken Uchida, Keio University
Yukiharu Uraoka, NAIST
Maud Vinet, Leti
Yee-Chia Yeo, TSMC
Grace Xing, University of Notre Dame

Secretary
Tsunaki Takahashi, Keio University

2015 Silicon Nanoelectronics Workshop

June 14-15, 2015

Rihga Royal Hotel Kyoto, Kyoto, Japan

Table of Contents

Sunday, June 14, 2015

8:30   Opening Remarks

Session 1: Plenary and Nonvolatile Memories

8:40   1-1   **(Plenary)** Advanced process technologies of 1S1R for High Density Cross 1 Point ReRAM, B. Y. Kim and H. S. Kim (SK Hynix Inc.)

9:10   1-2   Trade-off of Performance, Reliability and Cost of SCM/NAND Flash Hybrid 3 SSD, H. H. Takishita, S. Ning, and K. Takeuchi (Chuo University)

9:30   1-3   Additional Charge Trapping Layer SONOS Nonvolatile Memory Based on 5 Ultra-Thin Body Poly-Si Junctionless FinFET, Wei-Cheng Wang, Chien-Chih Chung, Ming-Hsien Chung, Cheng-Ping Wang, Yung-Chun Wu (National Tsing Hua University)

9:50   1-4   Understanding the Underlying Physics of Superior Endurance in Bi-layered 7 TaO_x-RRAM, Y. D. Zhao, P. Huang, Z. Chen, C. Liu, H. T. Li, W. J. Ma, B. Gao, X. Y. Liu, J. F. Kang (Peking University)

10:10   Break

Session 2: Novel Process Technology for Nanoelectronics

10:30   2-1   Variability Suppression of FinFETs by Smoothing Sidewall Roughness Using 9 Ion Beam Etching Technology, T. Matsukawa[1], K. Endo[1], H. Akasaka[2], Y. Kamiya[2], M. Ikeda[2], K. Tsunekawa[2], T. Nakagawa[2], Y.X. Liu[1], and M. Masahara[1] ([1]AIST, [2]Canon ANELVA)

10:50   2-2   sSOI Relaxation by BOX Creep Technique for Dual Strain CMOS Integration, 11 A. Bonnevialle[1,2], C. Le Royer[2], Y. Morand[1], S. Reboh[2], D. Rouchon[2], N. Bernier[2], B. Mathieu[2], C. Plantier[2], M. Vinet[2] ([1]STMicroelectronics, [2]CEA LETI)

11:10   2-3   Significance of Kinetic-linkage of Oxygen Vacancy with SiO_2/Si Interface for 13 SiO_2-IL Scavenging in HfO_2 Gate Stacks, X. Li, T. Yajima, T. Nishimura, and A. Toriumi (The University of Tokyo)

| 11:30 | 2-4 | Gate-stack engineering for self-aligned Ge-gate/SiO$_2$/SiGe-channel Insta-MOS devices, Wei-Ting Lai[1,2], Kuo-Ching Yang[2], Po-Hsiang Liao[2], Thomas George[2], and Pei-Wen Li[1,2] ([1]National Chiao Tung University, [2]National Central University) | 15 |

| 11:50 | 2-5 | Impact of H$_2$, O$_2$, and N$_2$ anneals on atomic-scale surface flattening for 3-D Ge channel architecture, Y. Morita, H. Ota, M. Masahara, and T. Maeda (AIST) | 17 |

| 12:10 | | Lunch | |

Session 3: Nanowire FETs

| 13:30 | 3-1 | **(Invited)** Experimental Study of Reliabilities in Tri-gate Nanowire Transistor ~What is Main Reliability Issue in 3D Transistor?~, K. Ota, C. Tanaka, D. Matsushita, T. Numata, and M. Saitoh (Toshiba Corporation) | 19 |

| 14:00 | 3-2 | Threshold Voltage and Current Variability of Extremely Narrow Silicon Nanowire MOSFETs with Width down to 2nm, T. Mizutani, Y. Tanahashi, R. Suzuki, T. Saraya, M. Kobayashi, and T. Hiramoto (University of Tokyo) | 21 |

| 14:20 | 3-3 | Performance of GAA Poly-Si Channel of Junctionless Field Effect Transistors with Ultra-Thin Body, Yan-Bo Liu, Yi-Ruei Jhan, Cheng-Ping Wang, and Yung-Chun Wu (National Tsing Hua University) | 23 |

| 14:40 | 3-4 | Investigation of Reconfigurable Silicon Nanowire Schottky Barrier Transistors-Based Logic Gate Circuits and SRAM Cell, J. Wang, G. Du, and X. Y. Liu (Peking University) | 25 |

| 15:00 | 3-5 | Impacts of Surface Roughness Scattering on Hole Mobility in Germanium Nanowires, H. Tanaka, J. Suda, and T. Kimoto (Kyoto University) | 27 |

| 15:20 | | Break | |

Session 4: Finite Dopants in Nanodevices

| 15:35 | 4-1 | **(Invited)** Single ion implantation of Ge donor impurity in silicon transistors, E. Prati [1], Y. Chiba[2], M. Yano[2], K. Kumagai[2], M. Hori[3], G. Ferrari[4], T. Shinada[5], and T. Tanii[2], [1]Consiglio Nazionale delle Ricerche, [2]Waseda University, [3]University of Toyama, [4]Politecnico di Milano, [5]Tohoku University) | 29 |

| 16:05 | 4-2 | Impact of Diffused Donor-Clusters near Lead/Channel Boundary on High-Temperature Single-Electron Tunneling in Narrow SOI-FETs, D. Moraru, A. Samanta[1], Y. Takasu[1], K. Tyszka[1,2], T. Mizuno[1], R. Jablonski[2] and M. Tabe[1] ([1]Shizuoka University, [2]Warsaw University of Technology) | 31 |

16:25   4-3   The impact of single donor and donor-acceptor pair on electronic and 33 transport properties of silicon nanostructures, L. T. Anh[1], D. Moraru[2], M. Manoharan[1], M.Tabe[2], and H. Mizuta[1,3] ([1]JAIST, [2]Shizuoka University, [3]Southampton University)

Session 5 Posters

16:45-18:30   Starting with Short Oral Presentation

5-1   Investigation of the Impact of Grain Boundary on Threshold Voltage of 3-D 35 MLC NAND Flash Memory, Z. Lun, L. Shen, Y. Cong, G. Du, X. Liu, Y, Wang (Peking University)

5-2   Resistive Switching Characteristics in HfOx Memory Devices with Local 37 Electrical Field Design, Tsung-Kuei Kang, Wei-Len Chen, Yu-Han Chen, Pei-Hsun Tsai (Feng-Chia University)

5-3   Position and Number Control of Donor-QD Potential by Pattern-doping in 39 SOI-FET Channels, K. Tyszka[1,2], D. Moraru[1], T. Mizuno[1], R. Jablonski[2], M. Tabe[1] (Shizuoka University, Warsaw University)

5-4   Nanodamascene metal-insulator-metal single electron transistor prepared 41 by atomic layer deposition of tunnel barrier and subsequent reduction of metal surface oxide, G. Karbasian, A. O. Orlov, and G. L. Snider (University of Notre Dame)

5-5   Impacts of Channel Constriction Dimensions of Graphene Single Carrier 43 Transistors on the Coulomb Diamond Characteristics, T. Iwasaki[1], M. Muruganathan[1], and H. Mizuta[1,2], ([1]JAIST, [2]University of Southampton)

5-6   Series-triple quantum dots fabricated under each control gate by the use of 45 thermal oxidation, T. Uchida[1], H. Sato[1], A. Tsurumaki-Fukuchi[1], M. Arita[1], A. Fujiwara[2], and Y. Takahashi[1] ([1]Hokkaido University, [2]NTT Corporation)

5-7   Fabrication and characterization of physically-defined double quantum dots 47 without unintentional localized states on highly-doped silicon substrate, Y. Yamaoka, T. Kodera, S. Oda (Tokyo Institute of Technology)

5-8   Characterization of carrier dynamics in Ge quantum dots through Ge 49 quantum-dot MOSFETs using pulsed voltage technique, Ming-Hao Kuo[1], Ho-Chane Chen[1], Wei-Ting Lai[1,2], and Pei-Wen Li[1,2] ([1]National Central University, [2]National Chiao Tung University)

5-9   Simultaneous two gate reflectometric spectroscopy of Si coupled donor-dot 51 system, X. Jehl[1], A. O. Orlov[2], R. Maurand[1], P. Fay[2], G. L. Snider[2], S. Barraud[3], and M. Sanquer[1] ([1]DSM-INAC, [2]University of Notre Dame, [3]DRT-Leti)

5-10 Fabrication of a highly controllable Si-MOS quantum dot device, T. Honda[1], 53
J. Yoneda[2,3], K, Takeda[2,3], T. Kodera[1,2], S. Tarucha[2,3], and S. Oda[1] ([1]Tokyo
Institute of Technology, [2]The University of Tokyo, [3]RIKEN)

5-11 Tunneling Field-Effect Transistor with a Grown Si Epitaxial Layer for 55
Boosting ON Current, J. Lee, J. H. Kim, D. W. Kwon, E. Park, T. Hyung Park,
and B.-G. Park (Seoul National University)

5-12 Short-Drain Effect of 5 nm Tunnel Field-Effect Transistors, Yu-Hsuan Chen[1], 57
Nguyen Dang Chien[2], Jr-Jie Tsai[1], Yan-Xiang Luo[3], and Chun-Hsing Shih[1]
([1]National Chi Nan University, [2]University of Da Lat, [3]National Tsing Hua
University)

5-13 Fabrication and Characterization of Silicon Nanowire Ultra-thin Channel 59
Poly-Si Junctionless Field Effect Transistors with a Trench Structure Ko-
Wei Lin, Mu-Shih Yeh, Min-Hsin Wu, Yung-Chun Wu (National Tsing Hua
University)

5-14 Hybrid Channel Poly-Si Junctionless Field-Effect Transistors with Trench 61
Structure Formed by Dry Etching Process, C.-P. Wang, Y.-R. Jhan, J.-J. Su,
Y.-C. Wu (National Tsing Hua University)

5-15 Built-in Effective Body-Bias Effect in UTBB Hetero-Channel MOSFETs and 63
Its Suppression, C.-H. Yu, P. Su (National Chiao Tung University)

5-16 Boolean logic circuit implementation using multi-input floating-body 65
MOSFET, Min-Woo Kwon, Hyungjin Kim, Jungjin Park, Byung-Gook Park
(Seoul National University)

5-17 Comparison of Electrical Characteristics of N-type Silicon Junctionless 67
Transistors with and without Film Profile Engineering by TCAD Simulation,
Jung-Ruey Tsai[1], Horng-Chih Lin[2], Hsiu-Fu Chang[1], Bo-Shiuan Shie[2], Ting-
Ting Wen[2], and Tiao-Yuan Huang[2] (Asia University, National Chiao Tung
University)

5-18 Thermodynamic stability of high phosphorus concentration in silicon 69
nanostructures, M. Perego[1], G. Seguini[1], E. Arduca[1,2], J. Frascaroli[1], D. De
Salvador[1,2], M. Mastromatteo[1,2], A. Carnera[1,2], G. Nicotra[1], M. Scuderi[1], C.
Spinella[1], G. Impellizzeri[1], C. Lenardi[2], and E. Napolitani[1,2] ([1]IMM-CNR,
[2]Università degli Studi di Padova)

5-19 3D-TCAD Simulation Study of the Novel T-FinFET Structure for Sub-14nm 71
Metal-Oxide-Semiconductor Field-Effect Transistor, Chen-Han Chou,
Chung-Chun Hsu, Steve S. Chung, Chao-Hsin Chien (National Chiao Tung
University)

5-20 Characteristics of Inversion, Accumulation and Junctionless mode Silicon N-Type and P-Type Bulk FinFETs with optimized 3-nm nano-fin structure, V. Thirunavukkarasu, Y.-R. Jhan, Y.-B. Liu, and Y.-C. Wu (National Tsing Hua University) 73

5-21 Bringing Physics to Device Design - a Fast and Predictive Device Simulation Framework, M. Karner, Z. Stanojevic, F. Mitterbauer, C. Kernstock, H. Demel (Global TCAD Solutions GmbH) 75

5-22 A Capacitance-Voltage model for DG-TFET, A. Biswas, A. M. Ionescu (Ecole Polytechnique Fédérale de Lausanne) 77

5-23 Physics-based Model for the Conductive Filament at the Low Resistance State of Thin SiO_2 Films, R. Yamaguchi, S. Sato, and Y. Omura (Kansai University) 79

5-24 Performance Evaluation of Si Ultra-Thin Body (1 nm) Junctionless FET with L_G = 1 nm and L_G = 3 nm, Yi-Ruei Jhan, Yan-Bo Liu, Yung-Chun Wu (National Tsing Hua University) 81

5-25 Design and analysis of electric-field-assisted nonlocal silicon-channel spin devices, D.Kitagata, T.Akushichi, Y.Takamura, Y.Shuto, S.Sugahara (Tokyo Institute of Technology) 83

5-26 Silicon-Compatible Resonant Plasma-Wave Transistor with 2D Silicene Channel for High-Performance Terahertz Electromagnetic Wave Emitters, Jong Yul Park, Sung-Ho Kim, and Kyung Rok Kim (Ulsan National Institute of Science and Technology) 85

5-27 Novel Trigate Field-Plated Poly-Si TFT with Improved Leakage Current and High On/Off Current Ratio, Yong-Hong Syu, Hsin-Hui Hu*, Jhen-Yu Tsai, Kai-Ming Wang, Jia-Jin Tsa (National Taipei University of Technology) 87

5-28 Frequency-Dependent Response of Nanoscale Thermocouples Using Temperature Oscillations Produced by Nanoscale Heaters, Gergo P. Szakmany, Alexei O. Orlov, Gary H. Bernstein, Wolfgang Porod (University of Notre Dame) 89

Monday, June 15, 2015

Session 6: New Low-Power Devices

8:30   6-1   New features in Planar SiGe Channel Tunnel FETs Performance and   91
Operation, C. Le Royer[1], L. Hutin[1], S. Martinie[1], P. Nguyen[1], S. Barraud[1], F. Glowacki[1], S. Cristoloveanu[2], M. Vinet[1] ([1]CEA LETI, [2]IMEP-LAHC)

8:50   6-2   Design of Complementary Tilt-gate TFETs with SiGe/Si and III-V Integrations   93
Feasible for Ultra-low-power Applications   E. R. Hsieh[1], Y. S. Lin[2], Y. B. Zhao[1], C. H. Liu[2], C. H. Chien[1], and Steve S. Chung[1] ([1]National Chiao Tung University, [2]National Taiwan Normal University)

9:10   6-3   Dopant-Assisted Tunnel-Current Enhancement in Two-Dimensional Esaki   95
Diodes, H.N. Tan[1], D. Moraru1, K. Tyszka[1,2], A. Sapteka[3], S. Purwiyanti[3], L.T. Anh[4], M. Manoharan[4], T. Mizuno[1], R. Jablonski[2], D. Hartanto[3], H. Mizuta[4,5], M. Tabe[1] ([1]Sizuoka University, [2]Warsaw University of Technology, [3]University of Indonesia, [4]JAIST, [5]University of Southampton)

9:30   6-4   Fabrication of high-quality $Co_2FeSi_{0.5}Al_{0.5}$/CoFe/MgO/Si spin injectors for Si-   97
channel spin devices, T. Kondo, Y. Kawame, Y. Takamura, Y. Shuto, S. Sugahara. (Tokyo Institute of Technology)

9:50   Break

Session 7: Quantum Computing and Electronics

10:10   7-1   **(Invited)** Spin-based Quantum Computing in Silicon, A. Dzurak (UNSW)   99

10:40   7-2   Variation of Coulomb diamonds and excited states caused by electric field in   101
Si single-electron transistor, H. Satoh[1], T. Uchida[1], A. Tsurumaki-Fukuchi[1], M. Arita[1], A. Fujiwara[2], and Y. Takahashi[1] ([1]Hokkaido University, [2]NTT Corporation)

11:00   7-3   Study of charged island formation in nanoscale Si single-electron transistors   103
using dual port reflectometric spectroscopy, A. O. Orlov[1], P. Fay[1], G. L. Snider[1], X. Jehl[2], R. Lavieville[3], S. Barraud[3], and M. Sanquer[2] ([1]University of Notre Dame, [2]DSM-INAC, [3]DRT-Leti)

11:20   7-4   Low Temperature Charge Pumping in SOI Gated PIN Diode, T. Watanabe[1],   105
M. Hori[1], T. Saruwatari[1], A. Fujiwara[2], and Y. Ono[1] ([1]University of Toyama, [2]NTT)

11:40   7-5   Charge sensing of p-channel double quantum dots fabricated on (110)   107
silicon substrate, K. Iwasaki, T. Kodera, and S. Oda (Tokyo Institute of Technology)

12:00   Lunch

Session 8:   Post Silicon Materials and Devices

13:30   8-1   **(Invited)** Fluctuations and Relaxation in Graphene, D. K. Ferry, B. Liu, and   109
              R. Akis (Arizona State University)

14:00   8-2   Low pull-in voltage graphene nanoelectromechanical switches, M.   111
              Manoharan[1], T. Chikuba[1], N. Kanetake[1], J. Sun[1], and H. Mizuta[1,2] ([1]JAIST,
              [2]University of Southampton)

14:20   8-3   Challenges of 3D VLSI-CoolCube[TM] process with p-Ge-OI and n-InGaAs-OI   113
              for Ultimate CMOS Nodes, F. Nemouchi[1], L. Hutin[1], H. Boutry[1], P.
              Rodriguez[1], E. Ghegin[1,2], J. Borrel[1,2], Y. Morand[2], S. Kerdiles[1], P. Batude[1],
              and M. Vinet[1] (CEA Leti, STMicroelectronics)

14:40   8-4   CMOS Roadmap Analysis from the Perspective of III-V technology using   115
              MASTAR, G. Hiblot[1,2], Q. Rafhay[2], G. Mugny[1], G. Ghibaudo[2], and F. Boeuf[1]
              ([1]STMicroelectronics, [2]IMEP-LAHC)

15:00   8-5   Effect of Free Carrier Accumulation or Depletion on Zone-center Vibrational   117
              Mode in Ge, S. Kabuyanagi, T. Nishimura, T. Yajima, and A. Toriumi (The
              University of Tokyo)

15:20   8-6   N[+]/P Shallow Junction with High Dopant Activation and Low Contact   119
              Resistivity Fabricated by Solid Phase Epitaxy Method for Ge Technology, P.
              Liu, M. Li, X. An, M. Lin, Y. Zhao, B. Zhang, X. Xia, R. Huang (Peking
              University)

15:40   Break

Session 9: 20th Anniversary Panel Session

16:00-17:30

Moderator: T. Irisawa (AIST)

Panelists: D. Ferry (Arizona State University)
                  H. Iwai (Tokyo Institute of Technology)
                  S. Oda (Tokyo Institute of Technology)
                  T. Hiramoto (The University of Tokyo)
                  K. De Meyer (imec)

Advanced process technologies of 1S1R for High Density Cross Point ReRAM

Beom Yong Kim and Hyeong Soo Kim

New Memory Process Team, R&D Division, SK Hynix Inc, Republic of Korea

Email: hyeongsoo.kim@sk.com

2Xnm cross point cell array of 1S1R has been investigated using advanced process technologies which include 1) a spacer to protect a selector material, 2) a selector to reduce sneak current, and 3) a resistor to have proper amount of oxygen vacancies (Vo). With such technologies optimization, the excellent bipolar 1S1R switching characteristics with low sneak current (300nA), low power operation (30μA), and high on/off ratio (>30) were acquired.

I. INTRODUCTION

The cross point array ($4F^2$) cell structure is mandatory for high density ReRAM production. The key for higher density cell array heavily depends on lowering the sneak current (I_{off}) which is the current level at the $1/2$ V_{set} when the resistor (1R) is in the low resistance state (LRS). In order to minimize the I_{off}, the selector (1S) for the ReRAM is required [1-3]. On the other hand, 1R should not only have high on/off current ratio but also need uniform distribution to operate 1S1R properly in terms of sensing margin. In this research, we demonstrated the improved switching properties (low I_{off} and high sensing margin) of 1S1R cell through the process development of 1S, spacer, and 1R.

II. EXPERIMENTAL

Test vehicles with 2Xnm ReRAM process technology were fabricated using 300mm wafers. TiN was applied to all electrodes of cell. The pillar type ReRAM cells were patterned between the two metal lines.

III. RESULTS AND DISCUSSION

Fig. 1 indicates the structure and process flow for the selector formation. NbO_2 having metal-insulator-transition (MIT) was used as the key material of selector device [4]. Fig. 2 and 3 show the importance of the process integration for the right threshold switching characteristics. The composition of NbO_2 can be degraded during the post sidewall spacer deposition, thus new spacer is needed to protect NbO_2 [5]. We tested the ReRAM friendly new Si_3N_4 spacer process to suppress additional oxidation or nitridation of NbO_2 (Fig. 4). As a result, the perfect bi-directional threshold switching (Fig. 5) and the excellent endurance (Fig. 6) were obtained at 300mm wafer level. We tried to minimize the I_{off} by increasing 1S cell resistance. However, I_{off} was not reduced despite the 1S cell shrinkage and the increase of its thickness

(Fig. 7). As an alternative, various barrier insertions were tested and as a result, clear reduction of I_{off} was observed (Fig. 8) Interestingly, the barrier having the highest standard Gibbs free energy ($| \Delta G |$) showed the lowest I_{off} (430nA) (Fig. 9). Note that all barrier stacks showed the forming process during the 1st sweep (Fig. 10) and NbO_2 without barrier had no forming. This clearly means that forming occurred in barriers. Therefore, barriers with higher ΔG can make the conductive filament size smaller, which allows I_{off} reduction (Fig. 11).

The structure and process flow for the resistor formation were shown in Fig.12. Resistance and V_o can be optimized by the process conditions (Fig. 13). As a result, the remarkable performance such as 10 years of retention, 1E8 of endurance, and 100nsec of switching speed was achieved. In terms of 1R statistics in array, "Program and verify (PNV)" scheme greatly improves the current distribution and pass rate of cells within a wafer (Fig. 14-16).

We finally fabricated 1S1R arrays using the optimized 1R and 1S (Fig. 17). The middle electrode (ME) was applied as the shared electrode for the independent operation of selector and resistor without disturbing each other. Ultimately, we have successfully demonstrated the excellent switching behaviors of 1S1R which exhibit ultra low I_{off} (300nA), large on/off current ratio (>30), low power operation (I_{sw}=30μA and V_{set}=2.2V), and perfect bidirectional and symmetric operation (Fig. 18). Moreover, the good switching uniformity of full cells and 97% yield at a 300mm wafer were achieved (Fig. 19)

VI. CONCLUSION

High performance of 1S1R was demonstrated through the optimization of the selector, spacer, and resistor processes at 2Xnm tech node. To find the way to further minimize I_{off} is the next challenge for a larger cell array.

REFERENCES

[1] S.M. Shapiro, et. al., Solid State Communications, 15, 377 (1974).

[2] David Bach, et. al., Microscopy and Microanalysis, 15, 524 (2009).

[3] Seonghyun Kim, et. al., VLSI, 155 (2012).

[4] Wootae Lee, et. al., VLSI, 37 (2012).

[5] Jiun-Jia Huang, et. al., IEDM, 733 (2011).

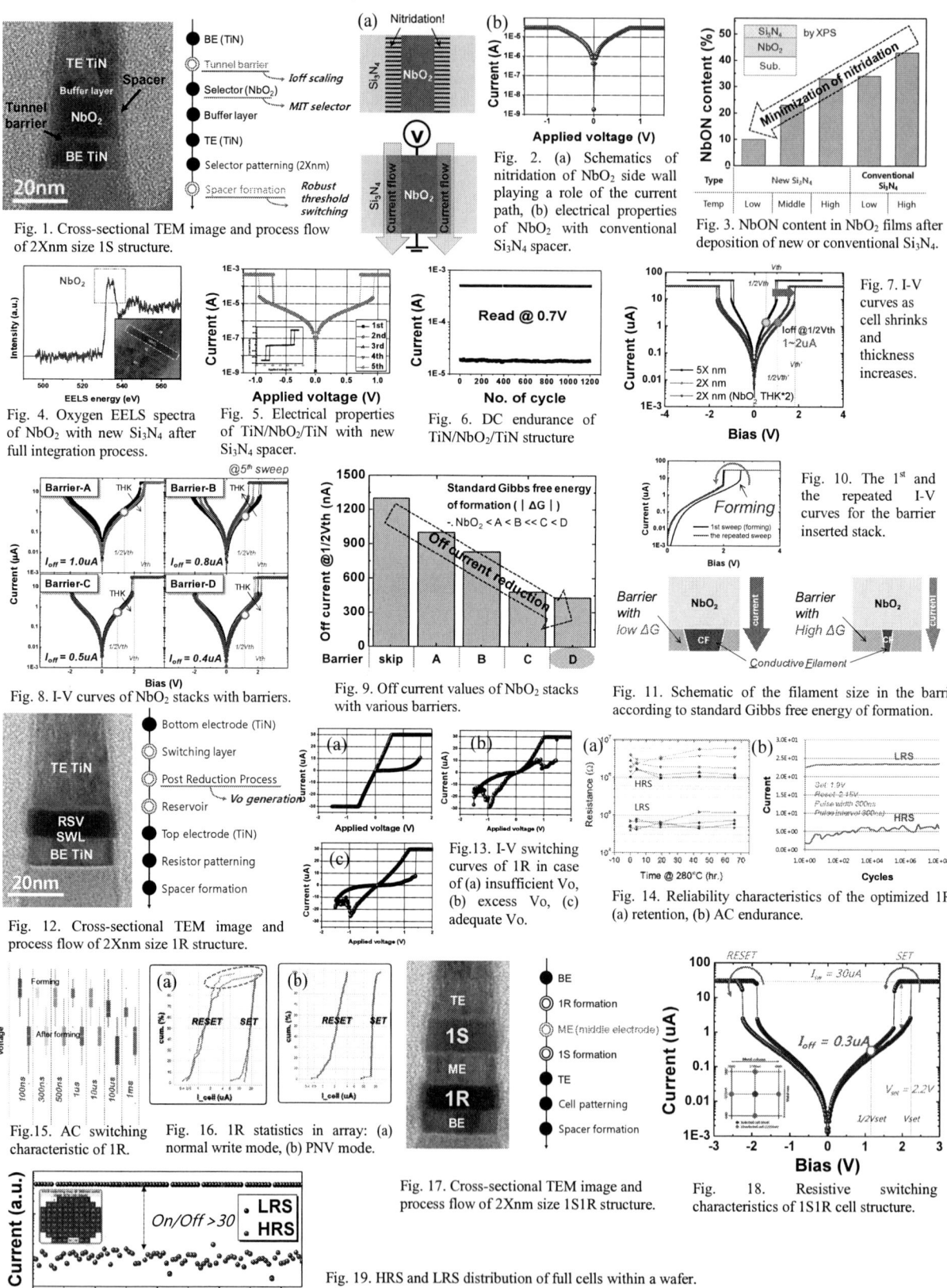

Fig. 1. Cross-sectional TEM image and process flow of 2Xnm size 1S structure.

Fig. 2. (a) Schematics of nitridation of NbO$_2$ side wall playing a role of the current path, (b) electrical properties of NbO$_2$ with conventional Si$_3$N$_4$ spacer.

Fig. 3. NbON content in NbO$_2$ films after deposition of new or conventional Si$_3$N$_4$.

Fig. 4. Oxygen EELS spectra of NbO$_2$ with new Si$_3$N$_4$ after full integration process.

Fig. 5. Electrical properties of TiN/NbO$_2$/TiN with new Si$_3$N$_4$ spacer.

Fig. 6. DC endurance of TiN/NbO$_2$/TiN structure

Fig. 7. I-V curves as cell shrinks and thickness increases.

Fig. 8. I-V curves of NbO$_2$ stacks with barriers.

Fig. 9. Off current values of NbO$_2$ stacks with various barriers.

Fig. 10. The 1st and the repeated I-V curves for the barrier inserted stack.

Fig. 11. Schematic of the filament size in the barrier according to standard Gibbs free energy of formation.

Fig. 12. Cross-sectional TEM image and process flow of 2Xnm size 1R structure.

Fig.13. I-V switching curves of 1R in case of (a) insufficient Vo, (b) excess Vo, (c) adequate Vo.

Fig. 14. Reliability characteristics of the optimized 1R: (a) retention, (b) AC endurance.

Fig.15. AC switching characteristic of 1R.

Fig. 16. 1R statistics in array: (a) normal write mode, (b) PNV mode.

Fig. 17. Cross-sectional TEM image and process flow of 2Xnm size 1S1R structure.

Fig. 18. Resistive switching characteristics of 1S1R cell structure.

Fig. 19. HRS and LRS distribution of full cells within a wafer.

Trade-off of Performance, Reliability and Cost of SCM/NAND Flash Hybrid SSD

Hirofumi Takishita, Sheyang Ning and Ken Takeuchi

Chuo University, Tokyo, Japan, E-mail: takishita@takeuchi-lab.org

Abstract— Storage-Class Memory (SCM), NAND flash hybrid Solid-State Drive (SSD) shows advantages of high performance and low power consumption compared with NAND flash only SSD. In this paper, first, three types of SCMs are analyzed respectively, with 0.1 μs, 1 μs and 10 μs read/write times. Then, their SCM/NAND flash capacity ratios are calculated to achieve the required SSD performance for the application. Next, by using these three types of SCMs, the acceptable SCM and NAND flash Bit Error Rates (BERs) are analyzed, which are limited by keeping specified SSD performance with error correction. Moreover, the acceptable BERs are also analyzed by considering the limitation of parity overhead size and SSD cost. At last, based on prior analyses, the acceptable BERs are summarized by considering the limitations of both SSD performance and cost.

I. INTRODUCTION

The architecture and data write of Storage-Class Memory (SCM)/NAND flash hybrid Solid-State Drive (SSD) are shown in Fig. 1 [1]. This hybrid structure is attractive because it largely increases SSD performance and reduces power consumption, compared with NAND flash only SSD [1]. In detail, SCM such as ReRAM, MRAM and PRAM [2] has over x10 higher performance than NAND flash. By adding small capacity of SCM, which equals to 6% of NAND flash capacity, the SSD performance boosts about 10-times [3]. On the other hand, using large SCM capacity is not cost effective because the SCM may have much higher bit-cost than NAND flash considering its early stage of development. Therefore, the hybrid SSD with small SCM capacity and large capacity NAND flash is the best solution to achieve both high SSD performance and low cost. In [4], the required SCM and NAND flash capacities are analyzed by considering both SSD performance and SCM bit-cost. However, the trade-off among SCM latency, SSD performance, cost and reliability is not analyzed. For example, NAND flash Bit Error Rate (BER) and required correctable bits increase along with write/erase cycling, as shown in Fig. 2 [5]. Therefore, large parity overhead and strong Error-Correcting Code (ECC) is needed to recover more errors and improve SSD reliability. On the other hand, strong ECC requires large parity overhead and high error correction complexity, which increases SSD cost and degrades the SSD performance. In this paper, the trade-off among SSD performance, cost and reliability is analyzed for the first time. The objective is to show the design guidelines of a high performance, high reliability and cost-effective SCM/NAND flash hybrid SSD. The workload of the financial server [6] and simulator in [3] are used in the following analyses.

II. ECC ENCODING AND DECODING

Bose-Chaudhuri-Hocquenghem (BCH) ECC is commonly used for NAND flash memories [7]. Fig. 3 shows the BCH ECC page structure, which is the write or read unit and is also the unit of the ECC codeword. To correct t bit errors, $13 \times t$ and $14 \times t$ parity overhead bits are needed, when user data sizes are 512 Bytes and 1 KBytes [8], respectively. Fig. 4 shows the ECC parity overhead versus acceptable raw BER [9]. The parity overhead is defined as the ratio of the redundant parity bit for ECC to the user data size. Though SCM has smaller capacity compared with NAND flash, the same amount of error bits per page is assumed because it stores hot data, and more write/erase cycles are applied. Fig. 5 shows data write and read flows by using BCH ECC. During the write, the ECC encoder latency can be ignored because the user data is bypassed and the parity is generated immediately [10]. In contrast, when the data is read from SCM/NAND flash, the large ECC decoding latency (t_{ECC}) is necessary which degrades the SSD performance. The relation of the large ECC correctable bits required and the decoding latency calculated based on [11] is shown in Fig. 6. Fig. 7 shows that the decoding also occurs during the write access of SCM/NAND flash hybrid SSD [1, 3], because the data read is needed even during the host write for the garbage collection.

III. DESIGN OF HYBRID SSD

A. Acceptable BERs limited by SSD performance

Fig. 8 depicts the SSD performance by using different types of SCMs with different write and read latencies (T_{W_SCM} and T_{R_SCM}) [12]. SSD performance is reduced if the SCM has larger read/write latencies. In case of the longer SCM latency, the required SSD performance (20MB/s) for the workload can be achieved by using larger SCM capacity [12]. Table. 1 shows the three different SCM configurations (very fast/fast/moderate) to achieve the target SSD performance. The

term "read/write latency" is defined to read or write the memory cell without ECC. In detail, for Case 1, small SCM capacity is sufficient because SCM write and read latencies are as small as 0.1 μs. For Case 2, though SCM write and read latencies are increased to 1 μs, the required SCM capacity does not increase significantly. In contrast, Case 3 uses the slowest SCM with 10 μs latencies. The SCM/NAND flash capacity ratio significantly increases to achieve the 20 MB/s SSD performance. Fig. 9 shows the calculated SSD performance with different NAND flash and SCM acceptable BERs at Case 2 [5]. In the discussion above, ECC calculation time is assumed to zero. However, the actual SSD performance degrades when the ECC decoding time is considered for the error correction. In the following analysis, 90% of the maximum performance is assumed to be acceptable when considering ECC. When NAND flash and SCM acceptable BERs increases, more errors should be corrected by using the stronger ECC. As a result, both the parity overhead (cost) and the ECC decoding time increases. In Fig. 10, the acceptable BERs are compared without considering the SSD cost constraint. Case 3 has 10 μs large write and read latencies. In Case 3, the larger acceptable BERs are acceptable because the SCM write and read are so slow that the ECC decoding can be operated simultaneously during the SCM read and write. The disadvantage of Case 3 is the larger SCM capacity and the parity overhead, which increases the SSD cost.

B. Acceptable BERs limited by SSD cost

The relation of acceptable BERs and SSD cost is also analyzed. Fig. 11 shows the cost (P) of one SSD sector. C and rC are NAND flash and SCM capacities, respectively. The parity overhead ratios are α and β for NAND flash and SCM, respectively. Assuming SCM bit cost is k-times of NAND flash, the total SSD cost (P) per sector is proportional to

$$P = (1+\alpha)C + k(1+\beta)rC \qquad (1)$$

In this assumption, the baseline SSD has ECC which corrects 40 errors per 1KByte page [13] and 29 errors per 512Byte page for NAND flash and SCM respectively. In this case, both NAND flash and SCM have the similar ECC strength. To make SCM sector-accessible, the user-data size of the ECC of SCM cannot be larger than the sector size (512Byte). In this situation, by calculating the required parity overhead, as shown in Fig. 4, the baseline SSD cost ($P_{Baseline}$) is derived as follows.

$$P_{Baseline} = 1.068C + 1.092krC \qquad (2)$$

Figs. 12(a) to 12(c) compare the acceptable BERs by using the three types SCMs as shown in Table 1. SCM bit cost is assumed 10 times of NAND flash bit cost (k=10). By assuming the constant SSD cost, when acceptable NAND BER is decreased, acceptable SCM BER can be increased, and vice versa. Assuming the sum of the parity overhead of NAND flash and SCM is the same, if less errors are corrected in NAND flash with weaker ECC, more errors are corrected in SCM with stronger ECC. In this analysis, SSD cost is assumed to be increased by 5% from the baseline. This extra cost increases the parity overhead capacity and recover more errors.

C. Summary of acceptable BERs

In Figs. 13(a) to 13(c), acceptable NAND flash and SCM BERs are limited by considering both SSD performance (Fig. 10) and SSD cost (Fig. 12). For both performance and cost, the 5% degradation compared with the baseline is allowed. Table. 2 summarizes the results. In Case 2 with balanced conditions, it has best acceptable SCM BER, with small SCM capacity and slightly-increased SSD cost. Case 1 has the large acceptable SCM BER, but 0.1 μs fast SCM read and write are required. Though Case 3 has best acceptable NAND BER, the SCM capacity is so large that the acceptable SCM BER is restricted by the SSD cost.

IV. CONCLUSION

The trade-off among SSD performance, cost and reliability is analyzed to show the design guidelines of a high performance, highly reliable and cost-effective SCM/NAND hybrid structure SSD.

REFERENCES

[1] H. Fujii, *et al.*, *Symp. VLSI Circ.*, pp. 134-135, 2012. [2] C. H Lam, *ICSICT.*, pp.1080-1083, 2010. [3] C. Sun, *et al.*, *TCAS-I*, vol. 61, no. 2, pp. 382-392, 2014. [4] C. Sun, *et al.*, *IMW*, pp. 64-67, 2013. [5] S. Tanakamaru, *et al.*, *IMW*, pp. 154-157, 2014. [6] http://traces.cs.umass.edu/index.php/Storage/Storage. [7] K. Prall, *et al.*, *IEDM*, pp. 102-105, 2010. [8] S. Tanakamaru, *et al.*, *SSE*, online, Dec. 2010. [9] N. Mielke, *et al.*, *IRPS*, pp.9-19, 2008. [10] Z. Jun, *et al.*, *IWVDVT*, pp.97-100, 2005. [11] Y. Lee, *et al.*, *ISSCC*, pp. 426-427, 2012. [12] C. Sun, *et al.*, *TCAS-I*, vol. 61, no. 8, pp. 2360-2369, 2014. [13] S. Nelson, *et al.*, *FMS*, 2011.

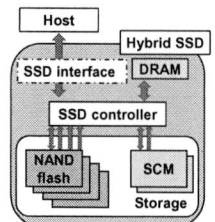

Fig. 1 SCM/NAND flash hybrid SSD [1]. Frequently accessed data is stored in SCM (faster). In contrast, rarely accessed data is stored in NAND flash (slower).

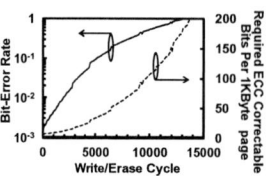

Fig. 2 Measured Bit Error Rate (BER) increment and required correctable bits on 2Xnm MLC NAND flash memory [5].

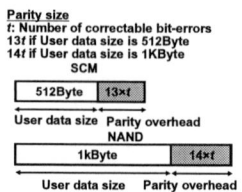

Fig. 3 BCH ECC page structure. When the user data size is increased [8], ECC correctable number is increased. However, SCM user data size cannot be larger than the sector size (512Byte).

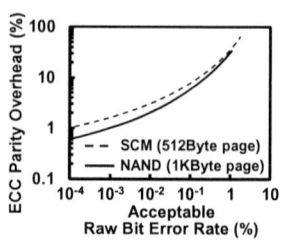

Fig. 4 Relation of the parity overhead and the acceptable raw BER [9].

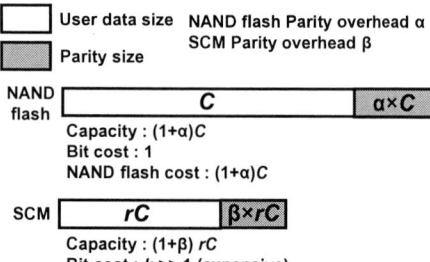

Fig. 5 BCH encoder and decoder. ECC encoding latency is negligibly small [5]. In contrast, ECC decoding latency t_{ECC} is large and reduces the SSD performance [11].

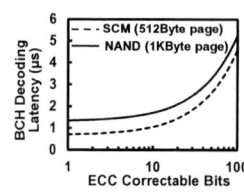

Fig. 6 ECC correctable bits versus BCH decoding latency based on [11]. 1KByte long BCH ECC structure increases the BCH decoding latency.

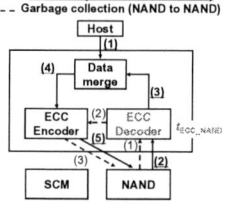

Fig. 7 Operation of SCM/NAND hybrid SSD during write access [3]. The operations during write access also require read from NAND flash or SCM, which needs ECC decoding time.

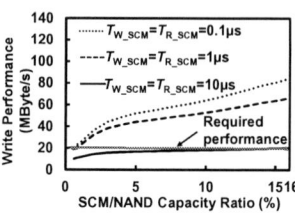

Fig. 8 SCM/NAND flash hybrid SSD performance by using SCMs with 0.1μs, 1μs and 10μs read and write times, respectively [12].

Table. 1 Three types of SCMs to keep the SSD performance above 20MB/s. Large SCM capacity is required if the SCM is slow and has large latencies.

	SCM write latency T_{W_SCM}	SCM read latency T_{R_SCM}	SCM/NAND capacity ratio r
Case 1	0.1μs	0.1μs	0.6%
Case 2	1μs	1μs	0.7%
Case 3	10μs	10μs	16%

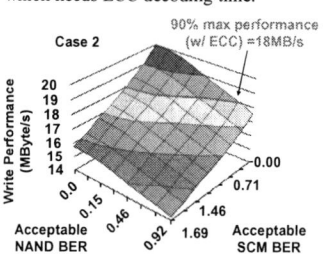

Fig. 9 SSD write performance versus acceptable NAND flash and SCM BERs [5]. Higher BER increases ECC complexity and reduces the write performance of SSD.

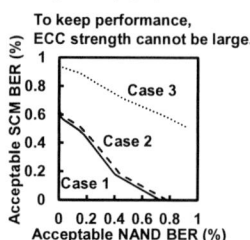

Fig. 10 Acceptable SCM and NAND flash BERs to keep SSD performance above 18MB/s with ECC.

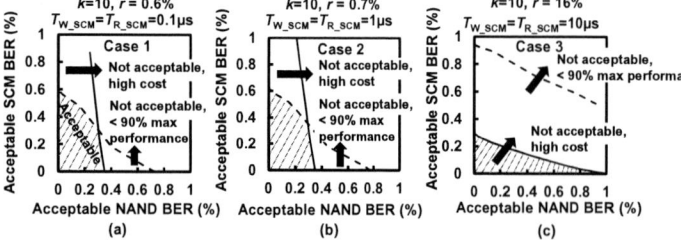

Fig. 11 Calculation of SSD cost. SCM bit cost is assumed k-times of NAND flash bit cost. Parity overhead is defined by correctable bits.

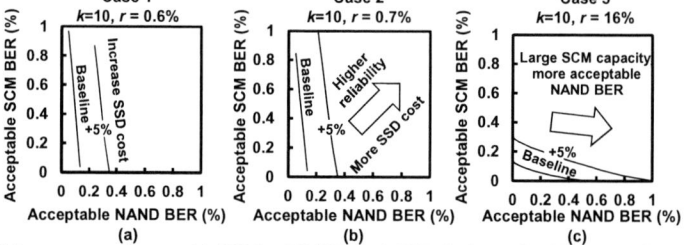

Fig. 12 Improvement of acceptable SCM and NAND flash BERs by increasing SSD cost and parity overhead. In baseline, the parity overheads of those ECCs are 6.8% and 9.2% for NAND flash and SCM, respectively. The baseline cost is defined as follows. $P_{Baseline} = 1.068C + 1.092krC$.

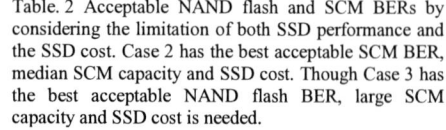

Fig. 13 Acceptable SCM and NAND flash BERs by considering limitations on both SSD performance (Fig. 10), and SSD cost (Fig. 12).

Table. 2 Acceptable NAND flash and SCM BERs by considering the limitation of both SSD performance and the SSD cost. Case 2 has the best acceptable SCM BER, median SCM capacity and SSD cost. Though Case 3 has the best acceptable NAND flash BER, large SCM capacity and SSD cost is needed.

	Case 1	Case 2	Case 3
Acceptable NAND BER	< 0.35%	< 0.35%	< 1.09%
Acceptable SCM BER	< 0.60%	< 0.62%	< 0.31%
SSD cost P (a.u.)	99	100	246

Additional Charge Trapping layer SONOS Nonvolatile Memory Based on Ultra-Thin Body Poly-Si Junctionless FinFET

Wei-Cheng Wang, Chien-Chih Chung, Ming-Hsien Chung, Cheng-Ping Wang and Yung-Chun Wu*

Department of Engineering and System Science, National Tsing Hua University, Hsinchu, Taiwan

Tel: +886-3-5715131 ext. 34287;Fax: +886-3-5720724;Email: tmwang04@gmail.com

Abstract–This work demonstrates ultra-thin body (UTB) trench structure Junctionless FinFET (trench JL-FinFET) with double stacked Si_3N_4 charge trapping layer (NN-CTL) $Si-SiO_2-Si_3N_4-Si_3N_4-SiO_2-Si$ (SONNOS) nonvolatile memory (NVM). It shows excellent memory characteristics, high program/erase (P/E) performance, good endurance ($>10^4$ cycles) and an excellent 10 years data retention with 99% electron remaining at 85℃.

I. Introduction

JL nanowire (NW) FET with high and uniform doping in channel and source/drain regions have much attraction [1]. JL-FET can avoid complex source and drain doping engineering. The proposed trench structure provides the hybrid FinFET and UTB [2][3] which reveals excellent performances of NVMs. Meanwhile, the poly-Si SONNOS NVM structure is suitable for 3D NAND, 3D stacked IC, AMLCD, and AMOLCD applications[4].We first provide a simple double and high performance stacked Si_3N_4 (NN-CTL) SONNOS NVM based on trench JL-FinFET to explore the future application.

II. Device Fabrication

Fig.1. (a) shows the structure of SONNOS NVM device. Fig.1. (b) presents the gate stacked; the double stacked Si_3N_4 (NN=3nm/3nm). Fig.1. (c) displays the process flow of fabrication in the trench-JL SONNOS and SONOS NVMs. The device was fabricated by growing a 400nm SiO_2 layer on 6 inch Si wafers and a 50nm undoped amorphous Si (a-Si) was deposited by low-pressure chemical vapor deposition (LPCVD) at 550℃.Then the poly-Si layer was formed by solid-phase recrystallization (SPC) method at 600℃ for 24hrs. The n-type poly-Si layer was implanted. Then they activated by furnace annealing at 600℃ for 4hrs. The active NWs channels were patterned e-beam lithography and transferred by reactive-ion etching (RIE). The trench structure were then defined by e-beam lithography and anisotropic etched by precisely time-controlled RIE. Thermal SiO_2 (3nm) / Si_3N_4 (3nm) / Si_3N_4 (3nm) / blocking SiO_2 (10nm) were formed by LPCVD for the gate stack. A 150nm in-situ doped n^+ poly-Si is for gate electrode. Finally, a 200nm SiO_2 layer passivation and a 300nm Al-Si-Cu metallization were performed and sintered.

III. Results and Discussion

Fig.2. (a) and (b) shows the FinFET and planar channel respectively, (c) and (d) present the trench structure by AFM images. Fig.3. (a) presents the NWs of devices, and (b) shows the gate structure of SONNOS device. Fig.4. presents the (a) program and (b) erase I_D-V_G hysteresis curves of the n-channel trench-JL FinFET SONNOS NVM by the FN tunneling. Fig.5. shows a comparison of the P/E performance of (a) the n-channel trench-JL planar SONNOS (NN-CTL) NVM device and (b) the n-channel trench-JL planar conventional SONOS NVMs. The SONNOS device has larger memory window (ΔV_{TH}) than SONOS.

Fig.6. indicates that the endurance characteristics of n-channel trench-JL FinFET and planar NVMs. The ΔV_{TH} exhibits an upward trend and memory window closing of planar device. This result reveals that every P/E cycle, there have no sufficient numbers of holes to compensate electrons in the trapping layer. In contrary, JL-FinFET SONNOS devices has large electrical field of NW corner, the sufficient numbers of holes can compensate electrons by electrical filed enhanced FN tunneling.

Fig.7. shows the retention characteristics for both of the JL-FinFET and JL-planar of SONNOS NVMs can almost maintain ~99% memory windows, owing to numerous and deep interface traps of the middle of double stacked Si_3N_4 (NN) layers. The interface traps artificially generated by the disorder double stacked Si_3N_4 layer might be played the deep traps which prevent electrons outflow.

IV. Conclusion

The trench-JL NW FinFET SONNOS NVM devices are demonstrated both high performances in NVM. The numerous and deep interface traps artificially generated by the double stacked Si_3N_4 (NN) layers improves the erasing and retention properties. The proposed trench-JL NW FinFET SONNOS NVM is suitable for 3D NAND flash memory application.

V. Reference

[1] J.-P. Coling et al., Nature Nanotechnol., vol. 5, p. 225–229, 2010.
[2] Mu-Shih Yeh et al., IEDM, p. 26-6, 2014.
[3] S.-Y. Kim et al., VLSI Tech. Dig., p.115, 2014.
[4] S. J. Choi et al., VLSI Tech. Dig., p.74, 2011.

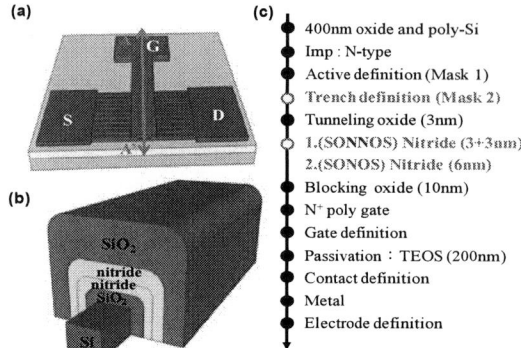

(c)
- 400nm oxide and poly-Si
- Imp : N-type
- Active definition (Mask 1)
- Trench definition (Mask 2)
- Tunneling oxide (3nm)
- 1.(SONNOS) Nitride (3+3nm)
- 2.(SONOS) Nitride (6nm)
- Blocking oxide (10nm)
- N+ poly gate
- Gate definition
- Passivation : TEOS (200nm)
- Contact definition
- Metal
- Electrode definition

Fig. 1.(a) The trench-JL FinFET SONNOS NVM device. (b) The cross section of the charge-trapping layer. (c) Details of the process flow chart of the fabricated devices.

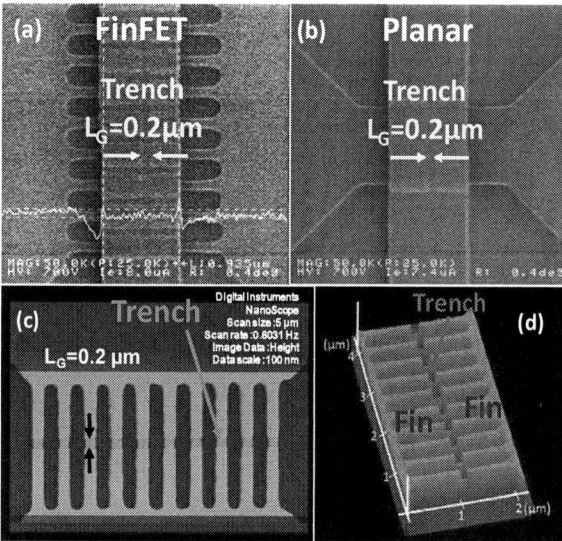

Fig. 2.(a)-(b) The SEM image of NW FinFET and planar structure. (c)-(d) Show the AFM images of trench-JL NW FinFET.

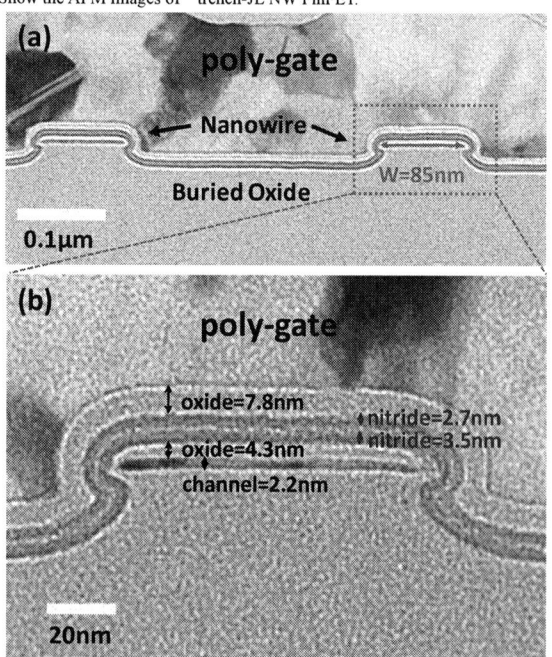

Fig. 3.(a) TEM image of NWs. (b) Magnified TEM image of the SONNOS with Poly-Si channel(2.2nm) SiO$_2$(4.3nm)/Si$_3$N$_4$ (2.7nm/3.5nm)/SiO$_2$(7.8nm).

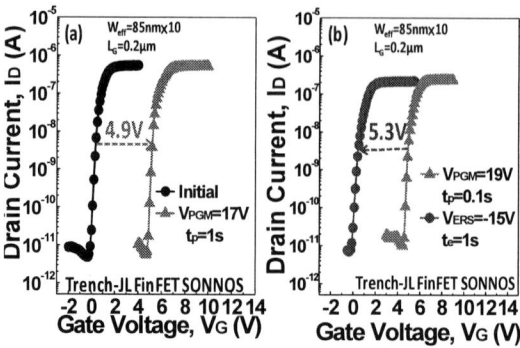

Fig.4. (a) FN programming and (b) FN erasing hysteresis curves of the n-channel trench-JL NW FinFET SONNOS NVM.

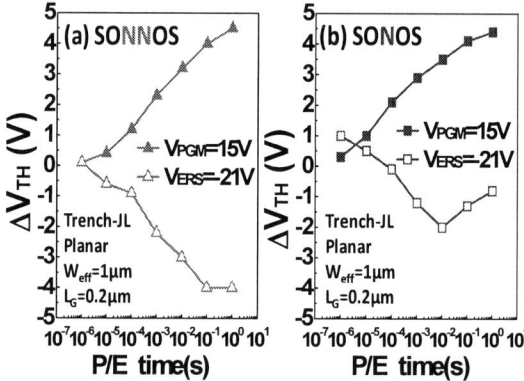

Fig.5. (a) Program/erase performance of trench-JL planar SONNOS NVM and (b) trench-JL planar SONOS NVM.

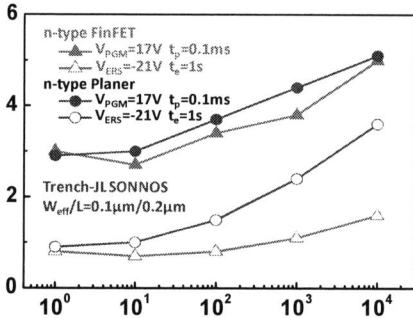

Fig.6. Endurance performance of n-ch. trench-JL NW FinFET and n-ch. trench-JL planar SONNOS NVMs. The FinFET devices show better endurance than planar.

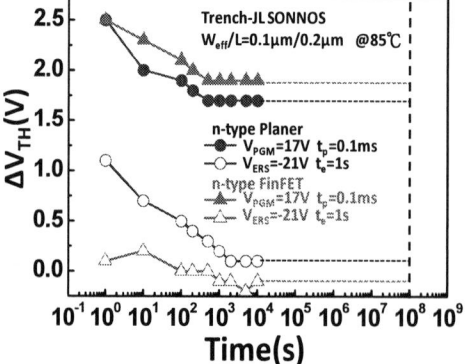

Fig. 7. Retention of trench-JL NW FinFET and trench-JL-planar SONNOS NVMs. Both of devices maintain almost 95%~99% memory windows for 10 years.

Understanding the Underlying Physics of Superior Endurance in Bi-layered TaO$_X$-RRAM

Y. D. Zhao, P. Huang, Z. Chen, C. Liu, H. T. Li, W. J. Ma, B. Gao, X. Y. Liu, J. F. Kang*

Institute of Microelectronics, Peking University, Beijing 100871, China *kangjf@pku.edu.cn

Abstract — A comprehensive physical model is introduced to account for the superior endurance characteristics of TaO$_X$-based RRAM with bi-layered Ta$_2$O$_{5-X}$/TaO$_{2-X}$ stack by considering the influence of oxygen content in TaO$_{2-X}$ layer. In the physical model, the TaO$_{2-X}$ layer acts as an oxygen reservoir to store and release oxygen ions (O^{2-}) during switching cycles and different oxygen contents in TaO$_{2-X}$ layer correspond to different capabilities to take redox reactions with O^{2-}. The proposed model is implemented into an atomic Monte-Carlo simulator involving the evolution of oxygen distribution in Ta$_2$O$_{5-X}$/TaO$_{2-X}$ stack to reproduce the impacts of different TaO$_{2-X}$ layers on endurance performance. Results reveal that the optimized TaO$_{2-X}$ layer with the feature of easily storing O^{2-} as lattice oxygen (LO) during SET and decomposing to release O^{2-} during RESET is beneficial to enhance the endurance property.

I. INTRODUCTION

TaO$_X$-based RRAM with asymmetric bi-layered Ta$_2$O$_{5-X}$/TaO$_{2-X}$ stack has drawn a lot of attention for its remarkable endurance characteristics of over 10^{12} cycles [1-4]. However, the physical origin of the excellent endurance characteristics in TaO$_X$-based RRAM is still unclear and insights into bi-layered Ta$_2$O$_{5-X}$/TaO$_{2-X}$ stack are particularly required. In this work, a new physical model to account for superior endurance property in bi-layered TaO$_X$-based RRAM is proposed by considering the oxygen content in TaO$_{2-X}$ layer which is related with the difficulty to take redox reactions with oxygen ions (O^{2-}). The proposed physical model is implemented into a developed Monte-Carlo simulator [5] involving the evolution of oxygen distribution in the bi-layered stack to simulate the influence of different TaO$_{2-X}$ layers on endurance performance. Results indicate that TaO$_{2-X}$ layer which can easily take redox reactions with O^{2-} is beneficial for endurance characteristics.

II. ENDURANCE MODEL

The typical TaO$_X$-based RRAM is a bi-layered structure composed of a stoichiometric Ta$_2$O$_{5-X}$ layer and an oxygen deficient TaO$_{2-X}$ layer [1]. Ta$_2$O$_{5-X}$ layer is the resistive switching layer (RSL), and TaO$_{2-X}$ base layer (TBL) helps to realize self-compliance during SET process [3]. TBL also acts as an oxygen reservoir to collect and release O^{2-} during switching cycles which is similar to the common oxygen reservoir such as TiN [5]. **Fig. 1** schematically shows the endurance model in bi-layered TaO$_X$-based RRAM. The oxygen reservoir in **Fig. 1(a)** can easily take redox reactions with O^{2-}. During SET, O^{2-} drift into TBL under electric field and a part of O^{2-} take redox reactions with TaO$_{2-X}$ and are stored as stable lattice oxygen (LO) in TBL near the conductive filaments (CF). During RESET, TBL decompose to release O^{2-} assisted by heat due to the relatively high temperature of RESET operation. The concentrated distribution of oxygen near CF guarantees the sufficient supply of O^{2-} to rupture CF during RESET and therefore the endurance of the device is highly enhanced. **Fig. 1(b)** shows the degraded endurance characteristics of TBL with difficulty to take redox reactions with O^{2-}. O^{2-} are difficult to be stored as stable LO during switching and the dissociative O^{2-} will diffuse in TBL laterally and vertically under electric field and thermal effects. After several cycles, the amount of O^{2-} near CF will be insufficient to rupture the filament, resulting in the degradation of endurance characteristics.

III. RESULTS AND DISCUSSIONS

The microscopic behaviors of TaO$_X$-based RRAM can be simulated by the atomic Monte-Carlo simulator involving generation & recombination of V$_O$ with O^{2-} as well as phase change between Ta$_2$O$_5$ and TaO$_2$ during switching [5]. In this study, the simulator is further developed to simulate the evolution of O^{2-} and LO distributions as well as the interaction between RSL and TBL. TaO$_X$-based RRAM with 20nm Ta$_2$O$_{5-X}$/20nm TaO$_{2-X}$ stack is simulated and compared with the fabricated Pt/Ta$_2$O$_{5-X}$/TaO$_{2-X}$/Ta RRAM devices. Electrical measurements were performed with Agilent-33250A pulse generator and Agilent-B1500 analyzer. **Fig. 2** shows the simulated and measured I-V characteristics. **Fig. 3** is the measured and simulated relationship between SET/RESET voltages and the pulse width under AC test. Excellent agreements are obtained between measured and simulated data. **Fig. 4** shows the simulated microscopic evolution of CF, O^{2-} and LO distributions in TaO$_X$-based RRAM during SET and RESET processes. The profile of TaO$_{2-X}$ is distributed with decreased 2-X by depth in TBL which accords with real fabrication process of implanting O into Ta. In order to compare the impacts of different TBL on endurance, two extreme conditions of different oxygen contents TaOx and TaOy in TBL are simulated. We assume TaOx can easily take redox reactions with O^{2-}, while TaOy is assumed to be extremely difficult to take redox reactions with O^{2-}. In the case of TaOx, the simulated microcosmic configuration of CF after FORMING is shown in **Fig. 5(a)**. **Fig. 5(b),(c)** are the oxygen distributions including both O^{2-} and LO after the 1st cycle and 10 cycles respectively. The corresponding figures in the case of TaOy are shown in **Fig. 6**. In the case of TaOx, a majority of O^{2-} are stored as LO in TBL and distribute near CF even after 10 cycles, assuring the supply of O^{2-} to rupture CF during RESET as shown in **Fig. 5(b)**. In case of TaOy, O^{2-} can hardly be stored as LO and dissociative O^{2-} will disperse in TBL as shown in **Fig. 6(c)**. **Fig. 5(d)** and **Fig. 6(d)** are simulated I-V characteristics of the 1st and 10th cycles in both cases. No obvious degradation of high resistance state is found after 10 cycles in the case of TaOx. While in the case of TaOy, the resistance window is reduced after 10 switching cycles. By considering the worst-case scenario of TaOy, the degradation of endurance can be accelerated and observed within 100 cycles as shown in the inset of **Fig. 6(d)**.

IV. CONCLUSION

A physical model accounting for the superior endurance characteristics of bi-layered TaO$_X$-based RRAM is proposed. Based on the model, a developed Monte-Carlo simulator is employed to simulate the impacts of different TaO$_{2-X}$ layers on endurance performance. Results reveal that optimized TaO$_{2-X}$ which can easily take redox reactions with O^{2-} is beneficial to enhance endurance property.

Acknowledgement: This work is supported in part by 973 Program (2011CBA00600), NSFC (61421005, 61404006).

Reference: [1] M-J. Lee et al. , Nat. Mat. 2011, vol. 10, pp. 625-630. [2] Z. Wei et al., IEDM 2011, pp. 721-724. [3] Young-Bae Kim et al., VLSI 2011, pp. 52-53. [4] L.Goux et al., VLSI 2014, pp. 162-163. [5] Y. D. Zhao et al., VLSI-TSA 2015 (accepted).

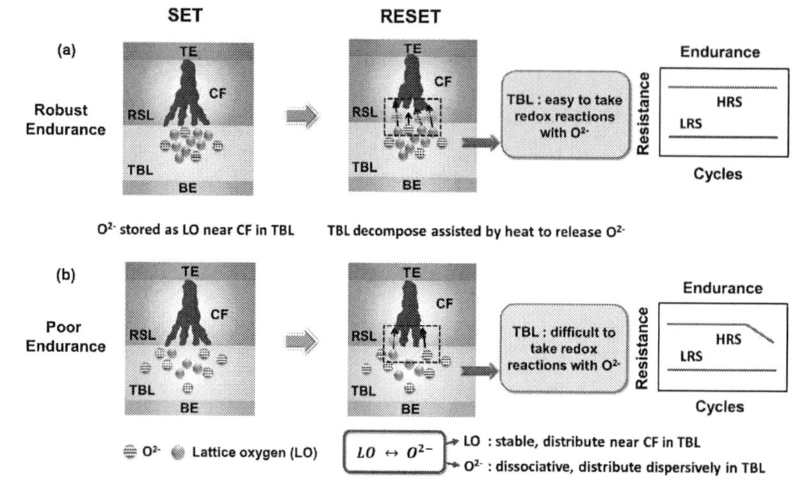

SET **RESET**

(a)
Robust Endurance

TE · CF · RSL · TBL · BE

TBL : easy to take redox reactions with O^{2-}

Endurance: HRS / LRS — Resistance vs Cycles

O^{2-} stored as LO near CF in TBL

TBL decompose assisted by heat to release O^{2-}

(b)
Poor Endurance

TE · CF · RSL · TBL · BE

TBL : difficult to take redox reactions with O^{2-}

Endurance: HRS / LRS — Resistance vs Cycles

⊖ O^{2-} ● Lattice oxygen (LO)

$LO \leftrightarrow O^{2-}$

→ LO : stable, distribute near CF in TBL
→ O^{2-} : dissociative, distribute dispersively in TBL

Fig. 1 Physical origin of superior endurance characteristics in bi-layered TaO$_X$-based RRAM. (a),(b) Schematic views of SET and RESET processes of deivces with TBL. TBL in (a) is easier to take redox reactions with O^{2-} than in (b), and (a) shows better endurance property. During SET, O^{2-} drift into TBL and a part of O^{2-} take redox reactions with TaO$_{2-X}$ in TBL to be stored as stable lattice oxygen (LO) near CF region. The rest dissociative O^{2-} will diffuse laterally and vertically under electric field and thermal effects. During RESET, TaO$_{2-X}$ decompose assisted by heat to release O^{2-} to rupture the CF. The RESET failure in (b) is due to lack of O^{2-} supplement to rupture the filament. TE and BE represent top electrode and bottom electrode, respectively.

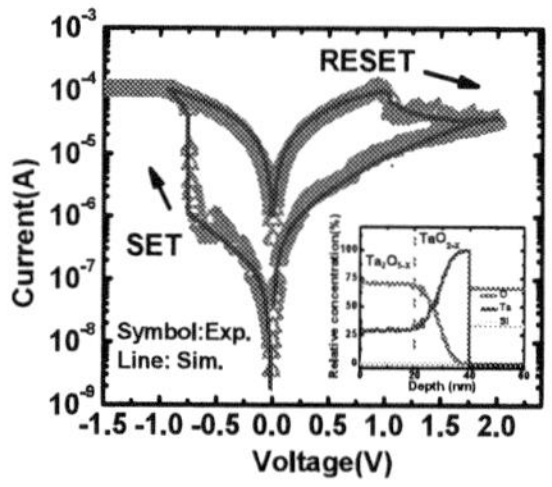

Fig.2 Measured and simulated I-V characteristics during SET and RESET processes of bi-layered TaO$_X$-based RRAM. Excellent agreements are got between measurements and simulations. Inset is the XPS depth profile of Ta$_2$O$_{5-X}$/TaO$_{2-X}$ stack.

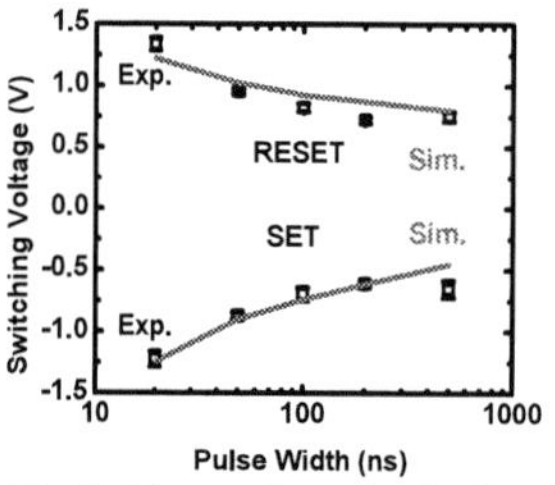

Fig.3 Measured and simulated relationship between SET/RESET voltages and the pulse width under AC test with pulse width of 20ns, 50ns, 100ns, 200ns, 500ns. The measured data are obtained from 20 cycles each pulse width.

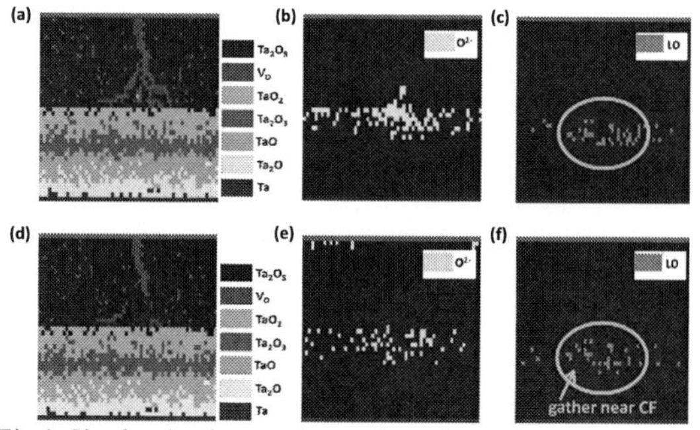

Fig.4 Simulated microscopic configurations of CF and oxygen distributions after SET (a-c) and RESET (d-f) processes. (a),(d) CF in RSL and TaO$_{2-X}$ distribution in TBL. (b),(e) Dissociative O^{2-} distributions in RSL and TBL. (c),(f) Lattice oxygen (LO) distributions in TBL.

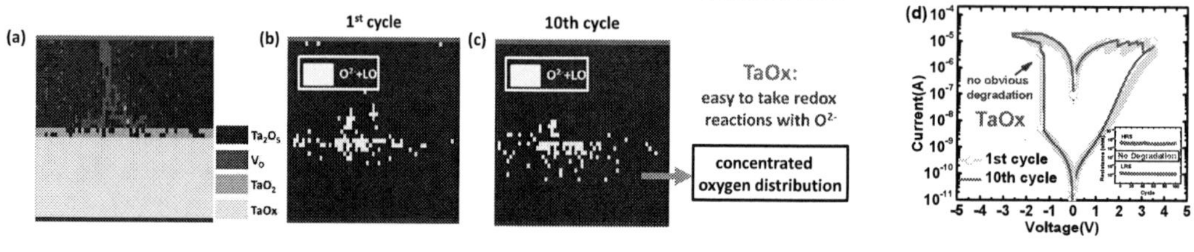

Fig.5 Simulated microscopic figures of (a) CF after FORMING and the corresponding oxygen (O^{2-} and LO) distributions with TBL composed of TaOx (b) after the 1st cycle and (c) after 10 cycles. (d) Simulated I-V characteristics with no obvious degradation after 10 cycles. Inset is the simulated endurance of 100 cycles.

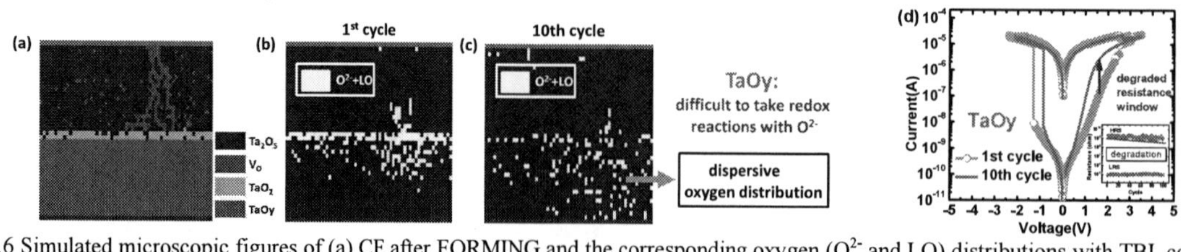

Fig.6 Simulated microscopic figures of (a) CF after FORMING and the corresponding oxygen (O^{2-} and LO) distributions with TBL composed of TaOy (b) after the 1st cycle and (c) after 10 cycles. (d) Simulated I-V characteristics with degradation after 10 cycles. Inset is the simulated endurance of 100 cycles with obvious degradation.

Variability Suppression of FinFETs by Smoothing Sidewall Roughness Using Ion Beam Etching Technology

T. Matsukawa[1], K. Endo[1], H. Akasaka[2], Y. Kamiya[2], M. Ikeda[2],
K. Tsunekawa[2], T. Nakagawa[2], Y.X. Liu[1], and M. Masahara[1]

[1]National Institute of Advanced Industrial Science and Technology (AIST), Japan
[2]Canon ANELVA Corporation, Japan
Email: t-matsu@aist.go.jp

Abstract — **Ion beam milling is successfully implemented for smoothing roughness of the fin sidewalls for the FinFETs with poly-crystalline TiN metal gate (MG). The V_t variability is improved significantly by smoothing the fin roughness without degradation of the carrier mobility. The suppressed V_t variability is interpreted as improved uniformity in the grain orientation of TiN which causes work function variation of the MG.**

I. INTRODUCTION

FinFETs introduced from 22 nm technology effectively suppress short channel effect and V_t variability caused by random dopant fluctuation (RDF) [1,2]. The remaining V_t variability for the FinFETs with suppressed RDF is dominantly caused by work function variation (WFV) of the poly-crystalline metal gate (MG) [3,4]. We have previously reported that smoothing the fin sidewall is effective to suppress the V_t variability [5]. Here, ideally flat fin sidewall is obtained by anisotropic wet etching of Si(111) and improves uniformity of crystal orientation of poly-crystalline MG. This technology, however, is not applicable for (110) oriented fin sidewalls commonly used in the actual FinFET platforms [1]. Ion beam milling is also reported as a possible technology option to improve smoothness of the silicon surface [6]. In this work, ion beam milling is utilized for improving smoothness of the (110) oriented fin sidewalls, and the effects on the electrical characteristics, i.e. V_t variability and the carrier mobility, are examined.

II. SAMPLE PREPARATION

Ion beam milling of the fin sidewall was implemented in a gate-first process flow of TiN-MG FinFETs (Fig. 1). Fin channels with (110) oriented sidewalls were fabricated from a nearly undoped SOI wafer using inductive-coupled plasma reactive ion etching (ICP-RIE). A part of the samples were processed by ion beam milling using the ion beam etching (IBE) apparatus [7] so that the 2 nm-thick Si is removed from each sidewalls. The IBE processed samples together with the control ones underwent gate stack formation with a thermal oxide and a poly-crystalline TiN-MG deposited by sputtering. After gate patterning and S/D doping, rapid thermal annealing (RTA) was carried out for dopant activation.

III. RESULTS OF CHARACTERIZATION

Morphology of the fin channels is compared between the IBE and control samples (Fig. 2). While the rough sidewall morphology is recognized for the control sample, the IBE sample shows significantly improved smoothness. Namely, the IBE is effective for smoothing the fin sidewalls as well as that reported for planar Si surface [6].

In the electrical characterization, we first compare I_d-V_g curves of the FinFETs having median V_t in a number of samples with identical design (Fig. 3). I_d-V_g curves, V_t and subthreshold slopes (SS) are almost identical for the control and IBE samples. Namely, the IBE process has no significant impact on the nominal characteristics. We next examine V_t variability by measuring standard deviation of V_t for various channel sizes as shown in the Pelgrom plot (Fig. 4), which is commonly used for benchmarking variability [2,3,5]. As clearly seen, slope of the Pelgrom plot is reduced by the IBE. Namely, ion beam milling of the fin sidewalls suppresses V_t variability effectively. Taking the morphology change (Fig. 2) into account, the improved smoothness of the fin sidewalls by the IBE contributes to improve uniformity of the TiN grain orientation, resulting in the suppressed V_t variability.

Finally, we examine the influence of the IBE process on the effective carrier mobility (μ_{eff}) which was evaluated by split-CV measurement (Fig. 4). The mobility curves are almost identical for the control and IBE processed samples. Namely, the IBE process does not degrade the carrier mobility.

IV. SUMMARY

Roughness of the fin sidewalls caused by the RIE for the fin formation is improved significantly by the ion milling process. Smoothing the fin sidewalls effectively suppresses V_t variability due to WFV without degradation of the carrier mobility.

REFERENCES

[1] C. Auth *et al.*, 2012 VLSI Tech., p.131.
[2] K. Kuhn *et al.*, IEEE T-ED, 58, p.2197, 2011.

[3] T. Matsukawa *et al.*, 2009 VLSI Tech., p.118.
[4] K. Endo *et al.*, IEEE EDL, 31, p.546, 2010.
[5] Y.X. Liu *et al.*, 2010 VLSI Tech., p.101.

[6] F. Frost *et al.*, Thin Solid Films 459, p.100, 2004.
[7] M. Ikeda *et al.*, 2014 IEEE Int'l Magnetics Conf., HQ-14.

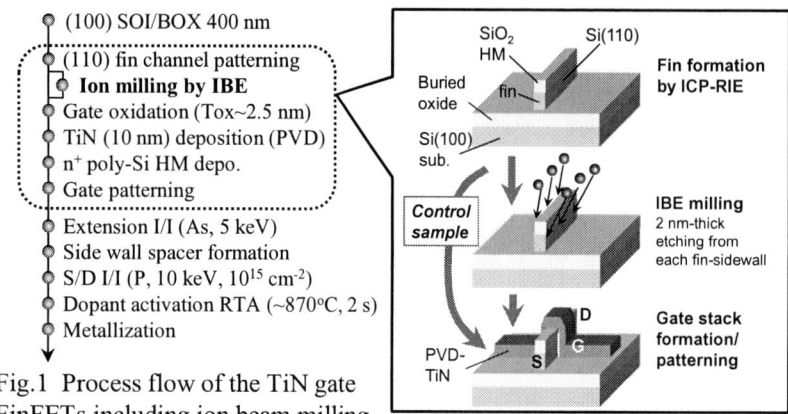

Fig.1 Process flow of the TiN gate FinFETs including ion beam milling.

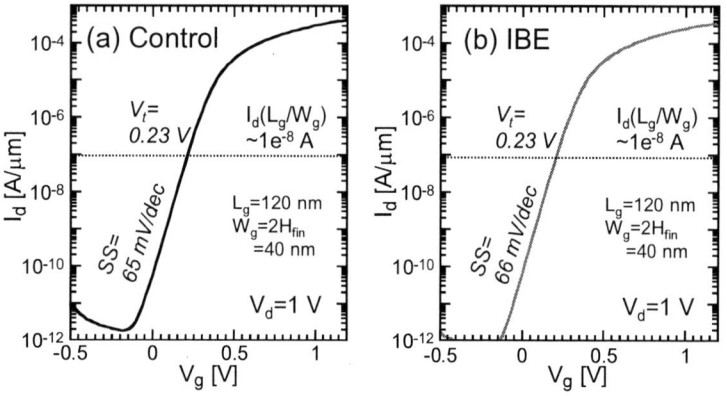

Fig.3 I_d-V_g characteristics of FinFETs with median V_t with designed L_g of 120 nm. (a) Control and (b) IBE sample. The IBE sample exhibits comparable V_t and subthreshold slope (SS) to the control sample.

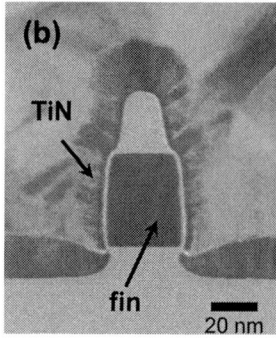

Fig.2 Cross sectional TEM of FinFETs with TiN metal gate. (a) Control and (b) IBE-processed sample. Morphology of the fin sidewall becomes smoother by the IBE milling.

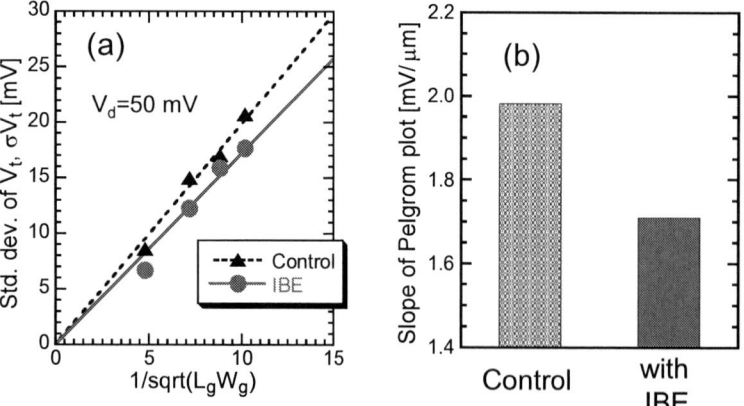

Fig.4 (a) Pelgrom plot and (b) its slope values representing the amount of V_t variability. The IBE milling process effectively reduces the V_t variability, thanks to smoothing of the fin-sidewall roughness caused by RIE for the fin formation.

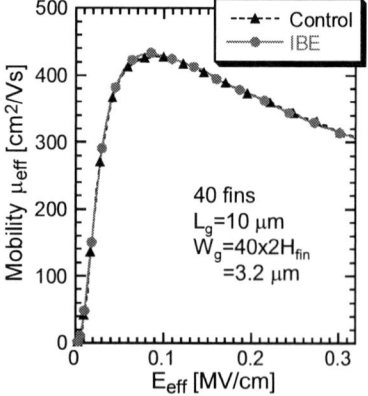

Fig.5 Electron mobility vs effective field obtained by split-CV measurement of the FinFETs. The mobility is kept identical through the IBE milling process.

sSOI Relaxation by BOX Creep Technique for Dual Strain CMOS Integration

A. Bonnevialle[1,2], C. Le Royer[2], Y. Morand[1], S. Reboh[2], D. Rouchon[2],
N. Bernier[2], B. Mathieu[2], C. Plantier[2] and M. Vinet[2]

[1]ST Microelectronics, 850 rue Jean Monnet 38926 Crolles, France
[2]CEA, LETI, MINATEC Campus, F-38054 Grenoble, France
Email: cyrille.leroyer@cea.fr

Abstract — **We report here on sSOI (strained Silicon On Insulator) strain relaxation at a local scale using a simple process based on BOX creep. This method consists in the transfer of SiN strain to a thin silicon layer during a high temperature anneal. This is allowed by the creep of the Buried Oxide (BOX). Thanks to Raman spectroscopy, the induced relaxation in the Si layer is determined. Using a tensile stressed SiN on + 1.4 GPa sSOI, we obtain around 60 % of strain relaxation.**

I. INTRODUCTION

The use of strained channels is an effective way to improve carriers mobility and furthermore advanced CMOS performances [1-3]. In the case of Fully Depleted SOI (FDSOI) for advanced technology nodes, strained channels can be obtained using sSOI substrate [4]. These wafers have been developed using SmartCut[TM] technology. The sSOI substrates exhibit a high level of stress (σ = + 1.4 GPa) offering significant nFETs performance boost: + 20% in the ON state current (I_{ON}) [5]. However, they are not suitable for pFETs which need compressive channels in order to improve holes mobility [3]. That is why a method able to locally relax the tensile strained SOI (for efficient dual strain CMOS) is required. In this work, we investigate the sSOI strain relaxation allowed by the creep of the buried oxide (BOX).

II. EXPERIMENTAL DETAILS

Starting from 200 mm sSOI wafers with a 14 nm thick strained Si on top of a 145 nm BOX, we have performed BOX creep [6] experiments with the process flow illustrated in **Fig.1.**: a) For the first step, a 10 nm thick layer of Silicon oxide deposition is performed. Then the + 0.9 GPa tensile silicon nitride is deposited by PECVD (plasma enhanced chemical vapor deposition); b) The second step consists in MESA patterning; c) Then a high temperature anneal at 1200 °C during few minutes under N_2 atmosphere is performed: the BOX creep occurs and the strain is transferred from the SiN to the underlying silicon layer; d) The final step consists in the selective SiN and SiO_2 layers removal using resp. H_3PO_4 and HF.

III. STRAIN CHARACTERIZATIONS

In this study, different recipe conditions are compared: SiN thickness (50 and 100 nm thick SiN with a stress of + 0.9 GPa) and the anneal thermal budget (2 and 4 min, 1200 °C). After BOX creep process, the 14 nm thick Si layer is characterized by Raman spectroscopy in order to extract the level of strain. Raman spectra are obtained using a Jobin Yvon T64000 triple monochromator. We have used UV-Raman spectroscopy using wavelength λ at 363 nm with an Ar+ laser [7, 8] to gain access to the strain and stress present in the Si layer. Strain and stress levels are obtained using the Raman frequency shift ($\Delta\omega$) extractions.

A. SiN thickness impact

Fig.2.a) shows a top view SEM micrograph of the 2x2 μm^2 MESA. **Fig.2.b)** compares both stress extractions (at the center of the pattern) with the two different SiN thicknesses after 1200 °C, 2 min anneal. The relaxation is around 15 % (σ = + 1.2 GPa) with a 50 nm thick SiN, whereas it reaches 35 % (that is to say σ = + 0.9 GPa) with 100 nm thick SiN: so twice more relaxation.

B. Annealing thermal budget impact

Fig.3 highlights the impact of the annealing thermal budget. 1200 °C 2 min anneal allows a 35% relaxation with a 100 nm nitride, whereas the Silicon layer is more relaxed after 3 min anneal at 1200 °C. With this recipe condition, the relaxation is around 60 % which corresponds to σ = + 0.6 GPa in the Si layer. This level of relaxation is confirmed by scanning Raman spectroscopy shown in **Fig.4.** Moreover, this stress profile highlights more relaxation at the edges than the center: around 70% (ie. σ = + 0.45 GPa). **Fig.5.** shows a) a Raman 2D mapping of patterns compared with b) simulations for a 100 nm nitride with 1200° C 2 min anneal: both are quite similar and exhibit the same edges effect. Finally, we have checked that the BOX creep process does not lead to crystalline defects generation: **Fig.6.** shows cross sectional STEM micrograph with no visible defects in the field.

CONCLUSIONS

We report on the successful local strain relaxation of sSOI using a new method based on BOX creep. The process steps have been detailed and analyzed using advanced characterizations for investigating the relaxation induced in the thin Si layer (Raman spectroscopy) and the crystallinity of the Si (STEM). With these very first experiments we were able to demonstrate a 60 % relaxation of sSOI tensile strain without any defects creation. This opens path for improved sSOI dual CMOS integration with tensile Si nFETs and relaxed Si or compressive SiGe pFETs.

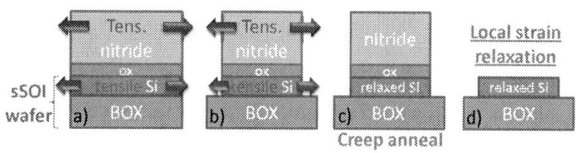

Fig.1: BOX creep process enabling local sSOI relaxation:
a) SiO₂ and tensile SiN deposition on sSOI; b) MESA patterning;
c) BOX creep anneal; d) SiN and SiO₂ selective etching.

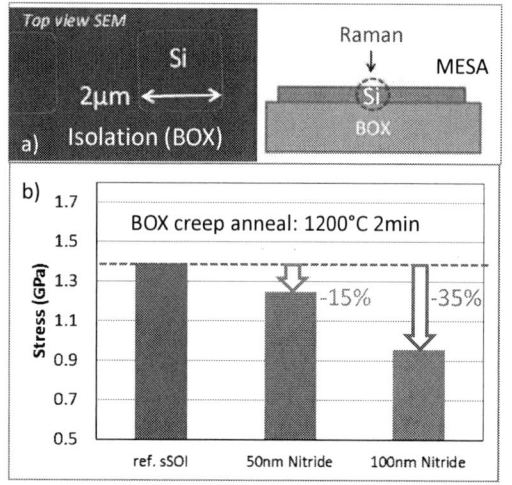

Fig.2: a) Top view SEM of the fabricated structures and corresponding schematic with the measurement site for Raman spectroscopy; b) Comparison of stress extractions (Raman) in the Si film after BOX creep process (Fig. 1) with 50 and 100 nm thick SiN.

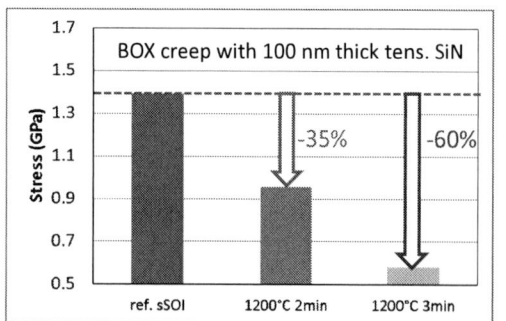

Fig.3: Impact of the thermal budget on the extracted stress (Raman) in the Si MESA after BOX creep process (with 100 nm tensile SiN).

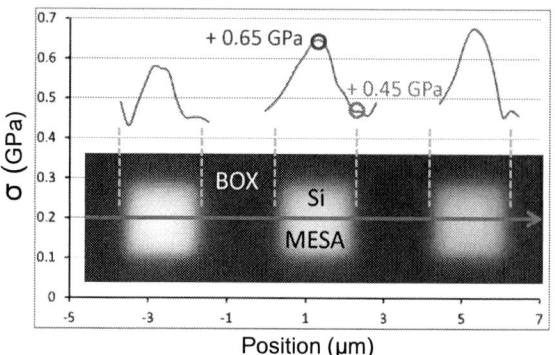

Fig.4: Scanning Raman results on 3 MESA patterns for 1200 °C, 2 min BOX creep process with a 100 nm nitride.

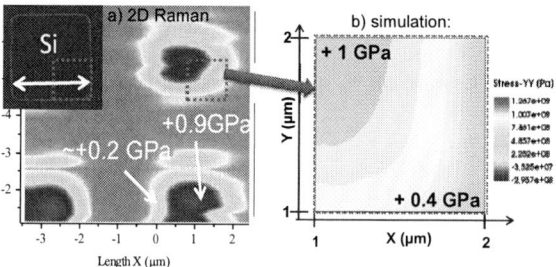

Fig.5: Comparison between a) experimental stress extractions (2D Raman mapping) and b) corresponding simulation (for BOX creep process with 100 nm nitride and a 1200 °C, 2 min anneal).

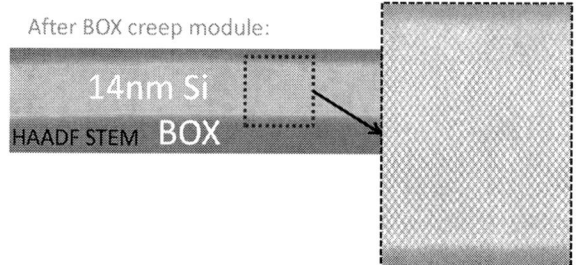

Fig.6: Cross-sectional HAADF STEM micrograph of 14 nm thick silicon layer on a scribe structure after sSOI BOX creep process.

ACKNOWLEDGMENT

This work has been pursued in the frame of the ST/IBM/LETI collaboration, with support of Reaching22 and Dynamic ULP projects. The ANR (French National Research Agency) is grandly acknowledged for its support in these developments in the frame of the Equipex FDSOI11 project.

REFERENCES

[1] N. Planes, *et al.*, VLSI tech. Symp, 2012, pp.133-134.
[2] O. Weber, *et al.*, VLSI tech. Symp, 2014, pp.16-17.
[3] B. de Salvo, *et al.*, IEDM 2014, pp. 172-175.
[4] W. Schwarzenbach, *et al.*, IEEE Int. Conference on Integrated Circuit Design and Technology (ICICDT), 2012, pp. 140-143.
[5] F. Andrieu, *et al.*, Proc. of ESSDERC, 2014, pp. 106-109.
[6] D. Chidambarrao, *et al.*, Patent US 2008/0169508 Al, 2008
[7] D. Rouchon, *et al.*, Journal of Crystal Growth 392 (2014) 66–7.
[8] J.-M. Hartmann, *et al.*, Thin Solid Films 516 (2008) 4238-4246.

Significance of Kinetic-linkage of Oxygen Vacancy with SiO$_2$/Si Interface for SiO$_2$-IL Scavenging in HfO$_2$ Gate Stacks

Xiuyan Li, Takeaki Yajima, Tomonori Nishimura, and Akira Toriumi

Department of Materials Engineering, The University of Tokyo, Tokyo 113-8656, Japan

Email: xiuyan@adam.t.u-tokyo.ac.jp

Abstract — **This work clarifies roles of substrate-Si in SiO$_2$-IL scavenging in HfO$_2$/SiO$_2$/Si gate stack experimentally, and points out a coupling effect of oxygen vacancy (V$_O$) with SiO$_2$/Si interface from thermodynamic viewpoint.**

I. INTRODUCTION

The mechanism of SiO$_2$-IL "scavenging" in HfO$_2$ gate stack needs to be understood for interface control as well as further EOT scaling [1]. The literatures so far reported have discussed it only from the viewpoint of V$_O$ transfer from HfO$_2$ into SiO$_2$ [2]. We recently found that substrate-"Si" also played a critical role in the scavenging experimentally [3, 4], as shown in **Fig. 1**. In fact, the scavenging does not occur on sapphire but does on SiC at higher temperature than on Si substrate [3]. The objective of this work is twofold. One is to clarify what the actual role of substrate-Si is as well as where Si disappears from SiO$_2$ in the scavenging. The other is to propose a physical model for the total process of scavenging.

II. RESULTS AND DISCUSSION

A. What is scavenging kinetics?

We experimentally reported V$_O$ diffusion from HfO$_2$ to SiO$_2$ [3]. The present work includes *two kinds of new experimental results*, concerning the role of Si in the substrate (Si^{sub}) and the movement of Si in SiO$_2$ (Si^{SiO2}).

i) What is the role of Si^{sub}?

To look into the possibility that Si^{SiO2} may grow on substrate or Si^{sub} be consumed in the scavenging, HfO$_2$/SiO$_2$ stacks on SiC was inspected. This is because 1) possibly grown Si^{SiO2} is distinguished from Si in SiC by XPS, 2) an inhomogeneous scavenging occurs on SiC at 1000°C with thicker HfO$_2$ and SiO$_2$ layers (**Fig. 2(a)/(c) and (b)/(d)**). The inhomogeneous scavenging is considered to be due to the facts that thicker HfO$_2$ is more easily crystallized [5], that the V$_O$ diffusion along inter-grains in polycrystallized HfO$_2$ is much faster than that through intra-grains [6], and that no silicidation occurs on SiC up to 1000°C [3]. TEM image in Fig. 2(d) shows that the bottom SiC interface keeps flat in both regions w/ and w/o SiO$_2$ scavenging. SiC is not consumed even at the region with SiO$_2$ scavenged completely. Moreover, after removing HfO$_2$ and SiO$_2$ by HF solution, SiC surface with the inhomogeneous scavenging (**Fig. 2(f)**) is as smooth as that of as-deposited one (**Fig. 2(e)**). No Si^{SiO2}

was detected on this surface by XPS. These results directly demonstrate that no consumption of the substrate and no growth of Si^{SiO2} on the substrate in scavenging. In other words, Si^{sub} has only a triggering effect on the scavenging without changing itself.

ii) Where do Si atoms go in the scavenging?

Now, the remaining issue is where Si^{SiO2} disappears. Since it does not grow on substrate, it should diffuse out from SiO$_2$-IL. However, we could not detect any Si-associated desorption by TDS (the mass spectroscopy in vacuum) measurement previously. This might be due to the fact that $m_{Si}=m_{CO}=m_{N2}$ and $m_{SiO}=m_{CO2}$, because both CO (CO$_2$) and N$_2$ in the chamber are likely to increase the background noise level. So, in this work all experiments were carried out using ^{29}Si and ^{18}O. **Fig. 3(a)** and **(b)** show SiO$_2$-IL scavenging results, and ^{29}SiO desorption peak from HfO$_2$/^{29}SiO$_2$/^{28}Si stack with sweeping the temperature. This is the first and direct evidence of out-diffusion of Si^{SiO2} from HfO$_2$ gate stacks. The fact that no Si^{18}O desorption is observed from Hf^{16}O$_2$/Si^{18}O$_2$/Si stack (**Fig. 3(c)**) suggests that atomic-Si^{SiO2} is diffused out from SiO$_2$ in the scavenging.

Thus, scavenging kinetics can be described as V$_O$ released from HfO$_2$ together with Si^{sub} triggers the SiO$_2$ scavenging, while generated Si^{SiO2} diffuses out form the HfO$_2$ surface, maintaining the initial substrate interface.

B. How is scavenging justified thermodynamically?

The decomposition of pure SiO$_2$ is not expected thermodynamically in the present condition [7] (**Fig. 4(a)**). In fact, no scavenging has been observed experimentally for ultra-thin SiO$_2$ on Si substrate without HfO$_2$ layer, even though Si-suboxide layer near SiO$_2$/Si interface might be energetically less stable than pure SiO$_2$. Note that no SiO$_2$ scavenging occurs without Si^{sub} as well. Thus, it is concluded that the kinetic linkage of V$_O$ from HfO$_2$ with SiO$_2$/Si interface provides the thermodynamic driving force for the scavenging, as schematically shown in **Fig. 4(b)**. The formula of scavenging reaction can be described as "SiO$_2$/Si + 2V$_O$ ↔ Si^{sub} + Si^{SiO2} ↑".

III. CONCLUSION

What occurs in SiO$_2$ scavenging in HfO$_2$ gate stack and how it occurs have been clarified experimentally and physically. The key is the kinetic-linkage of V$_O$ from HfO$_2$ with SiO$_2$/Si interface.

ACKNOWLEDGEMENT: This work was partly in collaboration with STARC in Japan. Xiuyan Li was grateful to the Marubun Research Fund and China Scholarship Council for financial support.
REFERENCES: [1] V. Misra *et al*, *Appl. Phys. Lett.*, **78**, 2001, 4166.

[2] T. Ando, *Materials,* **5,** 2012, 478. [3] X. Li *et al, Appl. Phys. Lett.* **105,** 2014, 182902. [4] X. Li *et al*, IEDM, 2014. 21.2. [5] S. Toyoda *et al*, J. Appl. Phys. **106,** 2009, 064103. [6] K. McKenna *et al*, Appl. Phys. Lett. **95,** 2009, 222111. [7] HSC Chemistry 6.

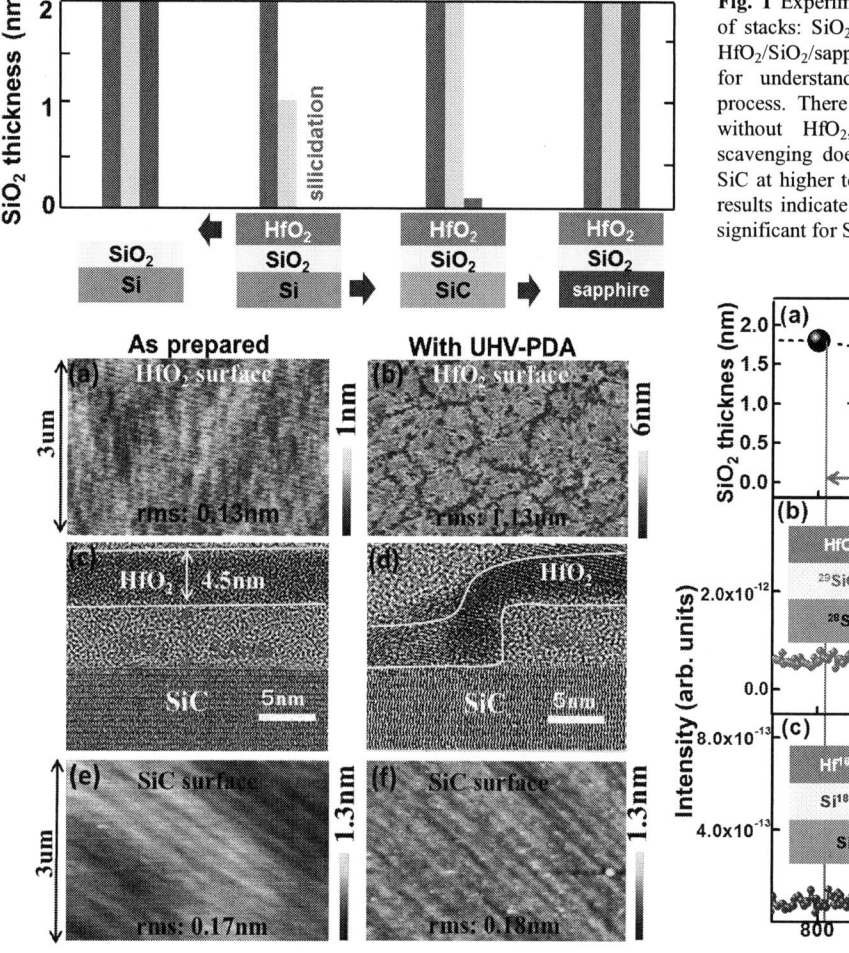

Fig. 1 Experiments of SiO_2 scavenging in four kinds of stacks: SiO_2/Si, $HfO_2/SiO_2/Si$, $HfO_2/SiO_2/SiC$ and $HfO_2/SiO_2/sapphire$ based on previous results [3, 4] for understanding the mechanism of scavenging process. There are two critical points to notice: 1) without HfO_2, no SiO_2 scavenging occurs; 2) scavenging does not occur on sapphire but does on SiC at higher temperature than on Si substrate. These results indicate both HfO_2 and Si in the substrate are significant for SiO_2 scavenging process.

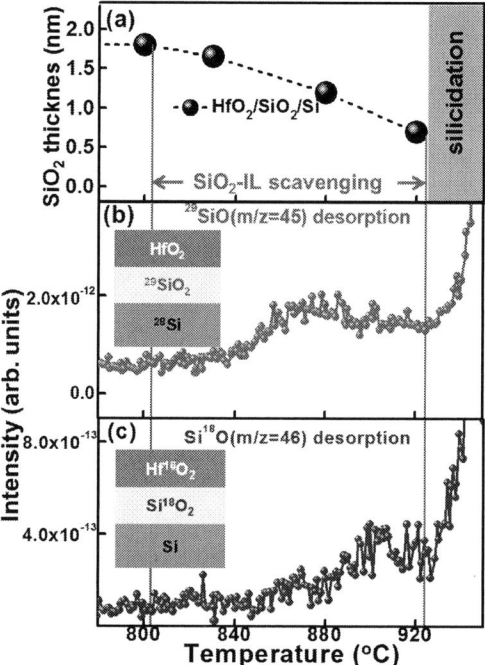

Fig. 2 (a)/(c) and **(b)/(d)** AFM/TEM images of $HfO_2/SiO_2/SiC$ stack as prepared and after UHV-PDA at 1000°C for 30min. **(e)** and **(f)** AFM images of SiC surface of the sample in Fig. 2(a)/(c) and (b)/(d) after removing HfO_2 and SiO_2 by HF solution. Inhomogeneous SiO_2 scavenging is observed while substrate is not affected by scavenging.

Fig. 3 (a) SiO_2-IL scavenging and **(b)** ^{29}SiO desorption of $HfO_2/^{29}SiO_2/^{28}Si$ stack. ^{29}SiO desorption peak was observed in temperature region of SiO_2 scavenging. **(c)** $Si^{18}O$ desorption of $Hf^{16}O_2/Si^{18}O_2/Si$ stack. No ^{18}SiO desorption was observed during SiO_2 scavenging.

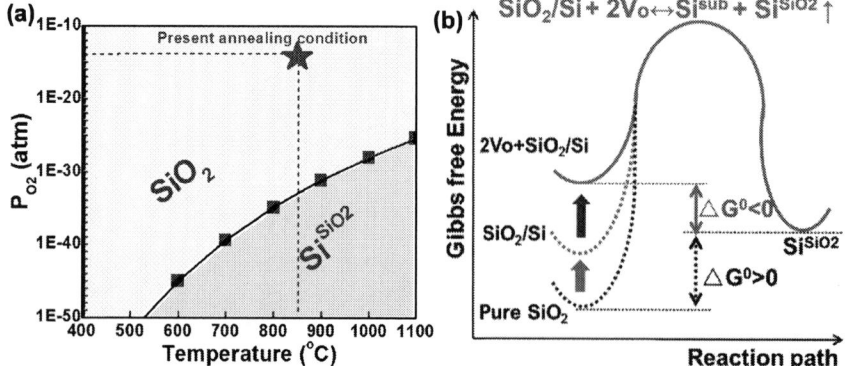

Fig. 4 (a) O_2 partial pressure as a function of temperature for conditions of SiO_2 decomposition [7]. Present condition is not enough to cause the decomposition of pure SiO_2. **(b)** Gibbs free energy description for the scavenging reaction. The kinetic-linkage of V_O from HfO_2 with SiO_2/Si interface should provide the driving force for SiO_2 scavenging.

14

Gate-stack engineering for self-aligned Ge-gate/SiO$_2$/SiGe-channel Insta-MOS devices

Wei-Ting Lai[1,2], Kuo-Ching Yang[2], Po-Hsiang Liao[2], Thomas George[3], and Pei-Wen Li[1,2]

[1] National ChiaoTung University, HsinChu, Taiwan, 300, Republic of China
[2] National Central University, ChungLi, Taiwan, 320, Republic of China
Email: pwli@nctu.edu.tw

Abstract — **We reported a first-of-its-kind, self-aligned gate-stack heterostructure of Ge-nanoshpere-gate/SiO$_2$/SiGe-channel on Si in a single-step approach through selective oxidation of a SiGe nano-patterned pillar over a Si$_3$N$_4$ buffer layer on Si substrate. Good tunability on the Ge-nanoshpere size, SiO$_2$ thickness, and SiGe-shell thickness provides a practically-achievable core building block for Ge-based metal-oxide-semiconductor (MOS) devices with size-tunable Ge gates, SiO$_2$ gate oxide, and SiGe channels. Detailed interfacial morphologies and structural properties between the Ge nanosphere/SiO$_2$ and SiO$_2$/SiGe-channel were examined using transmission electron microscopy, energy dispersive *x*-ray spectroscopy, and temperature-dependent high/low-frequency capacitance-voltage measurements. Both Al/SiO$_2$/Ge-nanospheres and NiGe/SiO$_2$/SiGe MOS capacitors exhibit quite low interface trap densities of 3–5×10^{11} cm^{-2}eV^{-1}, which is beneficial for advanced Ge MOS applications.**

I. INTRODUCTION

Ge continues to attract attentions in the arena of high-mobility materials for boosting the performance of CMOS transistors because of its superior intrinsic carrier mobility than Si. The key challenges for the fabrication of Ge CMOS transistors lie in the formation of gate dielectrics on Ge with satisfactory interfacial and electrical properties as well as the growth of Ge on Si heterostructures with sufficiently low defect densities [1-3]. Recently, we have demonstrated a unique approach to deliberately generate self-aligned SiO$_2$/Ge-nanosphere (NP)/SiO$_2$/SiGe-shell heterostructures over the Si substrate in a single fabrication step by means of the control available through lithographic patterning and selective oxidation of Si$_{1-x}$Ge$_x$ pillars over buffer layers of Si$_3$N$_4$ deposited over the Si substrates [4]. In this paper, we advanced this designer gate-stack structure for the production of Al-gate/SiO$_2$/Ge-channel and NiGe-gate/SiO$_2$/SiGe-channel MOS devices and demonstrated superior SiO$_2$/Ge and SiO$_2$/SiGe interface properties, which are beneficial and ready for Ge MOS applications.

Our unique, insta-MOS (i-MOS) gate-stack structure consists of SiO$_2$/Ge-NP/SiO$_2$/SiGe (Fig. 1), which is extendible for the fabrication of Al-gate/SiO$_2$/Ge-NP-channel or Ge-NP-gate/SiO$_2$/SiGe-channel MOS devices. In particular, the latter MOS structure is analogous to the prevailing poly-Si/SiO$_2$/Si one. Good size-tunablility over the Ge gates, SiO$_2$ gate oxide, and SiGe channel is achievable using the precise control available through our single-step oxidation process, which also effectively eliminates complicated surface treatment and cleaning processes. Detailed interfacial morphologies and structural properties between the Ge nanosphere and the Si substrate

were examined using high-resolution, cross-sectional transmission electron microscopy (CTEM) as well as energy dispersive *x*-ray (EDX) spectroscopy. Nickel germanidation process was also conducted on the Ge NP for the formation of self-aligned NiGe-gate/SiO$_2$/SiGe-channel capacitors.

II. EXPERIMENTAL WORKS

Process flow, schematic diagram and the corresponding TEM micrographs of Al/SiO$_2$/Ge-NP-channel and NiGe/SiO$_2$/SiGe-channel MOS capacitors are summarized in Fig. 1. Thermal oxidation converts a 30–230nm-diametered poly-SiGe nano-pillar over a buffer layer of Si$_3$N$_4$ into one single spherical, 20–90nm Ge NP penetrating Si substrate exactly below each oxidized pillar (Fig. 1). Between the Ge NP and the Si substrate there appears a 2.5 − 4nm-thick interfacial SiO$_2$ layer over a 3−15nm-thick Si$_{1-x}$Ge$_x$-shell ($x = 0.5−0.7$) with a "cup"-shape morphology that is conformal with the Ge NP and the Si substrate. The chemical purity of the Ge NP, thin interfacial layer of SiO$_2$ and Si$_{1-x}$Ge$_x$ shell has been verified by EDX examinations. Two gate metallization processes were conducted for forming Al gate and NiGe gate, respectively, either by a direct etching back the newly-formed SiO$_2$ layer over Ge NPs to suitable gate oxide thickness followed by the deposition of Al layers or by a completely removal of the newly-formed SiO$_2$ layer over Ge NPs followed by the deposition of Ni and rapid-thermal annealing at 400 °C. Thereby, two MOS configurations of Al-gate/SiO$_2$/Ge-NP-channel and NiGe-gate/SiO$_2$/SiGe-channel capacitors are produced (Fig. 1(b)).

III. RESULTS AND DISCUSSION

It is instructive to note that the thin, interfacial SiO$_2$ gate-oxide layer between the Ge QD and the SiGe shell is formed during this single oxidation step. In order to create this oxide, we successfully exploited our recently discovered phenomenon of controllable thermal oxidation of Si interstitials released from the Si substrate. The thickness of this interfacial SiO$_2$ layer is less than 4nm regardless of the size of the QD or the oxidation time because of an exquisitely-controlled dynamic balance between the fluxes of oxygen and silicon interstitials.

Figure 2 displays high/low-frequency capacitance-voltage (*C-V*) characteristics of , respectively, measured at 77–300K. For Al-gate/SiO$_2$/Ge-NP capacitors, a large negative flat-band voltage (V_{fb}) shift is observable for both n- and p-type devices, suggesting the presence of positive fixed charges at the interface of thermally grown SiO$_2$/Ge NPs. There appears to be no observable frequency dispersion and humps for p-MOS capacitors, whereas apparent weak-inversion humps for n-MOS devices are induced due to faster capture and emission of carriers in small-bandgap Ge [5]. Extracted interface trap density (D_{it}) exhibits a typical U-shaped distribution profile across the energy band gap with a minimal D_{it} of ~3×10^{11} cm^{-2}eV^{-1} for Al/SiO$_2$/Ge capacitors (Fig. 2(b)), suggesting a device-quality interface being obtainable for the SiO$_2$/Ge NP system.

In order to fully exploit competence of our insta-Ge MOS structures, Ge-NP-gate/SiO$_2$/SiGe-channel capacitors are favorable. Notably, a low D_{it} of ~4.5×10^{11} cm^{-2}eV^{-1} is measured on NiGe/SiO$_2$/SiGe capacitors, suggesting both high-quality SiO$_2$/Ge NPs and SiO$_2$/SiGe-shell interface properties are achievable based on our Insta-MOS structures. Further device structure design and process development including thermal oxidation, post-oxidation annealing, and etching back are undergoing for the optimal performance of Ge n- and p-MOSFETs.

IV. CONCLUSION

We reported the first-of-its-kind, unique CMOS-compatible approach for generating a self-aligned, gate-stacking MOS heterostructure based on an insta-MOS structure of SiO$_2$/Ge-NP/SiO$_2$/Ge-shell structure that it is nanofabricated in a single oxidation step of SiGe nano-pillars lithographically patterned over a buffer Si$_3$N$_4$ layer on the Si substrate. Good interfacial and electrical properties of low D_{it} 3–5×10^{11} cm^{-2}eV^{-1} are measured on both Al/SiO$_2$/Ge-NP and NiGe/SiO$_2$/SiGe-shell systems. We believe that our self-aligned SiO$_2$/Ge-QD/SiO$_2$/SiGe-shell heterostructure is a very promising, "foundational" candidate not only for its superior performance but also for the simplicity and elegance of its one-step fabrication process, and envisaged further scientific exploration of this heterostructure for the demonstration of advanced Ge-gate/SiO$_2$/Ge-channel Insta-MOS devices.

ACKNOWLEDGEMENT

This work was supported by the Ministry of Science and Technology of R. O. C. (MOST102-2221-E-009-195-MY3).

REFERENCES

[1] D. P. Brunco et al., *J. Electrochem. Soc.*, **155**, H552 (2008).
[2] A. Dimoulas et al., *Thin Solid Films*, **515**, 6337 (2007).
[3] M. Caymax et al., *Mater. Sci. and Eng. B*, **135**, 256 (2006).
[4] C. Y. Chien et al., *Nanotechnology*, **22**, 435602 (2011).
[5] K. Martens et al., *IEEE TED*, **55**, 547 (2008).

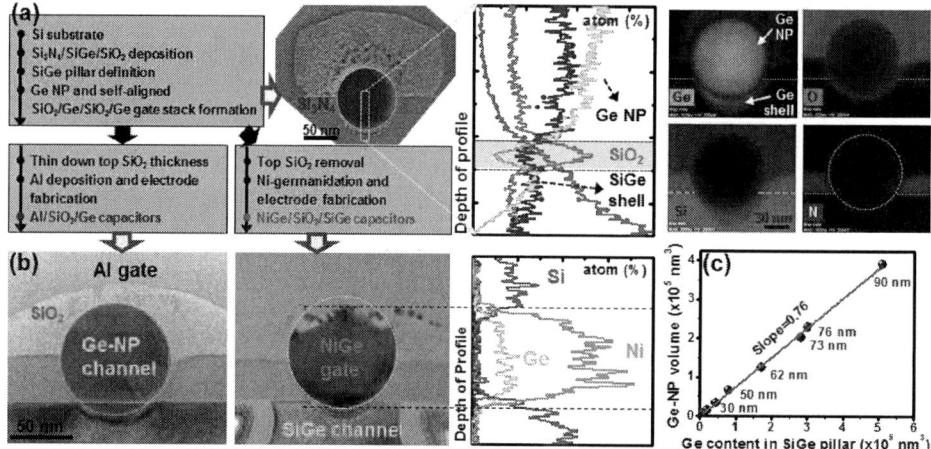

Figure 1 (a) Nano-fabrication procedures, TEM image and EDX results of a SiO$_2$/Ge-NP/SiO$_2$/SiGe-shell heterostructure on Si substrate. (b) Configurations of Al/SiO$_2$/Ge-NP-channel and NiGe/SiO$_2$/SiGe-channel MOS capacitors. (c) Controllability of Ge-NP/SiGe-shell size through total Ge content in SiGe pillar before thermal oxidation.

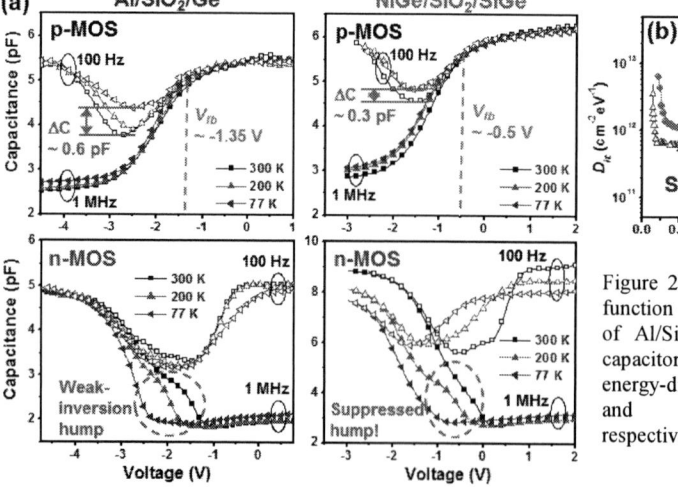

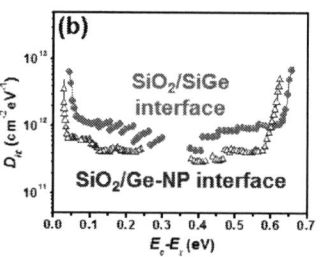

Figure 2 (a) *C-V* characteristics as a function of frequency and temperature of Al/SiO$_2$/Ge and NiGe/SiO$_2$/SiGe capacitors and (b) the corresponding energy-distributed D_{it} for SiO$_2$/Ge-NP and SiO$_2$/SiGe interfaces, respectively.

Impact of H_2, O_2, and N_2 anneals on atomic-scale surface flattening for 3-D Ge channel architecture

Yukinori Morita, Hiroyuki Ota, Meishoku Masahara, and Tatsuro Maeda

Nanoelectronics Research Institute (NERI), National Institute of Advanced Industrial Sciences and Technology (AIST)
Tsukuba West, 16-1 Onogawa, Tsukuba, Ibaraki 305-8569, Japan
Phone: +81-29-861-2433, E-mail: y.morita@aist.go.jp

Abstract

We evaluated atomic-scale surface morphology changes of Ge wafers after thermal flattening in H_2, O_2, and N_2 gas ambient. Analysis of atomic force microscopy results revealed that the atomic steps and terraces of Ge were evident after annealing at over 500°C for 160 s. The atomistic surface morphologies varied with the different annealing atmospheres. Nearly uniform step-terrace structure due to the suppression of terrace etching was obtained on N_2-annealed Ge surfaces.

Introduction

Due to the recent slowdown of the Si metal–oxide–semiconductor field-effect transistor (MOSFET) scaling trend, replacement of Si with high mobility channel materials is an attractive approach for increasing the MOSFET drive current. Ge is regarded as a candidate next-generation channel material because of its higher electron and hole mobilities compared to Si. Moreover, after many years of SiGe alloy studies, integration of Ge into the Si platform is easier than for the other novel channel materials such as the III–V semiconductors.

One of the most serious drawbacks of high mobility channel MOSFETs is the larger off current. In the case of the scaled Si MOSFET, 3-D channel architectures such as Fin- or nanowire-FETs can effectively suppress the off current that is caused by the "short channel effect." [1] Similar to the Si case, 3-D Ge MOSFET has been already considered to switch from 3-D Si channels to 3-D Ge channels.

In a 3-D Ge channel, the Ge surface prepared by reactive ion etching should be a channel interface. Hence, flattening of such a damaged and roughened surface is the key to reducing carrier scattering and enhancing the mobility for FET operation. Hydrogen annealing has been reported to reduce the surface roughness not only for Si but also for Ge surfaces. [2,3] In order to apply this type of thermal process to the nanometer-scale 3-D Ge channel, thermal etching or roughening behaviors by ambient gases should be considered together with the surface flattening.

In this study, we perform H_2, O_2, and N_2 anneals of Ge surfaces, and analyze the surface structures. Consequently, the atomistic morphologies vary for different annealing atmospheres. Based on the analysis, the origin of atomistic morphology changes of the Ge surfaces is discussed.

Experimental procedure

For flattening of planar surfaces, p-type Ge(001) and (111) wafers are used. The sacrificial oxide layers are etched using a low pH (HF:HCl=1:20) treatment. [4,5] After the wet treatment, the specimens are introduced into the vacuum chamber and annealed using a halogen-lamp-heater with rapid-thermal annealing (RTA) sequence, under purified gas flow. The pressures of purified gases are of the order of a few Pa for H_2 and N_2, and 2×10^{-4} Pa for O_2. The annealing duration is 160 s. After annealing, the surfaces are analyzed using stand-alone tapping-mode atomic force microscopy (AFM) in air atmosphere.

Results and discussion

Figure 1 summarizes the surface morphology changes due to thermal flattening in three different gas ambients. Panels (a) and (j) show the AFM topography of Ge (001) and (111) surfaces after the removal of the sacrificial oxide. No step-like features can be recognized. As shown in Figure 2, due to the thermal treatment at over 500°C, surface roughness values decrease. In addition, step-terrace features reflecting the crystallographic structure of Ge surface become more apparent. However, the atomistic features of the surface morphology vary with the different annealing atmospheres: in the case of the H_2 anneal [Fig. 1 (b)–(c), Ge(001)], the step-terrace structure is obvious but a clear "step-flow" feature represented by a uniform terrace width is not obtained. In the case of the O_2 anneal [Fig. 1 (d)–(f), Ge(001)], the terrace is relatively wide even for the 500°C anneal. However, the step edge shows a corrugated shape. The surfaces annealed in N_2 show step-flow features, and the step edges are relatively smooth in the 640°C anneal that seems to be the best surface of the three types of gas anneal. [Fig. 1 (g)–(i), Ge(001)] The clear step-terrace feature is also obtained on Ge(111) surface annealed at 500°C in N_2 [Fig. 1 (k)], but the surface is roughened over 640°C anneal.

The mechanism of the atomistic surface flattening can be explained on the basis of these AFM observations, as shown in Figure 3. Thermal activation of Ge migration is an intrinsic driving force for flattening. In addition, Ge sublimation and surface etching by ambient gases can also impact surface morphologies. The corrugated step edges in the O_2-annealed surfaces clearly indicate the effect of oxygen etching (active oxidation). Figure 4(a) shows an AFM image (air view) of the etch depth by O_2 anneal at 640°C, with the preparation procedure for making the etch depth evident by using a mask-rebate technique shown in Figure 4(b). [6] 0.5–1.0 nm/min etching by O_2 anneal is confirmed. This means that at a very low (2×10^{-4} Pa) O_2 pressure, O_2 etching can be controlled for the 3-D Ge channel etching process. In addition, at 500°C, O_2 etching effectively reduces surface roughness; this is promising for low temperature flattening. These process temperatures for flattening and etching can be lower compared with those for the Si nanochannel process. [7] Figure 5 shows an AFM image of Ge(001) surface annealed in O_2 pressure of 2×10^{-3} Pa at 640°C. Cluster-like islands grow on the Ge surface, indicating that the beginning of oxidation. Therefore, it is important to reduce ($< 10^{-3}$ Pa) oxygen partial pressure during the thermal flattening of Ge surface. For both H_2 and N_2 anneals, O_2 etching is suppressed but morphologies are different. Etching by H_2 molecules, or suppression of Ge migration could be suggested. N_2-annealed Ge(001) surface shows clear step-flow feature and no etch pit in the terraces and therefore appears to be advantageous for preparing an atomically-flat channel interface without Ge etching.

Summary

We evaluated atomic-scale surface morphology changes of Ge wafers after thermal flattening by H_2, O_2, and N_2 gas ambient. The AFM analysis of the surfaces revealed that the atomic steps and terraces of Ge were evident after annealing at over 500°C. The atomistic morphologies vary with different annealing ambient. At low O_2 pressure (2×10^{-4} Pa) conditions, the etching by O_2 is controllable and can be applied to the 3-D Ge channel process. Nearly uniform step-terrace structure was obtained on N_2-annealed Ge surfaces. This is realized by suppressing O_2-etching with a purified gas-flow condition.

Acknowledgement

The authors express our thanks to members of GNC high-mobility CMOS team for their useful discussions.

References

[1] S. Veeraraghavan, and J.G. Fossum, IEEE Trans. Electron Devices 36 (1989) 522.
[2] T. Nishimura, et al., Appl. Phys. Express 7 (2014) 051301.
[3] T. Nishimura, et al., Appl. Phys. Express 5 (2012) 121301.
[4] Y. Morita, and M. Nishizawa, Appl. Phys. Lett. 86 (2005) 171907.
[5] Y. Morita, and H. Ota, Appl. Phys. Lett. 100 (2012) 261605.
[6] Y. Morita, et al., Surf. Sci. 604 (2010) 1432.
[7] Y. Morita, et al., Silicon Nanoelectronics Workshop (SNW), 2010, 25.

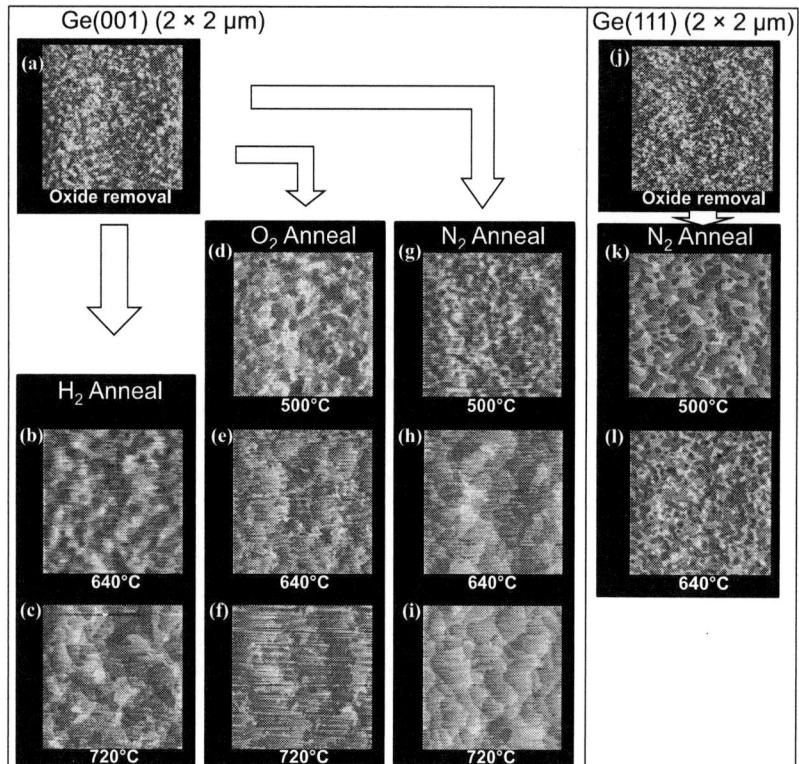

Fig. 1 Tapping-mode AFM images (2 × 2 μm) of Ge(001) [(a)–(i)] and (111) [(j)–(l)] surfaces obtained in air atmosphere. (a) and (j) show surfaces after removal of sacrificial oxide by HF:HCl =1:20 solution. (k) and (l): In the case of N_2-annealed Ge(111), the number of small islands and pits increased on the terrace for the 640°C anneal, while clear step-terrace is recognized for the 500°C anneal. Acceleration of sublimation of Ge adatoms on Ge(111) surface is suspected.

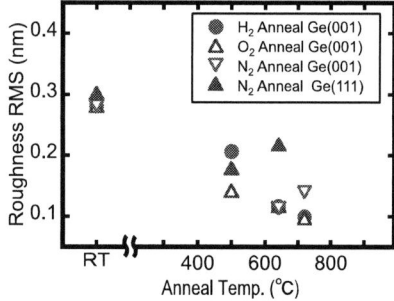

Fig. 2 Variation of RMS roughness of Ge surfaces measured by AFM.

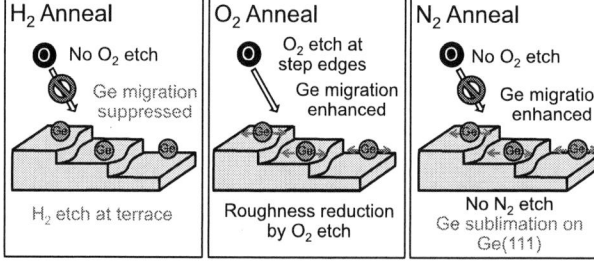

Fig. 3 Schematics explaining origins of atomistic surface morphologies for H_2, O_2, and N_2 annealed Ge surfaces.

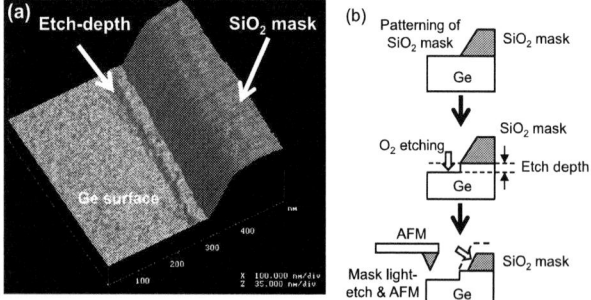

Fig. 4 (a) Tapping-mode AFM image (air view, 2 × 2 μm) of etch depth by O_2 etching in 2×10^{-4} Pa O_2 at 640°C. The surface is prepared by using mask-rebate technique. (b) Surface preparation procedure (mask-rebate technique) to make the etch-depth obvious. [6]

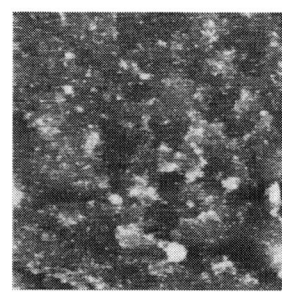

Fig. 5 Tapping-mode AFM image (2 × 2 μm) of Ge(001) surface annealed in 2×10^{-3} Pa O_2 at 640°C. Cluster-like oxide islands are formed on flat terraces. RMS roughness is 0.34 nm.

Experimental Study of Reliabilities in Tri-gate Nanowire Transistor
~What is Main Reliability Issue in 3D Transistor?~

Kensuke Ota, Chika Tanaka, Daisuke Matsushita, Toshinori Numata, and Masumi Saitoh

Advanced LSI Technology Laboratory, Corporate R&D Center, Toshiba Corporation, 1 Komukai-Toshiba-cho, Saiwai-ku,
Kawasaki 212-8582, Japan, Phone: +81-44-549-2192, E-mail: kensuke.ota@toshiba.co.jp

Abstract

We experimentally study the various reliabilities in tri-gate nanowire transistors (NW Tr.) such as negative bias temperature instability (NBTI), hot carrier injection (HCI), and random telegraph noise (RTN). Those are crucial reliability issues for low power applications. NBTI in narrower width was enhanced because of the NW corner effects, whereas RTN amplitude can be well fitted to the conventional size dependence reported in the planar Tr. HCI and NBTI induced additional carrier traps leading to the increase in the number of observed RTN. Moreover, RTN amplitude was enhanced by HCI and NBTI and enhancement becomes larger in narrower width.

1. Introduction

The progress of LSIs has been accomplished by the down-scaling of the components such as MOSFETs since the beginning of early 1970s. Recently, the conventional planar transistor has been encountered the limit of scaling because off-state leakage current (I_{off}) severely increased due to the short channel effect. Nanowire (NW) transistors have been attracted much attention as promising candidates to solve this problem. Since NW channel is surrounded by the gate, strong top gate controllability can be obtained which enables the suppression of I_{off}. In order to put NW Tr. into practical use, the reliability needs to be clarified. For low power applications, crucial reliability issues are NBTI, HCI, and RTN so that we experimentally investigated these reliabilities in NW Tr. [1,2,3]. Furthermore, RTN after HCI and NBTI were studied.

2. Enhanced degradation by NBTI in NW Tr.

Tri-gate NW Tr. was fabricated on a non-doped 300 mm SOI wafer. Fig. 1 shows the schematic and TEM images of NW Tr. Gate stacks consist of poly-Si gate and thermal SiO_2. In order to reduce the parasitic resistance, source and drain were elevated by Si-epi growth with thin sidewall spacer of 10 nm [4]. NBTI was investigated by measuring the threshold voltage shift (ΔV_{th}) in I_d-V_g characteristics after the constant stress gate bias was applied in pMOS NW Tr. Fig. 2 shows stress time dependence of ΔV_{th} in NW Tr. with various NW widths (W) and that in planar SOI Tr. fabricated in the same wafer. ΔV_{th} in NW Tr. is larger than that of planar SOI Tr., and ΔV_{th} becomes larger as W decreases. In order to clarify the origin of enhanced degradation by NBTI in NW Tr. with narrower W, side surface orientation dependence was measured, as shown in Fig. 3. Although degradation of NBTI in bulk-planar Tr. with (110) oriented channel was reported to be larger than that with (100) channel [5], ΔV_{th} in NW Tr. is independent of side surface orientation. Fig. 4 shows NW height (H) dependence of ΔV_{th}. As H decreases, ΔV_{th} increases. These results suggest that the enhanced degradation in NW Tr. with narrower W is not responsible for the NW side surface since side surface contribution decreases as H decreases. Fig. 5 shows ΔV_{th} plotted against circumference ($W_{eff}=W+2H$). Regardless of the ratio between W and H, ΔV_{th} is on a single curve suggesting that local degradation in NW channel was enhanced leading to the larger NBTI. It can be considered that electric field concentration at NW corner is the origin of the enhanced degradation. As shown in Fig. 6, simulation also supported that the electric field

concentration exists at NW corner.

3. Size dependence of RTN in NW Tr.

RTN is caused by the carrier trap and detrap resulting in the dynamic drain current fluctuations. Fig. 7 shows the typical observed RTN, histograms of RTN, and schematics of the capture/emission of carrier trap in case of one trap and two traps. We determined the amplitude of RTN (ΔI_d) with the histograms. Fig.8 shows the W dependence of threshold voltage fluctuation ($\Delta V_{th}=\Delta I_d/g_m$), where $g_m=dI_d/dV_g$. ΔV_{th} increase with the decrease in W. As shown in Fig.8(b), ΔV_{th} is inversely proportional to the one-half power of circumference ($W+2H$) rather than that of W, indicating that even in small diameter NW Tr. down to 12.5 nm, noise increase can be explained by the reduction of channel surface area. Fig.9 shows ΔV_{th} in <110> and <100> oriented NW Tr. Side surface orientation is not responsible for ΔV_{th} of RTN which is similar to $1/f$ noise [6]. In Fig.10, all measured ΔV_{th} with different W, L, H, and channel directions are plotted against $1/\{L(W+2H)\}^{0.5}$. ΔV_{th} is located on a single universal line irrespective of W, L, H, and side surface orientations. These results suggest that ΔV_{th} of RTN in NW Tr. even in small diameter down to 12.5nm can be well explained by conventional carrier number fluctuations.

4. RTN after HCI and NBTI in NW Tr.

We studied RTN after HCI in nFET and that after NBTI in pFET. Fig.11 shows an example of RTN before and after HCI. Clear RTN behavior was observed after RTN and the amplitude of observed RTN becomes larger as stress time increases. Fig.12 shows ΔV_{th} distributions before and after HCI. ΔV_{th} was enhanced by HCI. Not only the number of traps but also median ΔV_{th} increase as the stress time increases (Fig.13), suggesting that the traps induced by HCI generate larger RTN amplitude. Fig.14 shows W dependence of ΔV_{th} of RTN before and after HCI. ΔV_{th} enhancement by HCI increases with the decrease in W. RTN after NBTI showed the same behavior with that after HCI. In summary, as W decreases, along with enhanced degradation by HCI and NBTI, RTN induced by HCI or NBTI becomes quite significant. Since the larger degradation by HCI or NBTI in NW Tr. with narrower W is attributed to the electric field concentration at the channel corner, induced traps are supposed to be corrected at the channel corner which generates larger RTN.

5. Conclusion

NBTI and RTN were measured in NW Tr. Degradation by NBTI were found to be enhanced in NW Tr. with narrower W due to the corner effect. On the other hand, size dependence of RTN amplitude was inversely proportional to the one-half power of circumference suggesting the conventional carrier number fluctuations and no quantum confinement effect was observed. HCI and NBTI induced additional RTN and noise amplitude of induced trap was larger in smaller W due to the traps at channel corner. Therefore, not only gate dielectric improvement to suppress the initial RTN but also channel corner treatment to reduce RTN after HCI or NBTI degradation are needed in scaled NW Tr.

Acknowledgment This work was partly supported by NEDO's Development of Nanoelectronic Device Technology.

References [1] K. Ota et al., JJAP 51, 02BC08 (2012). [2] K. Ota et al., JJAP 53, 04EC13 (2014). [3] K. Ota et al., VLSI2014, p.144. [4] M. Saitoh et al., IEDM2010, p.780. [5] K. Ota et al., Appl. Phys. Let. 100, (2012), 212109. [6] M. Saitoh et al.,VLSI2013, p.228.

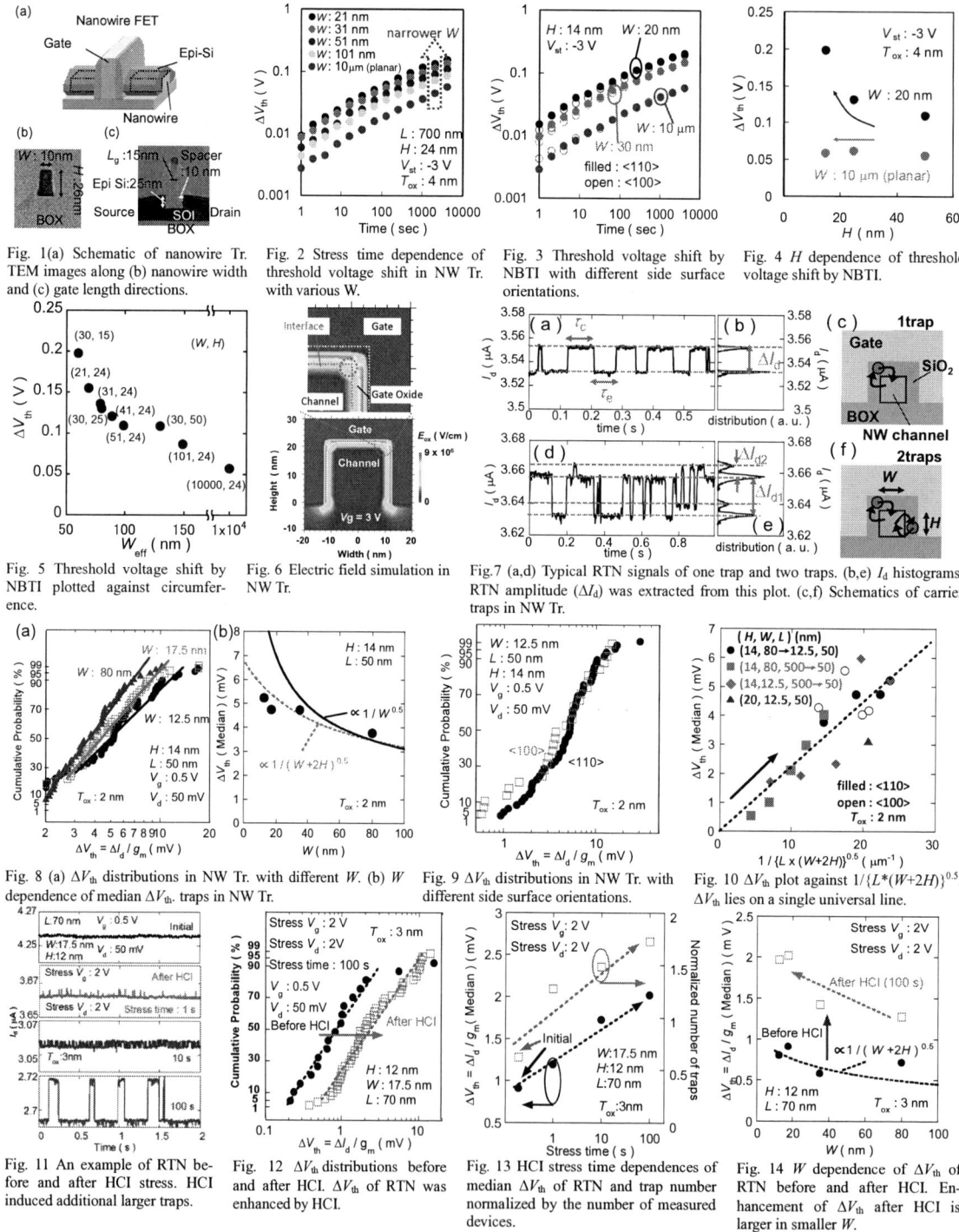

Fig. 1(a) Schematic of nanowire Tr. TEM images along (b) nanowire width and (c) gate length directions.

Fig. 2 Stress time dependence of threshold voltage shift in NW Tr. with various W.

Fig. 3 Threshold voltage shift by NBTI with different side surface orientations.

Fig. 4 H dependence of threshold voltage shift by NBTI.

Fig. 5 Threshold voltage shift by NBTI plotted against circumference.

Fig. 6 Electric field simulation in NW Tr.

Fig.7 (a,d) Typical RTN signals of one trap and two traps. (b,e) I_d histograms. RTN amplitude (ΔI_d) was extracted from this plot. (c,f) Schematics of carrier traps in NW Tr.

Fig. 8 (a) ΔV_{th} distributions in NW Tr. with different W. (b) W dependence of median ΔV_{th}. traps in NW Tr.

Fig. 9 ΔV_{th} distributions in NW Tr. with different side surface orientations.

Fig. 10 ΔV_{th} plot against $1/\{L*(W+2H)\}^{0.5}$. ΔV_{th} lies on a single universal line.

Fig. 11 An example of RTN before and after HCI stress. HCI induced additional larger traps.

Fig. 12 ΔV_{th} distributions before and after HCI. ΔV_{th} of RTN was enhanced by HCI.

Fig. 13 HCI stress time dependences of median ΔV_{th} of RTN and trap number normalized by the number of measured devices.

Fig. 14 W dependence of ΔV_{th} of RTN before and after HCI. Enhancement of ΔV_{th} after HCI is larger in smaller W.

Threshold Voltage and Current Variability of Extremely Narrow Silicon Nanowire MOSFETs with Width down to 2nm

T. Mizutani, Y. Tanahashi, R. Suzuki, T. Saraya, M. Kobayashi, and T. Hiramoto

Institute of Industrial Science, The University of Tokyo

4-6-1 Komaba, Meguro-ku, Tokyo 153-8505, Japan, Phone: +81-3-5452-6264, E-mail: mizutani@nano.iis.u-tokyo.ac.jp

Abstract

The threshold voltage (V_{TH}) and on-current (I_{ON}) variability of extremely narrow silicon nanowire channel FETs is intensively measured and statistically analyzed. It is found that the Pelgrom coefficient (A_{VT}) of 7nm-wide nanowire FETs is much smaller than that of FDSOI FETs, while A_{VT} rapidly increases as the nanowire width decreases down to 2nm. The increase in variability is ascribed to V_{TH} fluctuations due to the quantum confinement effect induced by nanowire width fluctuations.

Introduction

A silicon nanowire channel FET has attracted attention as a promising device for near-future LSIs [1]. The short-channel effect can be suppressed in a nanowire channel FET with a three-dimensional structure due to the superior electric field controllability by the gate electrode. However, nanosize transistor scaling causes the severe increase in the characteristic variability [2-4]. Although, the variability of 10nm-wide-scale nanowire FETs has been reported [2-4], the detailed statistical analysis of nanowire FETs with width less than 5nm has not done yet.

In this work, the characteristic variability in extremely narrow silicon nanowire FETs is intensively measured and statistically analyzed. The minimum nanowire width is 2nm. It is clarified that V_{TH} and I_{ON} variability increase rapidly as the width decreases due to the quantum confinement effect.

Measurements

Intrinsic channel gate-all-around (GAA) silicon nanowire nFETs with (110)-directed channel were fabricated on (001) SOI substrate. Detailed fabrication process is described in [5]. The nanowire channel height (H) is 3nm, and the channel length (L) is 100nm. The estimated channel width (W_e) is varied from 2nm to 7nm. In this study, 170 nanowire nFETs for each W_e were measured. Fig.1 compares measured I_{ds}-V_{gs} characteristics of nanowire nFETs with W_e=2~7nm at V_{ds}=50mV. As W_e decreases, V_{TH} variability as well as drain current variability increases.

V_{TH} Variability

Figs.2-3 show cumulative distributions of V_{THC} and sub-threshold swing (SS) with various W_e, respectively, where V_{THC} is V_{TH} defined by subthreshold constant current (I_0=1×10^{-9}A) and SS is defined at I_0. As W_e decreases, V_{THC} variability as well as SS variability increases. V_{THC} distribution for W_e>4nm almost lies on a straight line and hence follows a normal distribution, while V_{THC} distributions for narrower W_e apparently deviate from the normal distribution.

Fig.4 shows average V_{THC} and σV_{THC} as function of W_e. The average V_{THC} increases as W_e decreases. This V_{THC} increase originates from by the quantum confinement effect [7].

V_{THC} is very sensitive to nanowire width especially in nanowire FETs, indicating that slight line width roughness (LWR) will cause large V_{THC} fluctuations in extremely narrow nanowire FETs.

Fig.5 shows the Pelgrom plot of σV_{THC}. Here, the effective width (W_{eff}=2W_e+2H) is defined since the device has GAA structure, and the horizontal axis is (LW_{eff})$^{-1/2}$. σV_{THC} in bulk, FDSOI, nanowire FETs in the literature [3-4, 6] is also plotted. A_{VT} for W_e=7nm is 0.69mVμm and is as small as that of nanowire FETs in [3]. It is interesting to note that nanowire FETs has smaller A_{VT} than FDSOI due to the aligned grain boundary formation of poly-gate in nanowire FETs [3].

However, nanowires for W_e<7nm deviate from the straight line. If we assume that the deviation from the straight line of A_{VT}=0.69mVμm is caused by the LWR-induced quantum confinement effect only, the channel width fluctuation is estimated to be σ~ 0.25nm.

I_{ON} variability

Fig.6 shows cumulative distributions of I_{ON} at V_{gs}=1.5V and V_{ds}=50mV with various W_e and correlation between I_{ON} and V_{THC}. I_{ON} distributions almost follow a normal distribution. The correlation between I_{ON} and V_{THC} is poor in wider FETs, while I_{ON} apparently depends on V_{THC} in narrower FETs. In Fig.7, the I_{ON} distributions and the correlation at constant overdrive voltage (V_{over}=V_{gs}−V_{THC}=1.0V) are plotted in order to remove the effect of V_{THC} variability. Both I_{ON} variability and V_{THC} dependence still remain especially for W_e=2nm.

Fig.8 shows Pelgrom plot of $\sigma I_{ON}/\mu I_{ON}$ at V_{over}=1.0V. It is found that, contrary to V_{THC}, $\sigma I_{ON}/\mu I_{ON}$ of FETs with $W_e \geq$4nm lie on a straight line and the slope (0.2%μm) is much smaller than that in [3] (0.6%μm) and as small as that in FDSOI [8]. On the other hand, $\sigma I_{ON}/\mu I_{ON}$ rapidly increases as W_e decreases for W_e<4nm. The V_{THC} dependence of I_{ON} in Fig.7 and the increase in $\sigma I_{ON}/\mu I_{ON}$ in Fig.8 for $W_e \leq$4nm is caused by the LWR-induced quantum confinement effect.

Conclusion

The characteristic variability in silicon nanowire FETs is intensively investigated. It is found that relatively wide nanowire FETs (W_e=7nm) show significantly small V_{TH} and I_{ON} variability compared with bulk and FDSOI FETs. It is also clarified that the increase in V_{TH} variability in extremely narrow FETs (W_e=2nm) is caused by the quantum confinement effect. W_e fluctuations should be suppressed to reduce V_{TH} variability in extremely narrow nanowire FETs.

Acknowledgement

The author would like to thank Dr. K. Takeuchi for helpful discussion. This work was partly supported by a Grant-in-Aid for Scientific Research and by Project for Developing Innovation Systems of MEXT, Japan.

References

[1] K. J. Kuhn, IEEE TED, vol. 59, p. 1813, 2012.
[2] S. D. Suk et al., VLSI Tech. Symp., p. 142, 2009.
[3] M. Saitoh et al., VLSI Tech. Symp., p. 132, 2011.
[4] K. Mao et al., JJAP, vol. 51, 02BC06, 2012.
[5] R. Suzuki et al., JJAP, vol. 52, 104001, 2013.
[6] Y. Yamamoto et al., VLSI Tech. Symp., p. 212, 2013.
[7] H. Majima et al., IEEE EDL, vol. 21, p. 396, 2000
[8] T. Mizutani et al., Silicon Nanoelectronics Workshop, p. 71, 2012.

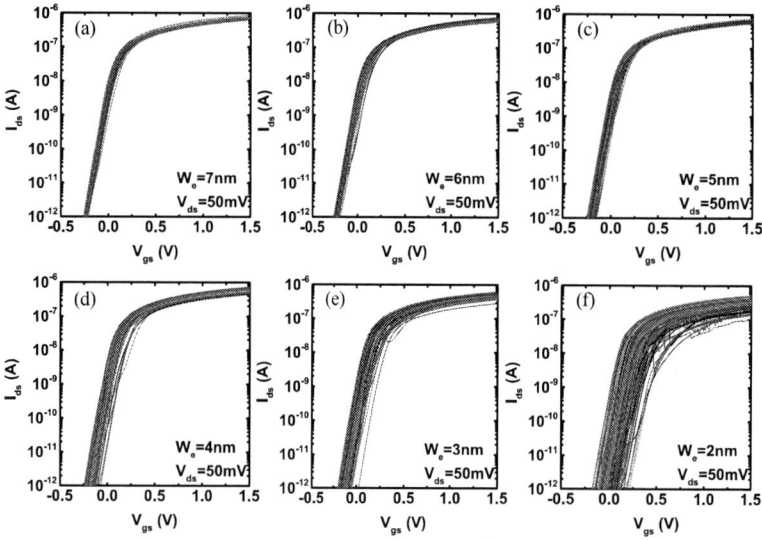

Fig.1. Measured I_d-V_g characteristics of nanowire nFETs at V_{ds}=50mV. (a) W_e=7nm, (b) W_e=6nm, (c) W_e=5nm, (d) W_e=4nm, (e) W_e=3nm, and (f) W_e=2nm.

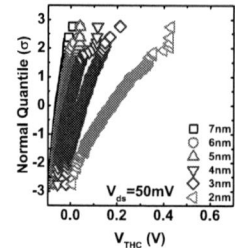

Fig.2. V_{THC} distribution of nanowire nFETs with various W_e at V_{ds}=50mV.

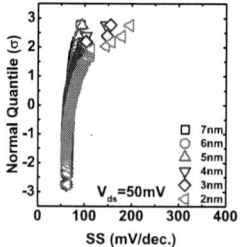

Fig.3. SS distribution of nanowire nFETs with various W_e at V_{ds}=50mV.

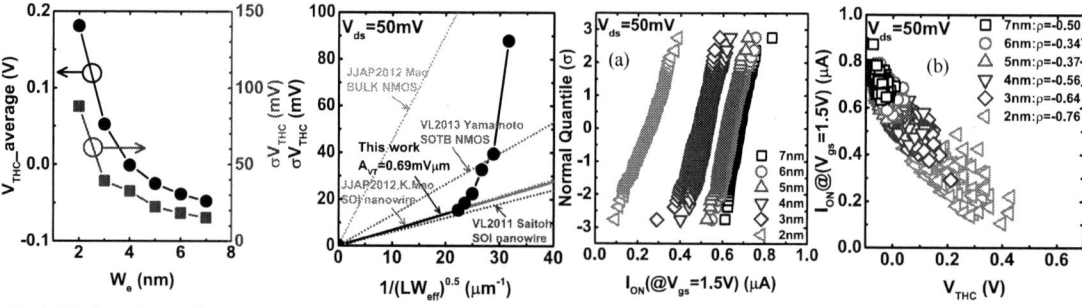

Fig.4. W_e dependence of average V_{THC} and σV_{THC} of nanowire nFETs at V_{ds}=50mV.

Fig.5. Pelgrom plot of σV_{THC} of nanowire nFETs at V_{ds}=50mV.

Fig.6. (a) Cumulative distributions of I_{ON} and (b) correlations between V_{THC} and I_{ON} of nanowire nFETs with various W_e at V_{gs}=1.5V at V_{ds}=50mV.

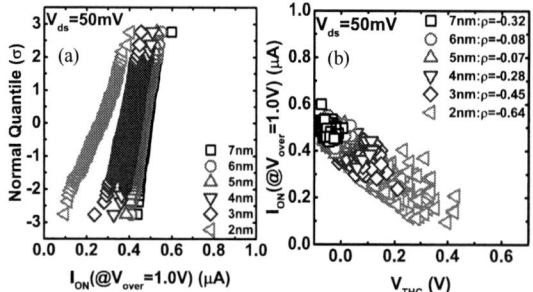

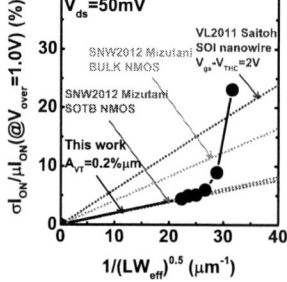

Fig.7. (a) Cumulative distributions of I_{ON} and (b) correlations between V_{THC} and I_{ON} of nanowire nFETs with various W_e at V_{over}=1.0V at V_{ds}=50mV.

Fig.8. Pelgrom plot of $\sigma I_{ON}/\mu I_{ON}$ of nanowire nFETs at V_{over}=1.0V at V_{ds}=50mV.

Performance of GAA Poly-Si Channel of Junctionless Field Effect Transistors with Ultra-Thin Body

Yan-Bo Liu, Yi-Ruei Jhan, Cheng-Ping Wang and Yung-Chun Wu*

Department of Engineering and System Science, National Tsing Hua University, Hsinchu, Taiwan

Mail Address: R408 ESS Building, 101, section 2 Kuang Fu Road, Hsinchu, 300, Taiwan

Tel: +886-3-5715131 ext. 35802; Fax: +886-3-5720724; Email: turbon48@gmail.com

Abstract–Agate-all-around (GAA) with trench structure poly-Si channel junctionless Field-Effect Transistor (JL-FET) has been successfully demonstrated. This JL-FET shows excellent performance in a low drain-induced barrier lowering (DIBL), a steep Sub-threshold Swing (SS) ~70 mV/decade and high I_{ON}/I_{OFF} (>10^8) ratio.

I. Introduction

JL silicon devices with high doping concentrations in their channel and source/drain (S/D) regions have attracted much interest.These devices are characterized by their (1) avoidance of the formation of ultra-steep S/D profiles (2) low thermal budgets owing to elimination of implant activation anneal after gate stack formation, and (3) large buried current which transport in the bulk of the device, which reduces the impact of imperfect interfaces between a semiconductor and insulator [1] [2]. Such a JL feature is a promising candidate for poly-Si TFTs [3]. In this work, GAA with trench structure JL-FET was demonstrated.

II. Device Fabrication

Figure 1(a) presents the structure of the GAA JL-FET. Figure 1(b) shows the key process flow of fabricating GAA and planar JL-FET by different channel width in layout designs. A 400 nm Tetraethyl orthosilicate (TEOS) layer was deposited on 6 inch silicon wafers. Then, a 40 nm undoped amorphous silicon (a-Si) layer was deposited by low-pressure chemical vapor deposition (LPCVD) at 550°C. Following, the a-Si layer was treated by solid-phase recrystallized (SPC) method which is annealed at 600 °C for 24 hrs in a nitrogen ambient atmosphere. And the silicon layer was implanted by Boron ions at a dose of 2×10^{14} cm^{-2}, followed by furnace annealing at 600°C for 4 hrs. The active of the device were defined by e-beam lithography and etched by RIE. The trench structure with 25nm channel thickness was defined by e-beam lithography and anisotropic etched by RIE.The GAA structure was suspended by BOE dip. Subsequently, a 8 nm-thick gate oxide using thermal oxide which consumed around 8 nm poly-Si on both top side and bottom side of channel that remained 9 nm-thick channel. The value of L_G was determined by trench structure. Additionally, 250 nm in-situ doped n$^+$ poly-Siwas depositedfor poly gate, and which was patterned by e-beam lithography and RIE. Finally, a standard passivation and metallization was performed.

III. Results and Discussion

Figure 2(a) and 2(b) present the devices structure of the GAA JL-FET and planar JL-FET with trench structure. Figure 2(c) and 2(d) show the cross-section along source side to drain side.

Figure 3(a) and 3(b) display a top-view scanning electron microscopic (SEM) image of the active region of the GAA and planar JL-FET before gate deposition, respectively. Then, figure 3(c) and 3(d) show a top-view SEM image of the GAA and planar JL-FET after gate deposition, respectively. The width of each channel in GAA JL-FET and planar JL-FET are 0.3 μm and 1 μm, respectively.

Figure 4 plots the I_D-V_G curves of GAA and planar JL-FET at V_D = -1V. As compared to planar JL-FET, GAA JL-FET has a remarked reduction of OFF current (I_{OFF}) and a steep SS value that is close to 70 mV/decade, owing to the gate controllability through gate-all-around structure and ultra-thin body. Because of the fully depleted channel, the GAA JL-FET has lower V_{TH} than the planar JL-FET.

Figure 5 plots I_D-V_D curves at fixed overdrive voltage. The planar JL-FET has higher saturation current than GAA JL-FET, because it has largercross-sectional area of channel than the GAA JL-FET.

Figure 6 shows the I_D-V_G curves of GAA and planar JL-FET at V_D = -0.5V and -3V. The GAA NWs JL-FET has lower V_{TH} (-1.67 V) and DIBL (3.6 mV/V) than the planar JL-FET. Where V_{TH} refers to the gate voltage at I_D = 10^{-8} A. The DIBL is defined as the difference in V_{TH} value between V_D = -0.5 V and -3 V. The DIBL value is normalized by the difference in voltage bias between V_D = -0.5 V and -3 V.

Figure 7 displays DIBL values with various gate lengths (L_G) from 0.2 μm to 0.5 μm. A nearly negligible DIBL of 3.6 mV/V is obtained at L_G = 0.2 μm for GAA JL-FET. The fully depleted channel results in this superior short-channel control ability.

Figure 8 shows the cumulative distribution of V_{TH} and SS in both devices. The SS variationof the GAA device (70~90mV/decade) is smaller than that of the planar device (184~205mV/decade). Otherwise, the V_{TH}variationof the GAA device (-1.65 V ~ -1.55 V) is lower than that of the planar device(0.19 V ~ 0.37 V).

IV. Conclusion

The trench combine with GAA structure for JL-FET was demonstrated successfully in this work. The trench structure is easily integrated into the JL-FET device. Additionally, the JL-FET with trench and GAA structure has excellent electrical characteristics such as low I_{OFF} and SS, negligible DIBL and high I_{ON}/I_{OFF} ratio. This JL-FET is potential candidate for 3D-IC applications.

V. Reference

[1] J. P. Colinge et al., Nature Nanotech., vol. 5, pp. 225, 2010.
[2] J. P. Colinge et al., APL, vol.96, pp. 073510, 2010.
[3]H.C.Linetal.,EDL,vol.33, pp. 53, 2012.

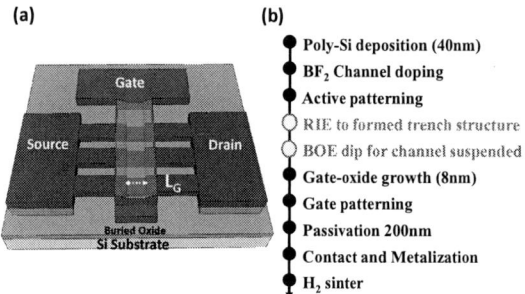

(a)

(b)
- Poly-Si deposition (40nm)
- BF₂ Channel doping
- Active patterning
- RIE to formed trench structure
- BOE dip for channel suspended
- Gate-oxide growth (8nm)
- Gate patterning
- Passivation 200nm
- Contact and Metalization
- H₂ sinter

Fig. 1. (a) GAA Trench JL-FET device. (b) Details of the process flow chart of the fabricated devices.

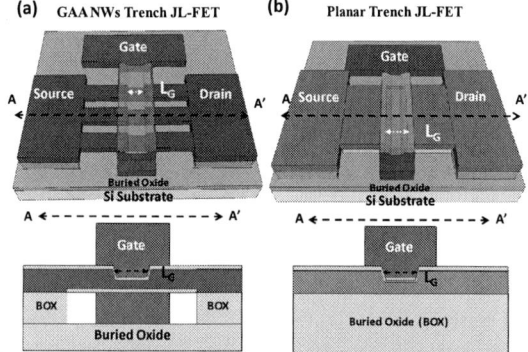

Fig. 2. (a)-(b) Cross-section of the channel structure of GAA JL-FET and Planar JL-FET. A and A' (B and B') indicate cross-section form source to drain.

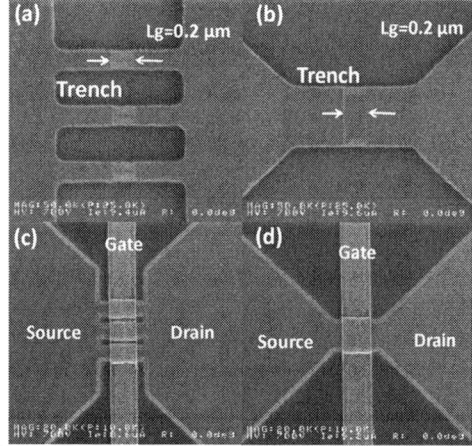

Fig. 3. (a)-(b) Top-view SEM image of GAA and Planar JL-FET before gate deposition. (c)-(d) Top-view SEM image of GAA and Planar JL-FET after gate deposition

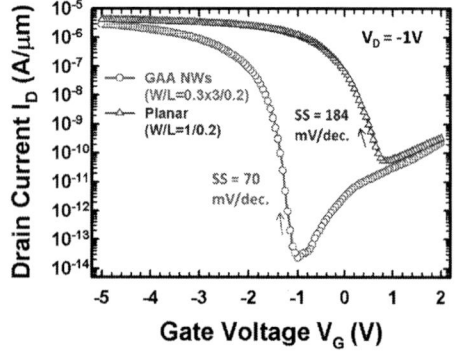

Fig. 4. Comparison of the I_D-V_G curves of Planar and GAA JL-FET with L_G=0.2μm at V_d= -1V. The GAA JL-FET displays higher I_{ON}/I_{OFF} ratio (10^8) than the Planar JL-FET(8×10^4).

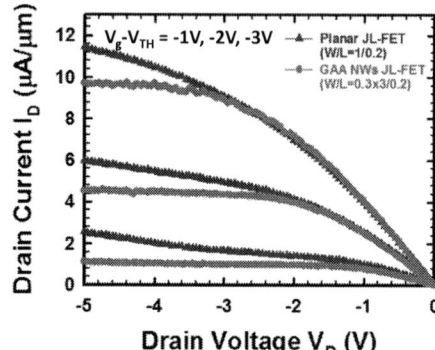

Fig.5. I_D-V_D curves of Planar and GAA JL-FET with L_G = 0.2 μm.

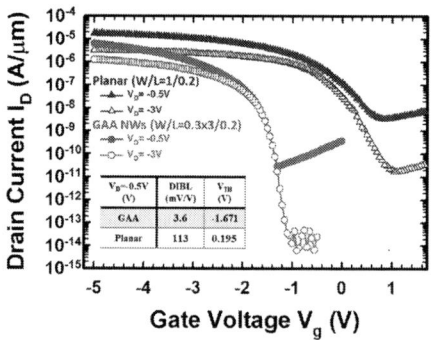

Fig. 6. Comparison of the I_D-V_G curves of Planar and GAA JL-FET with L_G = 0.2 μm at V_d = -0.5V and -3V.

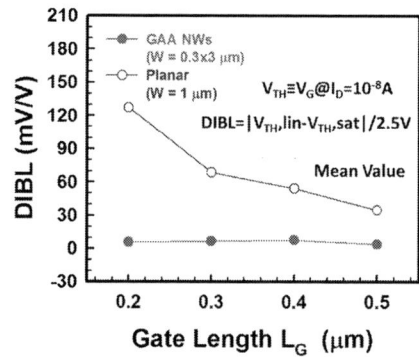

Fig. 7. The DIBL value of GAA JL-FET is negligible from L_G = 0.5 μm to L_G = 0.2 μm (< 8 mV/V). The mean value is obtained from results of five device.

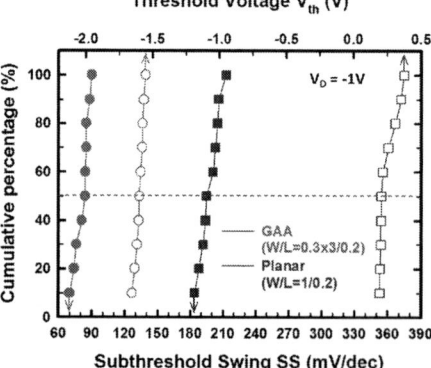

Fig. 8. The cumulative distribution of V_{TH} and SS for GAA JL-FET (V_{TH} = -1.65V, SS = 70mV/dec.) and Planar JL-FET (V_{TH} = 0.19V, SS=184mV/dec.).

Invetigation of Reconfigurable Silicon Nanowire Schottky Barrier Transistors-Based Logic Gate Circuits and SRAM Cell

Juncheng Wang, Gang Du*, and Xiaoyan Liu*

Institute of Microelectronics, Peking University, Beijing, 100871, China

Email: gangdu@pku.edu.cn, xyliu@ime.pku.edu.cn

Abstract — **Reconfigurable Silicon nanowire Schottky Barrier transistors (RFETs) with configurability to be programmed as n/p-type polarity are promising for future integrated circuits. In this work, the tunable polarity characteristics of RFETs are investigated. TCAD simulations have been performed for RFETs-based INV, NOR, NAND logic gates and SRAM cell. 4-terminal RFETs presented show the potential of programmable circuits and high density integration.**

I. INTRODUCTION

The different Reconfigurable Silicon nanowire Schottky Barrier transistors (RFETs) have been reported recently with tunable polarity and high on/off current ratio [1-7]. In contrast to Schottky barrier transistors with metal/silicide as source/drain [6-7], there are two separate gates Control Gate (CG) and Program Gate (PG) in RFETs. PG of RFETs in [3-4] controls both two Schottky junctions, CG acts within the active region by switching on/off the device, which lead to a limited scalability and degrade the transient behaviour of the device [5]. However RFETs [1-2], of which PG and CG are located at two Schottky junctions, show the potential of high density integration. The performance of RFETs-based INV, NOR, NAND logic gates and SRAM cell are investigated with simulation method in the work. Simulation results of basic logic gates can give insights into the in-circuit behaviour of RFETs in the future.

II. RFET-BASED LOGIC GATES AND SRAM CELL

The structures of RFETs are plotted in **Fig. 1**. PG (Vg2) in RFETs is to select n/p-type program, and CG (Vg1) is to control the injection of the desired carriers into the channel. 2-RFETs INV, 4-RFETs NOR and NAND gates are shown in **Fig. 2**(a-c) and when the n-/p-type PG voltage swapped and supply voltage Vdd and Gnd switched, NOR gate is modified to NAND gate. RFETs-based SRAM cell schematic is shown in **Fig. 3**. Besides the configurability in circuits, RFETs also have the potential of high density integration [8]. **Fig. 4** shows the high density integration crossbar architecture with n-/p-type program controller and address decoder made up of high density RFETs array.

II. SIMULATION METHOD

The performance of RFET and its circuits is simulated with TCAD tool [9]. We considered drift diffusion transport within the Si region, thermionic emission and quantum tunnelling at Schottky junctions. **Fig. 5** shows fitted characteristics of RFETs with Ref. [1]. Good agreement shows the validity of the simulated model and parameters. The parameters of RFETs used in the following simulations are given in **Table I**. Silicide Schottky barrier height is chosen to be 0.66eV [1], so that the current characteristics of the n/p-type program devices can be well matched (**Fig. 6**).

II. RESULTS AND DISCUSSION

Fig. 6 and **Fig. 7** show the source-drain current as a function of Vg1 and Vds in n-/p-type program RFETs. RFETs show high on/off ratio if properly programmed. However they cannot work correctly due to ambipolar conduction if PG and CG are interchanged. To give insights into RFETs circuit behaviours, we carried out simulations for RFETs binary logic operation circuits. **Fig. 8** shows voltage transfer characteristics of RFETs inverter at different supply voltages and gives the transient analysis of INV gate with 0.05 fF load capacitance. 4-RFETs NOR and NAND gate logic operation characteristic with transient circuit simulations are shown in **Fig. 9** and **Fig. 10**, which present full swing output in programmable circuits. Delay and energy per switch of RFET-based INV, NOR and NAND gate circuit are shown in **Table II**. The simulation results show that RFETs are promising for programmable circuits even in lower Vdd. 6-T SRAM cell read butterfly based RFETs is shown in **Fig. 11**. The read static noise margins at different supply voltages are achieved.

II. CONCLUSIONS

The tunable polarity characteristics of the RFETs are investigated by TCAD simulations. We demonstrate the binary logic operations such as INV, NOR, NAND Gate and SRAM cell based RFETs. With the potential of high density integration, RFETs are suitable for the programmable circuits in the future.

ACKNOWLEDGEMENTS

This work is supported by the National Fundamental Basic Research Program of China (Grant No. 2011CBA00604).

REFERENCES

[1] A. Heinzig, et al. *Nano Lett.*, 12(1): 119-124, 2012.
[2] A. Heinzig, et al. *Nano Lett.*, 13(9): 4176-4181,
2013. [3] M. De Marchi, et al. *IEDM Tech. Dig.*, 8.4.1-
8.4.4, 2012. [4] M. Mongillo, et al. *Nano Lett.*, 12(6):
3074-3079, 2012. [5] W. M. Weber, et al. *IEEE Proc.*

ESSDERC, 246-251, 2013. [6] J. Wang, et al, *IEEE
ICSICT*, Xi'an, 2012. [7] J. Wang, et al, *IWCE*, Nara,
2013. [8] D. Sacchetto, et al, *Proc. IEEE*, 100(6):
2008, 2012. [9] *TCAD Sentaurus Device Users
Manual, Synopsys*, 2012.

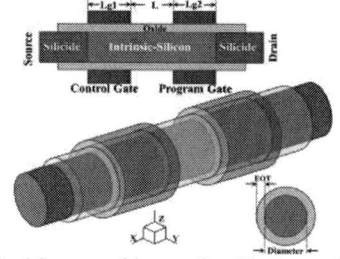

Fig. 1 Structure of the reconfigurable Si nanowire Schottky barrier transistor for TCAD simulation.

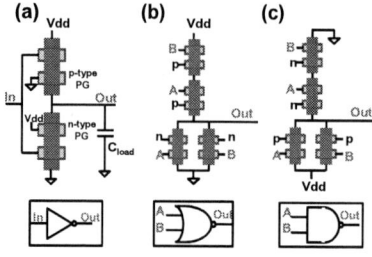

Fig. 2 (a) INV, (b) NOR and (c) NAND logic gates with n/p-type program RFETs.

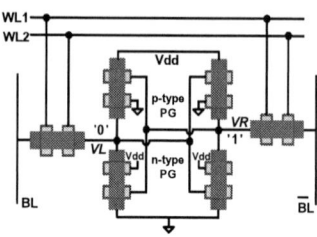

Fig. 3 Schematic of a conventional 6-T SRAM cell based RFETs.

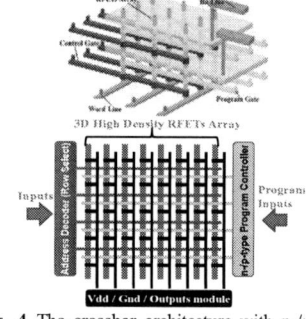

Fig. 4 The crossbar architecture with n-/p-type program controller and address decoder made with high density RFETs array.

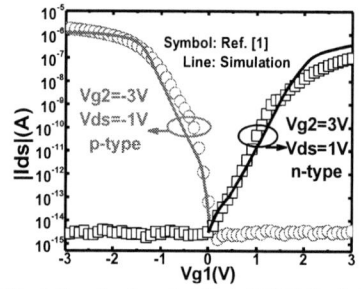

Fig. 5 Transfer characteristics of RFET. Symbols are from Ref. [1], lines are our simulation results. RFET gate length is 225 nm, NWT Diameter is 30nm and EOT is 10nm.

Table I Device Parameters in Simulation

PARAMETERS	VALUE
Gate Length (Lg/Lg1/Lg2)	30 nm
Length Between Two Gates (L)	30 nm
EOT (Tox)	1 nm
NWT Diameter	16 nm
Silicide Schottky Barrier Height	0.66eV

Table II Simulation Results of RFETs logic gates

Gate	INV			NOR	NAND
Vdd(V)	1	1.2	1.5	1.5	1.5
Delay(ps)	207	93	63	134	129
Energy per switch(fJ)	0.033	0.053	0.076	0.051	0.044

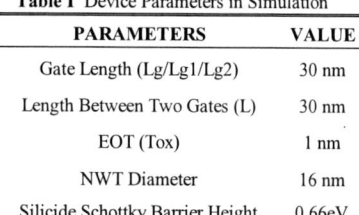

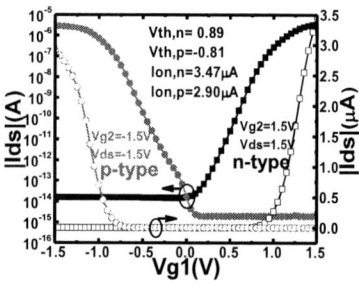

Fig. 6 Well matched transfer characteristics of n-/p-type program RFETs.

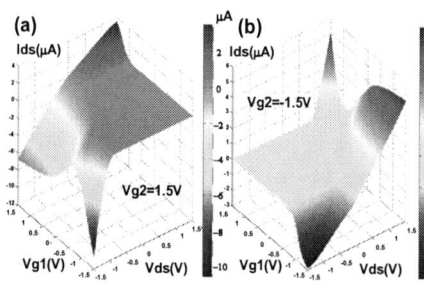

Fig. 7 The source-drain current as a function of Vg1 and Vds in (a) n-type and (b) p-type program RFETs.

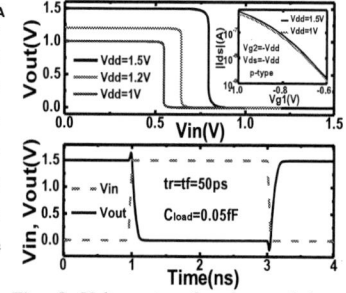

Fig. 8 Voltage transfer characteristic at different supply voltages and Transient analysis of RFETs inverter.

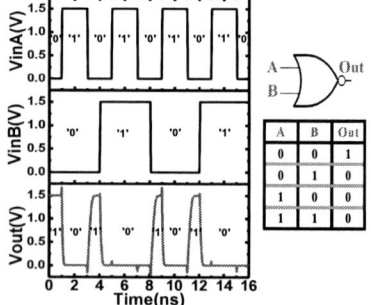

Fig. 9 The 4-RFETs-based NOR gate binary logic operation characteristic with transient circuit simulations when n-/p-type PG voltage is 1.5V/-1.5V, Vdd is 1.5V.

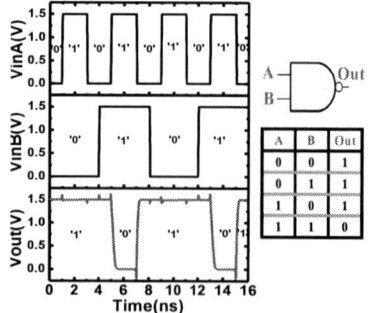

Fig. 10 The 4-RFETs-based NAND gate binary logic operation characteristic with transient circuit simulations when n-/p-type PG voltage is 1.5V/-1.5V, Vdd is 1.5V.

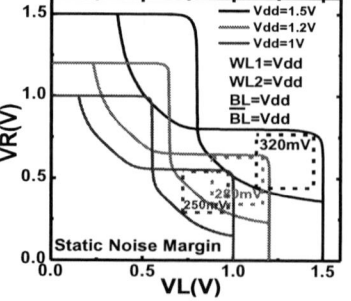

Fig. 11 6-T SRAM cell read butterfly based RFETs. The read static noise margins at different supply voltages are achieved.

Impacts of Surface Roughness Scattering on Hole Mobility in Germanium Nanowires

Hajime Tanaka, Jun Suda, and Tsunenobu Kimoto

Department of Electronic Science and Engineering, Kyoto University, Kyoto 615-8510, Japan
Email: tanaka@semicon.kuee.kyoto-u.ac.jp   Phone/Fax: +81-75-383-2302/2303

Abstract — **The hole mobility in rectangular cross-sectional germanium nanowires was calculated taking into account phonon and surface roughness scattering (SRS). SRS was modeled based on atomistic description of electronic states, and the impacts of SRS on hole mobility were analyzed.**

I. INTRODUCTION

Germanium nanowires (Ge NWs) are a promising candidate as a p-channel material of future MOSFETs owing to their high hole mobility and electrostatic controllability. However, thin NWs may suffer strong surface roughness scattering (SRS) due to strong quantum confinement on carriers [1]. Thus, the impacts of SRS on hole transport in Ge NWs should be investigated to realize high-performance Ge NW pMOSFETs.

In this study, we calculated the hole mobility μ in rectangular cross-sectional [001], [110], [111], and [112]-oriented Ge NWs based on tight-binding (TB) approximation. Considered scattering mechanisms are phonon scattering and SRS, and the impacts of SRS were analyzed.

II. CALCULATION METHOD

A. Modeling of surface roughness scattering

Various models for SRS in NWs have been proposed based on continuum [2,3] or atomistic [1,4,5] description of electronic states. In this study, we developed a new model that can take into account the 2-dimentional surface roughness on the sidewalls of a rectangular NW directly based on atomistic wave functions (Fig. 1). Using this model, the scattering probability corresponding to a stochastic surface roughness was calculated by the Fermi's golden rule. The surface roughness on a sidewall is assumed to be uncorrelated with that on the other sidewalls. For the surface roughness on each face, an exponential autocorrelation function with the root mean square (RMS) of 0.48 nm and correlation length of 1.3 nm [6] was assumed.

B. Mobility calculation

The valence band structures of Ge NWs were computed by an $sp^3d^5s^*$ TB approximation [7] taking a spin-orbit coupling into account. The phonons were calculated by a Valence Force Field model [8] with free boundary condition at the surface. The rate of phonon scattering was calculated by the Fermi's golden rule considering the wavenumber and energy conservation laws [9]. Using the scattering probabilities by SRS and phonon derived as above, the low-field hole μ was computed by solving Boltzmann's transport equation [9,10]. The temperature was assumed to be 300 K.

III. RESULTS AND DISCUSSION

A. Mobility at low hole density

At low hole density, SRS is caused through fluctuation of quantum confinement and is strongly dependent on the geometry of NWs (Fig. 2). SRS is strong when the NW height is small, because the holes are strongly confined. Even when the cross-sectional size is the same, NWs with larger confinement-induced energy shift (Fig. 3) is more affected by SRS. The negative correlation between the effective mass along transport (Fig. 4) and confinement directions leads to the tendency that NWs with high phonon-limited μ (light transport mass) have small energy shift (heavy confinement mass) and thus they are insensitive to SRS. As a result, [110]/(001) NWs with large height, which have high phonon-limited μ, exhibit high μ even when SRS is considered.

B. Mobility at high hole density

Fig. 5 shows the calculated hole density dependence of μ in a [110]/(001) Ge NW. The mobilities limited by SRS, phonon, and both phonon and SRS are plotted. Self-consistent (SC) calculation of the band structure and the electrostatic potential was also performed. The degradation of μ at high hole density is stronger for SRS-limited μ than for phonon-limited μ. This originates from the strong quantum confinement and large energy shift of higher-order subbands. At the same time, the effect of vertical field, which is included by SC, on SRS becomes significant and further degrade μ at high hole density. This weakens the geometry dependence of μ at high hole density compared to that at low hole density.

REFERENCES

[1] N. Neophytou *et al.*, *Phys. Rev. B*, vol. 84, 085313, 2011.
[2] S. Jin *et al.*, *J. Appl. Phys.*, vol. 102, 083715, 2007.
[3] Z. Stanojevic *et al.*, *Proc. 18th SISPAD*, p. 352, 2013.
[4] R. Kotlyar et al., *J. Appl. Phys.*, vol. 111, 123718, 2012.
[5] A. Svizhenko *et al.*, *Phys. Rev. B*, vol. 75, 125417, 2007.
[6] S. M. Goodnick *et al.*, *Phys. Rev. B*, vol. 32, 8171, 1985.
[7] Y.-M. Niquet *et al.*, *Phys. Rev. B*, vol. 79, 245201, 2009.
[8] Z. Sui *et al.*, *Phys. Rev. B*, vol. 48, 17938, 2004.
[9] W. Zhang *et al.*, *Phys. Rev. B*, vol. 82, 115319, 2010.
[10] Y. Yamada *et al.*, *J. Appl. Phys.*, vol. 111, 063720, 2012.
[11] Y.-M. Niquet *et al.*, *J. Appl. Phys.*, vol. 112, 084301, 2012.

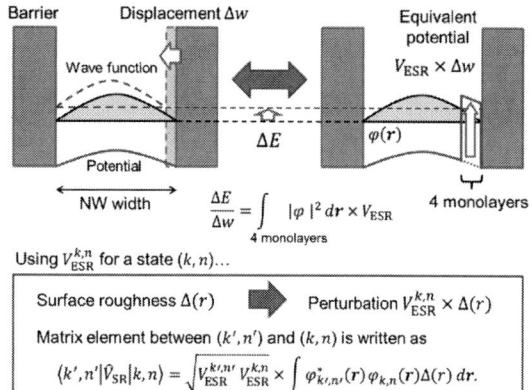

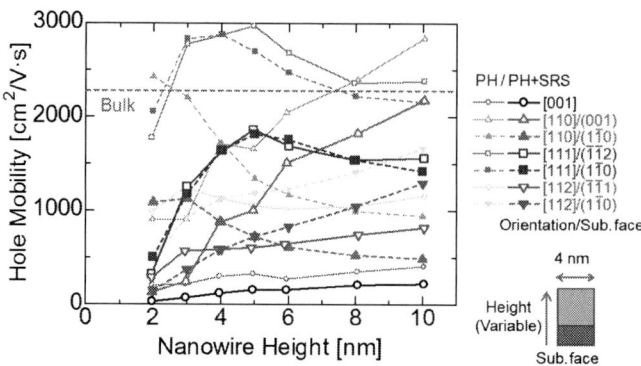

Fig. 1: Schematic image of the model of SRS used in this work. First, the energy variation of subbands due to a small displacement of a sidewall was calculated by TB, and this was converted to an equivalent potential energy in 4 atomic layers near the sidewall. Then the matrix element corresponding to a surface roughness is expressed using this potential energy.

Fig. 2: Height dependence of hole mobility in 4-nm-wide Ge NWs with various orientations and substrate faces considering phonon scattering and SRS (PH+SRS). For comparison, the mobility calculated considering only phonon scattering is shown by pale colored lines (PH). The gray horizontal dashed line denotes the bulk mobility calculated by a similar method [11]. Here, the Fermi level is set at the mid-gap and electrostatic potential is assumed to be flat in the cross-section of NWs.

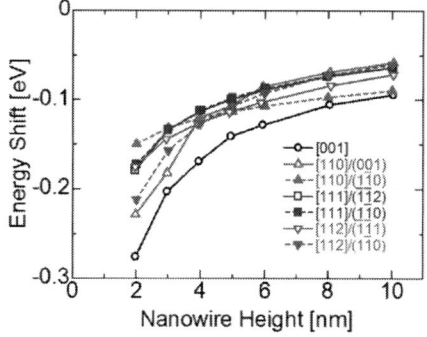

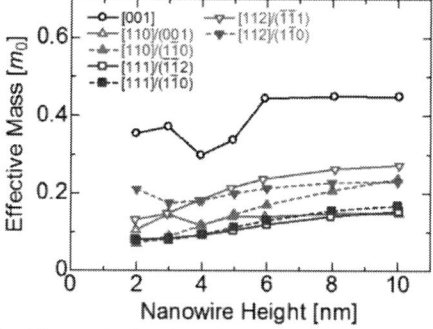

Fig. 3: The calculated confinement-induced energy shift at the Γ point. The energy is referenced from the valence band maximum of bulk Ge. A large absolute value of energy shift means light mass along the confinement directions and gives large perturbation due to roughness.

Fig. 4: The calculated hole transport effective mass averaged by distribution function. NWs with heavy transport mass tend to have light confinement mass, which leads to a large confinement-induced energy shift (Fig. 3).

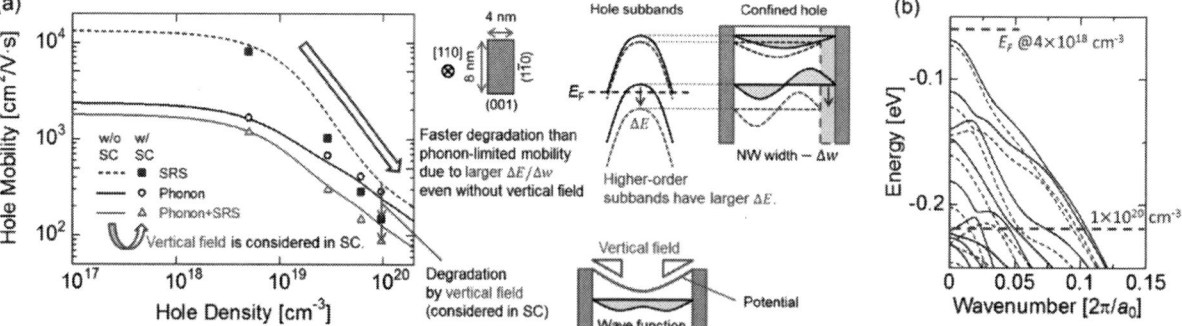

Fig. 5: (a) Hole density dependence of hole mobility in a [110]/(001) Ge NW with 4-nm width and 8-nm height. Lines and symbols show the results without and with self-consistent (SC) calculation, respectively. In the case with SC, gate-all-around structure with metal gate and 0.6-nm-thick SiO$_2$ as a gate oxide were assumed. Large impact of SRS compared with phonon scattering at high hole density originates from contribution of higher-order subbands and the vertical field. (b) The valence band structure and the Fermi level in the NW without SC. Dashed lines show the valence band structure when the width is reduced by 2 monolayers. Larger energy variation of higher-order subbands due to the NW width reduction can be confirmed.

Single ion implantation of Ge donor impurity in silicon transistors

E. Prati [1], Y. Chiba[2], M. Yano[2], K. Kumagai[2], M. Hori[3], G. Ferrari[4], T. Shinada[5], and T. Tanii[6]

[1]Istituto di Fotonica e Nanotecnologia, Consiglio Nazionale delle Ricerche, Italy
[2]School of Fundamental Science and Engineering, Waseda University Japan
[3]Graduate School of Science and Engineering, University of Toyama, Japan
[4]Dipartimento di Elettronica, Informazione e Bioingegneria, Politecnico di Milano, Italy
[5] Center for Innovative Integrated Electronics System, Tohoku University, Sendai, Japan
[6] Faculty of Science and Engineering, Waseda University, Tokyo, Japan

Email: enrico.prati@cnr.it

Abstract — Ge impurities in silicon generate deep donor states in the silicon bandgap. We demonstrate the single ion implantation of Ge ions in the channel of silicon transistors and their electrical activation. Because of the deep donor ground state of Ge, we realize room temperature impurity bands. Our method enables us to create atomic scale conductive paths in silicon with no need of external gate voltages.

I. INTRODUCTION

With the gate length of MOSFETs approaching 7 nm, the channel region contains only one or a few dopant atoms. The random dopant distribution will cause fatal fluctuation of the device performances, such as threshold voltage or transconductance [1-2]. On the other hand, the positive outcome of ultimately scaled-down doped devices has been demonstrated [2-6] in previous work. We have fabricated transistors with deterministically implanted few arsenic (As) or phosphorus (P) atoms and have revealed the quantum transport properties at low temperature for the first time and Hubbard impurity band formation [7-9]. Conventional impurity atoms such as As and P are fully ionized at room temperature, so they are in principle unable to provide conductive paths within the lattice. On the contrary, sufficiently deep donor states are supposed to be able to maintain bound electrons in even at room temperature, by providing thus conductive path if they are suitably arranged as arrays.

Here we report on the fabrication devices with arrays of germanium (Ge) atoms using single-ion implantation (SII) method, and investigation of the electron transport. The results of electrical measurement highlight the value of deterministic doping towards the doped-channel device limits.

II. GERMANIUM DOPED SILICON

Germanium is isoelectronic to silicon and shows donor levels after annealing at low temperatures around 400-500 °C [10-12]. As the levels disappears after annealing at higher temperatures, we applied 550 °C annealing for 60 s. Such states are caused by a non-thermal state of incorporation left over from ion implantation. It might be incorporation on an interstitial site or a more complex configuration with vacancies which breaks up at higher temperatures. Ge exhibits two donor states at 0.5 eV and 0.27 eV below the conduction band respectively. [9,10]

III. EXPERIMENT

Transistors were fabricated on n-type (100) silicon-on-insulator substrates and patterned using standard photolithography. The channel widths were 100 nm; we report on samples with the lengths and thicknesses of 100 nm and 90 nm, respectively. The drain current (I_d) was controlled by the gate bias (V_g) from the substrate through the 125-nm-thick buried oxide. Two doubly charged Ge ions were implanted at each site in the active channel region at 10 and 20 nm pitch using SII (Figure 1). Rapid thermal annealing for dopant activation was carried out at the temperature of 550 °C for 60 sec. The device shows accumulation-mode n-type transistor operation. Figure 2 shows the electron transport property of a transistor with an array of Ge dots (2 Ge ions) at room temperature. The Hubbard band [13] originates from the implanted atoms and diffusion of electrons along the path.

IV. CONCLUSIONS

According to our finding, it is possible to extend to deep donors such as Ge ions the SII technique in silicon. Thanks to the deep energy ground states of Ge in the silicon bandgap, we were able to achieve sub-threshold transport in silicon transistors by exploiting the impurity band created by an array of germanium atoms at room temperature.

ACKNOWLEDGEMENTS

The authors thank the Italy-Japan PEST 2010-2012, the Short Mobility Term grant 2013 of CNR, the JSPS Fellowship Program 2014 and Grant-in-Aid for Scientific Research (nos 23226009 and 25289109) from MEXT.

REFERENCES

[1] T. Shinada, S. Okamoto, T. Kobayashi and I. Ohdomari, "Enhancing semiconductor device performance using ordered dopant arrays," *Nature*, vol. 437, pp. 1128-1131, 2005.

[2] E. Prati and T. Shinada, *"Single-Atom Nanoelectronics,"* 1st ed. Pan Stanford, 2013.

[3] I. Ohdomari," Single-ion irradiation: physics, technology and applications", *J. Phys. D:Appl. Phys.*, vol. 41, no. 043001, pp. 27, 2008.

[4] M. Hori, T. Shinada, Y. Ono, A. Komatsubara, K. , Kumagai, T. Tanii, T. Endoh, and I. Ohdomari "Impact of a few dopant positions controlled by deterministic single-ion doping on the transconductance of field-effect transistors", *Appl. Phys. Lett.*, vol. 99, no. 062103, 2011.

[5] M. Tabe, D. Moraru, M. Ligowski, M. Anwar, R. Jablonski, Y. Ono and T. Mizuno, "Single-Electron Transport through Single Dopants in a Dopant-Rich Environment", *Phys. Rev. Lett.*, vol. 105, pp. 016803, 2010.

[6] Y. Shimizu, H. Takamizawa, K. Inoue, F. Yano, Y. Nagai, L. Lamagna, M. Perego, G. Mazzeo, E. Prati, "Behavior of phosphorous and contaminants from molecular doping combined with a conventional spike annealing method," *Nanoscale*, vol. 6, no. 2, pp. 706-710, 2014.

[7] E. Prati, M. Hori, F. Guagliardo, G. Ferrari and T. Shinada, "Anderson–Mott transition in arrays of a few dopant atoms in a silicon transistor," *Nature Nanotechnology.*, vol. 7, pp. 443-447, 2012.

[8] T. Shinada, M. Hori, F. Guagliardo, G. Ferrari, A. Komatsubara, K. Kumagai, T. Tanii, T. Endo, Y. Ono and E. Prati, "Quantum transport in deterministically implanted single-donors in Si FETs", *Tech. Dig. of Int. Electron Device Meeting*, pp. 30.4.1-30.4.4, December 2011.

[9] M. Hori, T. Shinada, F. Guagliardo, G. Ferrari, and E. Prati, "Quantum Transport Property in FETs with Deterministically Implanted Single-Arsenic Ions Using Single-ion Implantation.", *Abstract of IEEE Silicon Nanoelectronics Workshop*, Honolulu, pp. 1-2, June 2012.

[10] M. Schulz, "Determination of Deep Trap Levels in Silicon Using Ion-Implantation and CV-Measurements," *Appl. Phys.*, vol. 4, pp. 225-236, 1974.

[11] M. Schulz, "Deep trap levels of ion - implanted germanium in silicon measured by Schottky contact techniques," *Appl. Phys. Lett.*, vol. 23, pp. 31-33, 1973.

[12] Y. R. Suprun-Belevich and L. Palmetshofer, "Deep defect levels and mechanical strain in Ge+-

implanted silicon," *Nucl. Instr. and Meth. in Phys. Res. Sect. B*, vol. 96, no. 1, pp. 245-248, 1995.

[13] P. Norton, "Formation of the Upper Hubbard Band from Negative-Donor-Ion States in Silicon", Phys. Rev. Lett. vol. 37, no. 3, pp. 164, 1976

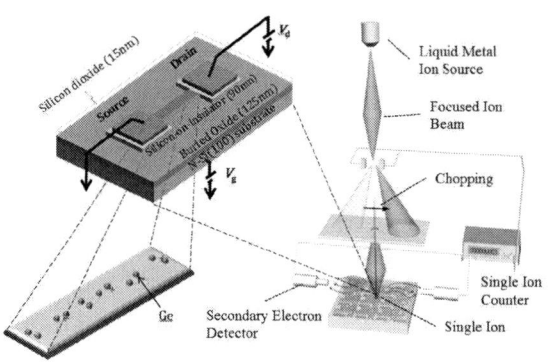

Fig.1. Illustration of device structure and the single-ion implantation method. Germaium ions are implanted two-by-two by using single ion implanter. The channel length is 100 nm and the width is 100 nm.

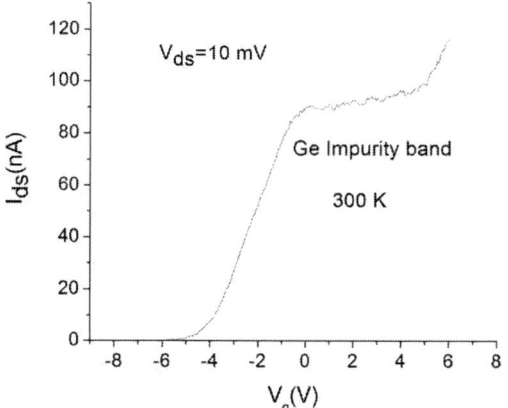

Fig.2. *I-V* characteristic for transistor with array of dots (2 donor ions per implanted site). Below the conduction band of the transistor a flat band appears at room temperature.

Impact of Diffused Donor-Clusters near Lead/Channel Boundary on High-Temperature Single-Electron Tunneling in Narrow SOI-FETs

D. Moraru[1], A. Samanta[2], Y. Takasu[2], K. Tyszka[2,3], T. Mizuno[2], R. Jablonski[3] and M. Tabe[2]

[1]Faculty of Engineering, Shizuoka University, 3-5-1 Johoku Hamamatsu 432-8011, Japan

[2]Research Institute of Electronics, Shizuoka University, 3-5-1 Johoku Hamamatsu 432-8011, Japan

[3]Institute of Metrology and Biomedical Engineering, Warsaw University of Technology, Poland

Email: moraru.daniel@shizuoka.ac.jp

Abstract — We study the effect of phosphorus donors diffused as clusters near the boundary between lead and channel of silicon-on-insulator field-effect transistors (SOI-FETs). In the narrowest-channel devices, one such P-donor-cluster can fully modulate single-electron tunneling transport up to elevated temperatures (>120 K). These results suggest the importance of properly designing the lead edges to enhance the tunneling-operation temperature of dopant-based nano-devices.

I. INTRODUCTION

In recent years, individual donors or clusters of a few donors have been demonstrated to operate as quantum dots (QDs) at low temperatures [1-5]. However, at higher temperatures single-electron tunneling (SET) cannot be observed, mainly due to the small barrier height. In our previous work, we showed that barrier height can be enhanced by channel design to promote dielectric confinement effect [6], leading to operation around T=100 K. However, a more practical approach may be to use donor-clusters [7].

In this work, we comparatively study SOI-FETs with narrowest (~10 nm) and wider (~50 nm) channels. As a result, we find SET transport dominated by clusters of phosphorus (P) donors diffused from the P-doped heavily-doped source edge. In the narrowest-channel devices, it is found that tunneling-transport through such a diffused multiple-donor cluster survives up to temperatures well above 100 K. This suggests that a significantly large barrier height of donor clusters near the channel and source/drain leads boundary.

II. SOI-FETs AND CHANNEL DESIGN

We fabricated nanoscale SOI-FETs, as shown in **Fig. 1**. The channels are designed as long and thin Si constrictions by an electron-beam lithography technique. Doping with P-donors was done by thermal diffusion with high concentration ($N_D \approx 2 \times 10^{19}$ cm^{-3}) in the source/drain leads. The channel was either left nominally non-doped or doped with low concentration ($N_D \approx 1 \times 10^{17}$ cm^{-3}). Channel width was used as a parameter, as shown in **Figs. 2(a)** and **2(b)**. P-donors are expected to diffuse with a gradient of concentration away from the lead edge (as schematically illustrated in the insets).

III. LOW-TEMPERATURE I-V CHARACTERISTICS

First, we study I_D-V_G characteristics at low temperatures (T=5.5 K). In **Figs. 3(a)** and **3(b)**, I_D-V_G curves are shown for wider-channel FETs ($W_{ch} \geq 50$ nm) with non-doped and weakly-doped channels, respectively. As a general observation, these wider-channel devices behave almost independently of channel doping, although slight current modulations as peaks are observed in the subthreshold region. These peaks are ascribed to tunneling via diffused P-donors near source (commonly found in both doped and non-doped devices). These features are quickly masked by the FET current that readily flows through the channel.

In **Figs. 3(c)** and **3(d)**, I_D-V_G curves are shown for devices with the narrowest channels ($W_{ch} \approx 10$ nm), again for non-doped and weakly-doped conditions. The behavior of these devices is strikingly different from **Figs. 3(a)** and **3(b)**, exhibiting a large number of fine current peaks and their characteristic "envelopes" (as marked in the graphs). From the period estimation, the fine peaks are ascribed to SET via a large island (approximately 100-200 nm length). More importantly, the fine and quasi-periodic peaks are strongly modulated in several "envelopes". A single modulation can be ascribed to tunneling-interaction of the SET island with a P-donor [8]. Hence, it can be expected that our multiple modulations can be ascribed to clusters of many P-donors (see **Fig. 4**).

IV. HIGH-TEMPERATURE SET OBSERVATION

For a number of SOI-FETs, we evaluated the temperature evolution (**Fig. 5**) for T<150 K. It can be observed that the current peaks vanish at $T \approx 40$ K for wider-channel FETs [**Fig. 5(a)**]. For narrow-channel FETs, as shown in **Fig. 5(b)**, fine features also vanish rapidly, which suggests a small charging energy of the SET island. However, envelopes and new broad peaks at lower V_G's survive even for T>100 K. These features are marked by arrows in **Fig. 5(b)**. From the period of these broader peaks, they can be ascribed to QDs with radius of 9.0 ± 2.0 nm. This is consistent with the behavior of a P-donor cluster.

V. CONCLUSIONS

We studied the impact of donor-clusters near the leads on tunneling-transport characteristics of nanoscale SOI-FETs. High-temperature operation is enhaced in narrow-channels in which a single P-donor-cluster may control the transport. These findings suggest that the design of the lead regionis critical in enhancing the operation temperature of dopant-devices.

ACKNOWLEDGMENTS

This work was supported by MEXT Kakenhi (23226009, 25630144) & Coop. Res. Project (RIE, Shizuoka Univ.).

REFERENCES

[1] H. Sellier *et al.*, Phys. Rev. Lett. **97**, 206805 (2008).
[2] G.P. Lansbergen *et al.*, Nature Phys. **4**, 656 (2008).
[3] M. Tabe *et al.*, Phys. Rev. Lett. **105**, 016803 (2010).
[4] M. Pierre *et al.*, Nature Nanotechnol. **5**, 133 (2010).
[5] E. Prati *et al.*, Appl. Phys. Lett. **98**, 053109 (2011).
[6] E. Hamid *et al.*, Phys. Rev. B **87**, 085420 (2013).
[7] D. Moraru *et al.*, Sci. Rep. **4**, 6219 (2014).
[8] V.N. Golovach *et al.*, Phys. Rev. B **83**, 075401 (2011).

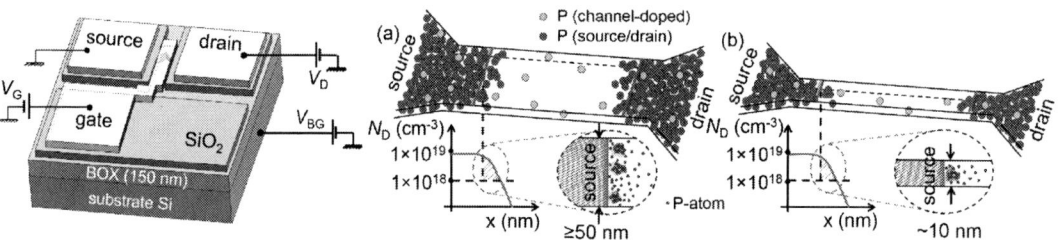

Fig. 1. SOI-FET structure and I-V measurement circuit. A top Al gate covers a nanoscale thin and long Si channel.

Fig. 2. Possible P-donor distributions in SOI-FETs, in the source/drain extensions, near the leads and in the channel, for devices with wider channels, $W_{ch} \geq 50$ nm [(a)] and narrowest channels, $W_{ch} \approx 10$ nm [(b)]. Effects of side diffusion are illustrated in the lower insets.

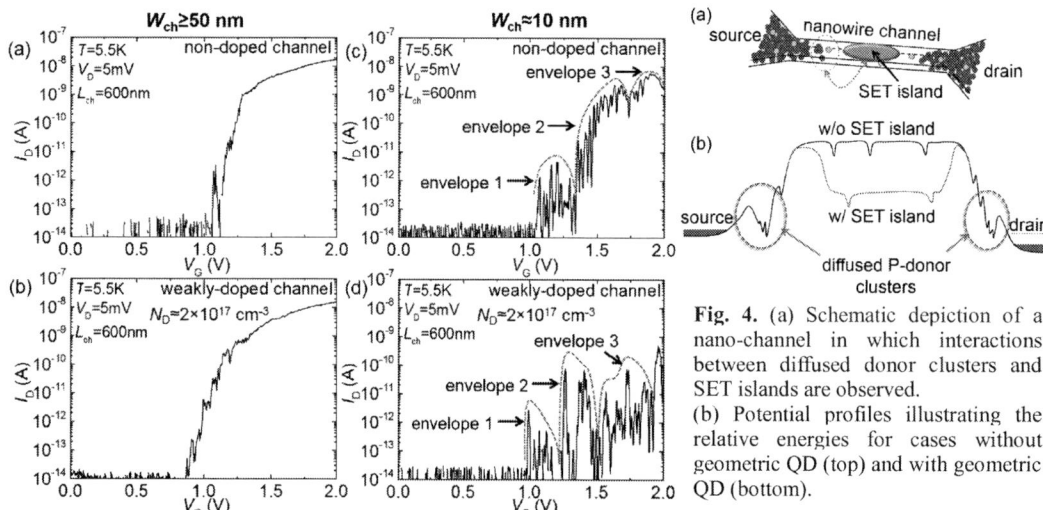

Fig. 4. (a) Schematic depiction of a nano-channel in which interactions between diffused donor clusters and SET islands are observed.
(b) Potential profiles illustrating the relative energies for cases without geometric QD (top) and with geometric QD (bottom).

Fig. 3. Low-temperature (5.5 K) I_D-V_G characteristics for devices with same nominal channel length (distance between heavily-doped source and drain), for: (a)-(b) devices with wider channels ($W_{ch} \geq 50$ nm); (c)-(d) devices with narrowest channels ($W_{ch} \approx 10$ nm). Top: nominally non-doped channels. Bottom: weakly-doped channels ($N_D \approx 2 \times 10^{17}$ cm^{-3}).

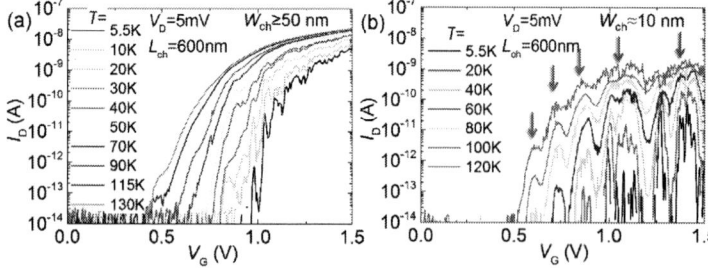

Fig. 5. I_D-V_G characteristics as a function of temperature for: (a) a device with wider channel ($W_{ch} \geq 50$ nm); (b) a device with narrowest channel ($W_{ch} \approx 10$ nm). Single-electron tunneling current peaks are preserved up to $T \approx 120$ K. (blue arrows: peak envelopes from low temperature; red arrows: newly emerged current peaks).

The impact of single donor and donor-acceptor pair on electronic and transport properties of silicon nanostructures

L. T. Anh[1*], D. Moraru[2], M. Manoharan[1], M. Tabe[2] and H. Mizuta[1,3]

[1] School of Materials Science, Japan Advanced Institute of Science and Technology, Japan

[2] Research Institute of Electronics, Shizuoka University, Japan

[3] Nanoelectronics and Nanotechnologies Research Group, Electronics and Computer Science, Faculty of Physical Sciences and Engineering, University of Southampton, United Kingdom

Email: letheanh@jaist.ac.jp

Abstract — **We present first-principles calculations of electronic and transport properties of silicon (Si) nanorods with sizes smaller than the Bohr radius, doped with single phosphorus (P) or co-doped with P and B (boron). For P-doped nanorods, we found that the first conductive energy state (1st CS) remains nearly unchanged by reducing the nanorod size, resulting in an electron binding energy (E_b) practically independent on size. This is due to the counterbalance between quantum confinement and attractive interaction from the P donor. We show, however, that E_b depends on the P position in Si cross-shaped nanostructures. Finally, the impact of P-B interaction on electronic and transport properties of Si nanostructures is investigated.**

I. INTRODUCTION

The impact of individual dopant on the tunneling-transport of nano-scale Si devices becomes significant [1, 2]. Therefore, a detailed theoretical understanding of the impact of dopants on the electronic and transport properties of nanostructures is essential. In this study, first, we calculate E_b for a single P donor in Si nanorods (Fig. 1(a)). Then, we study the dependence of E_b on the P position in Si cross-shaped nanostructures (Fig. 1(b)). Energy states of P donors are also expected to be modified by the presence of B acceptors nearby. Therefore, the interactions of P donors and B acceptors are studied to clarify their impact on Si nanostructures (Fig. 1(c)).

II. ELECTRON BINDING ENERGY CALCULATION

Binding energy is evaluated based on our method that combines projected-density-of-states (PDOS) and 3D wavefunction (3D WF) analysis. Fig. 2(a) shows PDOS at centered P atom, together with the molecular energy spectrum of the Si nanorod. The WFs of quantum states, resulting from hybridization between P and Si atomic orbitals [5], start to fully spread through the structure at the hybrid state labeled H^{28} (Figs. 2(a)-(b)). The ratio PDOS/DOS represents the contribution of the P atom to the hybrid state. For energy states higher than H^{28}, this ratio becomes negligibly small

(Fig. 2(c)). This confirms that the state H^{28} is the 1st CS. The E_b is the difference in energy between H^0 and H^{28}. In the past [3], this energy was calculated by: $E_b = I_d - A_p$; where I_d, A_p is ionization and affinity energy for doped and un-doped nanostructure, respectively. This method does not include interactions between P donor and the 1st CS. Our results show that E_b remains ~1.5 eV for nanorod radii, R_{avg}, in the range 0.35~1.44 nm (Fig. 3(a)). Fig. 3(b) shows also that the WF amplitude at center for $R_{avg} \approx 0.6$ nm is higher than that for $R_{avg} \approx 1.0$ nm. Hence, the overlap of the attractive potential of the P donor and electron WF in smaller nanorods is larger [4]. When reducing the size, a counterbalance between quantum confinement and attractive interaction makes the energy of the 1st CS to remain nearly unchanged and, thus, E_b is also only weakly dependent on size. In larger-size nanostructures, E_b is known to decrease [6].

III. POSITIONAL DEPENDENCE OF BINDING ENERGY

Fig. 4(a) shows comparatively E_b estimated with the conventional method [3] and our method, calculated for the P atom in cross-shaped nanostructures at: center (x+0), near the edge (x+2), and at the edge (x+4). Our method shows that E_b increases by moving the P atom toward the edge. Fig. 4(b) shows that the value of LUMO state at (x+0) is larger than that at (x+4). Thus, in this case, the overlap between attractive potential of P atom and the LUMO is smaller for P at center, resulting in larger E_b when the donor is at the edge.

IV. EFFECTS OF P-B CO-DOPING

In order to clarify the impact of P-B interaction on electronic and transport properties, we study the P-B co-doped Si nanostructures. We find that the ground state WFs of P donor and B acceptor interact strongly in a destructive manner when the dopant atoms get close to the nearest-neighbor sites (Fig. 5(a)). This destructive interaction results in a repulsion of their ground state energies (Fig. 5(b)) and also a remarkable reduction of the atomic built-in potential with decreasing the P-B atomic distance.

V. CONCLUSIONS

Electron binding energy of single-P-doped Si nanorods is ~1.5 eV almost independently of size, because of the counterbalance between quantum confinement and attractive interaction of P donor. However, E_b increases when P moves towards the edge of Si nano cross-shape. Finally, we found that P donor and B acceptor interact destructively in Si nanorods.

ACKNOWLEDGMENTS

This work is supported by MEXT Grants-in-Aid for Scientific Research (KAKENHI No. 23226009).

REFERENCES

[1] M. Tabe *et al.*, Phys. Rev. Lett., **105**, 016803 (2010).
[2] E. Hamid *et al.*, Phys. Rev. B **87**, 085420 (2013).
[3] D. V. Melnikov *et al.*, Phys. Rev. Lett., **92**(4), 046802 (2004).
[4] H. Mizuta, Jpn. J. Appl. Phys. **35** 2012 (1996).
[5] L. T. Anh *et al.*, J. Appl. Phys. **116**, 063705 (2014).
[6] M. Diarra *et al.*, J. Appl. Phys. **103**, 073703 (2008).

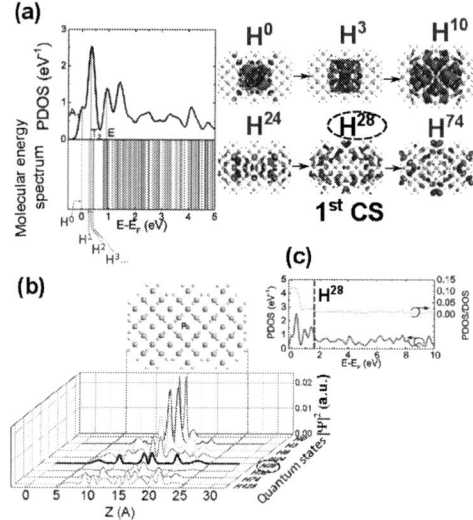

Fig. 2. (a) PDOS at the P atom and molecular energy spectrum of a Si nanorod (left); 3D wavefunctions of quantum states (right). (b) Projections of wavefunction square along Z<100> direction. (c) PDOS at the P atom (red) and the ratio PDOS/total DOS (green).

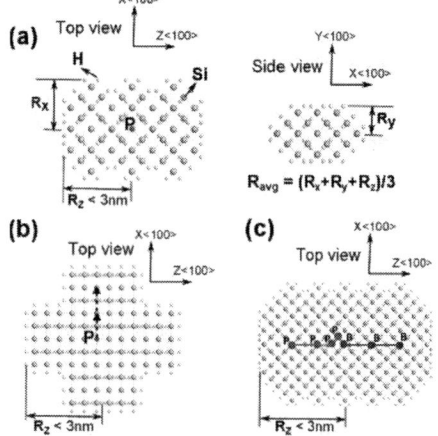

Fig. 1. Top and side views of: (a) single-P-doped Si nanorod; (b) single-P-doped Si nano cross-shape, with the P atom moved from center to the edge; (c) P-B co-doped nanostructures with P-B separation decreased to nearest-neighboring sites.

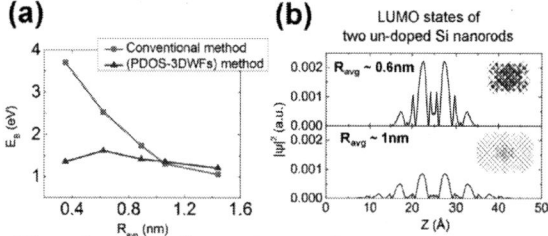

Fig. 3. (a) Comparison of E_b between the conventional method and our method (combining PDOS-3DWFs analysis) for Si nanorods. (b) Projections of wavefunction square of the LUMO states for two un-doped Si nanorods. The insets show the 3D wavefunction pictures.

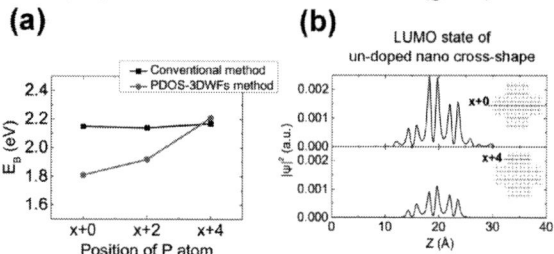

Fig. 4. (a) Comparison of E_b between the conventional method and our method (combining PDOS-3DWFs analysis) for Si cross-shape. (b) Projections of wavefunction square of the LUMO state of an undoped nano cross-shape along (x+0) and (x+4) line. The insets show the positions of the shown lines.

Fig. 5. (a) PDOS on B atom (blue) and on P atom (red) with the wavefunctions of B acceptor and P donor at the ground state. (b) Ground-state energies with respect to the Fermi level for B acceptor and P donor for different P-B separations.

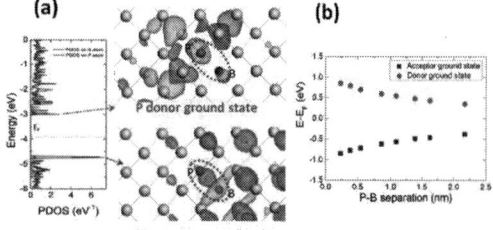

Investigation of the Impact of Grain Boundary on Threshold Voltage of 3-D MLC NAND Flash Memory

Zhiyuan Lun, Lei Shen, Yingying Cong, Gang Du*, Xiaoyan Liu[†], Yi Wang

Institute of Microelectronics, Peking University, Beijing 100871, China

* gangdu@pku.edu.cn   † xyliu@ime.pku.edu.cn

Introduction

Three-dimensional (3-D) NAND flash technology has been attracting much attention in recent years, for it overcomes the challenges of scaling limitation faced by 2-D NAND flash and meets the demand of higher storage density and lower bit cost. Several 3-D NAND flash architectures has been proposed recently. [1-3] However, most of these structures have adopted poly-crystalline silicon (poly-Si) as the channel material, which has interface traps distributing along the random grain boundary (GB). Previous study has shown that these GB traps can cause the fluctuations in electrical characteristics and result in a wider distribution of threshold voltage (V_{th}) for NAND flash memory [4], [5], where only the intrinsic V_{th} of the memory structure is analyzed. In this paper, we utilize 3-D simulation to examine the impact of GB with various positions on V_{th} of different storage states in a multi-level cell (MLC) NAND flash memory.

Simulation Methodology

Fig. 1 shows the schematics of the simulated structure, which features a string of 3D vertical channel NAND flash memory with cylindrical poly-Si channel. The parameters used in the simulation are summarized in Table I. To investigate the impact of GB on the memory cell characteristics, we implement a single grain boundary perpendicular to the memory string with various positions from the source-side adjacent cell to the bitline-side adjacent cell. The distribution of acceptor-type and donor-type interface states across the band gap of the GB is given by fitting the experimental data [6] with the formula in Appendix I. To model the programming/erasing (P/E) process of the memory cell and calculate the trapped charge in the storage layer, in-house numerical simulator for NAND flash [7] is used with the consideration of cylindrical geometry. The stored charge density is then used as input of 3D TCAD tools [8] to simulate the effect of GB in poly-Si channel and extract the V_{th} for the 2-bit memory cell. V_{th} is defined as the control gate voltage of the investigated central cell when the bitline current is 200nA with 5.5V read-pass voltage and 0.5V bitline voltage. For the poly-Si channel, mobility models considering mobility degradation caused by impurity scattering and influence of electrical field normal to the interface are included in the drift-diffusion transport.

Results and Discussion

1. Impact of GB on P/E operations

By using the proposed simulation method, **Fig. 2** plots the P/E characteristics of 150 samples with randomly distributed GB positions. It can be concluded that the GB does not affect P/E operations. This is because during P/E, despite the trapped charge along GB, the relatively high P/E voltage dominates the band profile of the gate stack and along the channel, which directly determines the electron/hole tunneling current. In this investigation,

for the GB-free cases, we assume the following V_{th} levels in MLC: -1.3V, 1.5V, 3V, and 4.5V, which are denoted by "00", "01", "10", and "11" respectively. The same fixed P/E pulses are used to calculate the stored charge density in different memory states for cells with GB.

2. Impact of GB on read operation

Fig. 3 shows the dispersion of transfer characteristics of 150 samples for each state with randomly distributed GB positions. Since the stored charge has hardly any difference in each state after P/E, the variation in these curves has to be ascribed to the impact of GB on read operation. **Fig .4(a)** compares the V_{th} variation due to the GB at different channel positions in "00" and "11" states. It should be noticed that, for "00" state, the maximum variation of V_{th} occurs when the GB is located at the inter-cell region between the investigated cell and the source-side adjacent cell, while for "11" state, a GB under its control gate near the source side has the largest impact on V_{th}. The V_{th} variation is directly correlated to the conduction band profile along channel, which can be raised by the charged GB, as shown in **Fig. 4(b)**. In both states, the positions of GB with the maximum V_{th} variation are related to the peak positions of the potential barrier. **Fig. 4(c)** shows the distributions of electron density in "00" and "11" states for GB-free cell and two GB positions, with the control gate voltage to be -1.5V and 4.5V respectively. **Fig. 5(a)** shows the V_{th} variation in different programmed states, where the maximum V_{th} variation increases with higher charge storage. This results from the fact that for higher state, the total trapped charge density along GB is larger, as shown in **Fig. 5(b)**.

3. Threshold voltage distribution

Fig. 6 shows the cumulative probability of V_{th} for four states in 2-bit MLC. It can be seen that higher state has larger decrease of read margins and wider V_{th} distribution due to the GB in the poly-Si channel.

Conclusion

The impact of a single GB with various positions along the poly-Si channel on V_{th} variation of 3-D cylindrical MLC NAND flash memory is investigated. The GB traps are found to have little effect on the P/E characteristics. During read operation, GB that accounts for the maximum V_{th} variation is located at the inter-cell region in the erased state and underneath the control gate in the programmed states. This study can provide better understanding of the physics for the impact of GB in 3-D MLC NAND flash memory.

Acknowledgement

This work was supported by NKBRP 2011CBA00604.

Reference:

[1] H. Tanaka *et al.*, VLSI, 2007, p.14. [2] J. Kim *et al.*, VLSI, 2009, p.186, [3] H.-T. Lue *et al.*, VLSI, 2010, p131. [4] C.-W. Yang and P. Su, TED, 61(4), 2014, p.1211. [5] Y.-H, Hsiao *et al.*, TED, 61(6), 2014, p.2064. [6] Z. Lun *et al.*, IWCE, 2014. [7] H. Sehil *et al.*, Synth. Met. 90(3), 1997, p.181. [8] TCAD Sentaurus vG2012-06, Synopsys, 2012.

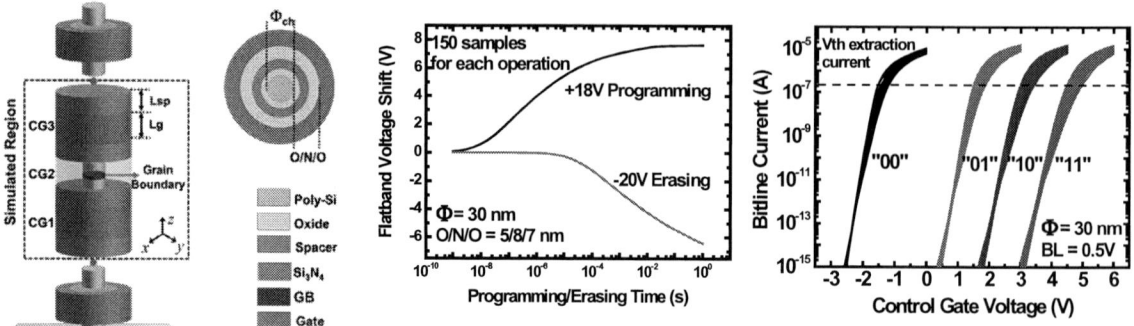

Fig. 1. Schematic illustration of simulated 3-D NAND flash memory. The position of the grain boundary is calculated from the center of central control gate (CG2).

Fig. 2. The P/E characteristics of 150 samples for each operation with randomly distributed GB position.

Fig. 3. Dispersion of transfer characteristics of 150 samples for each state with randomly distributed GB position. State "00", "01", "10", and "11" are defined using fixed P/E pulses: 10ms at -20V, 0.1µs at 18V, 2.5µm at 18V, and 400µs at 18V respectively.

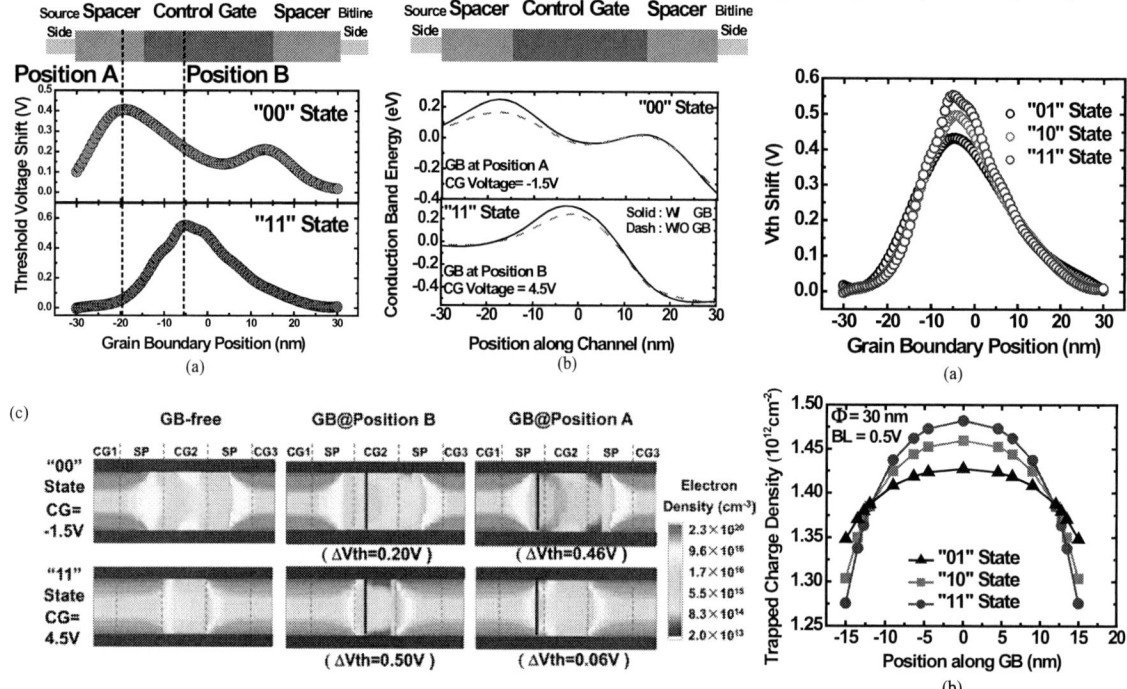

Fig. 4. (a) The V_{th} variation due to the GB at different channel positions in "00" and "11" states. Position A and B are the GB positions which correspond to the maximum V_{th} variation in "00" and "11" states respectively. (b) The conduction band energy profile along poly-Si channel at the semiconductor interface for cells with and without GB. (c) Distributions of electron density in the poly-Si channel in "00" and "11" states for GB-free cell and two GB positions. The control gate voltage for "00" and "11" states are -1.5V and 4.5V respectively.

Fig. 5. (a)The V_{th} variation due to GB at different channel positions in different programmed states. (b)The interface trapped charge density along GB in different programmed states before the "on" state of the memory string. The 0 point of the position is the center of the channel.

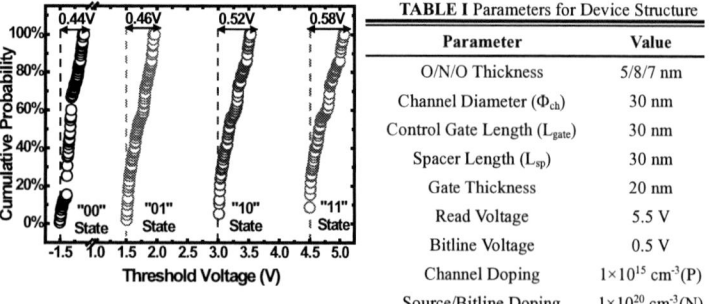

Fig. 6. Cumulative probability of V_{th} distribution for four states in 2-bit MLC NAND flash memory.

TABLE I Parameters for Device Structure

Parameter	Value
O/N/O Thickness	5/8/7 nm
Channel Diameter (Φ_{ch})	30 nm
Control Gate Length (L_{gate})	30 nm
Spacer Length (L_{sp})	30 nm
Gate Thickness	20 nm
Read Voltage	5.5 V
Bitline Voltage	0.5 V
Channel Doping	1×10^{15} cm^{-3}(P)
Source/Bitline Doping	1×10^{20} cm^{-3}(N)

Appendix I Formula used to calculate trap density in GB and related parameters

$$g(E) = N_{TA}\exp(\frac{E-E_c}{E_{TA}}) + N_{TD}\exp(\frac{E_V-E}{E_{TD}})$$
$$+ N_{GA}\exp\left[-(\frac{E_{GA}-E}{W_{GA}})^2\right]$$
$$+ N_{GD}\exp\left[-(\frac{E-E_{GD}}{W_{GD}})^2\right]$$

$N_{TA}=1\times10^{14}$ cm^{-2}, E_{TA}=0.07eV,
$N_{TD}=8\times10^{13}$ cm^{-2}, E_{TD}=0.08eV,
$N_{GA}=5\times10^{12}$ cm^{-2}, E_{GA}=0.65eV, W_{GA}=0.1eV,
$N_{GD}=5\times10^{12}$ cm^{-2}, E_{GD}=0.45eV, W_{GD}= 0.1eV

36

Resistive Switching Characteristics in HfOx Memory Devices with Local Electrical Field Design

Tsung-Kuei Kang, Wei-Len Chen, Yu-Han Chen and Pei-Hsun Tsai
Department of Electronic Engineering, Feng-Chia University,
Taichung, Taiwan, R.O.C.
TEL: 886-4-24517250 ext. 4958, FAX: 886-4-24510405, kangtk@fcu.edu.tw

Abstract

The switching characteristics of the resistive switching memory with local electrical field are investigated. The local electrical fields in the HfOx film are created by some sharp regions of bottom electrode. Compared with planar structures, this design can improve the yield and uniformity of set/reset voltage for Pd/HfOx/TiN structure. Results indicate that the sample with spacer and raised structures can improve the switching characteristics.

Introduction

Recently, resistive random access memories (RRAMs) have been paid much attention [1-2]. For the metal/insulator/metal (MIM) structure, its memory characteristic shows the reversible and reproducible resistive switching between high/low resistance states. In electrical characteristics, the set/reset uniformity, memory window, yield, and endurance are very important. Previous work showed that the memory embedded with nanocrystals (NCs) had been fabricated and it improved resistive switching characteristics in HfOx film owing to the enhancement of the local electric field [3-4]. In this study, the local electrical fields in the HfOx film are created by some sharp regions of bottom electrode and their electrical characteristics will be compared with the one without the design of local electrical field.

Device Fabrication

Metal/HfO$_x$/metal (MIM) stack were fabricated on Si$_3$N$_4$/SiO$_2$/Si and Si spacer structure. Si spacer structure is the same as previous work [5]. After the substrate is ready, a 200-nm-thick titanium nitride layer was deposited as a bottom electrode. Subsequently, a 10 nm thick HfOx was deposited as the resistance switching layer. Finally, a 50-nm-thick Pd layer was evaporated on the HfOx film as a top electrode. The Schematic diagrams were shown in Figs. 1(a)-(e).

Results and Discussion

There are three structures, including 50 nm Pd/10 nm HfO$_x$/200 nm TiN /Si$_3$N$_4$/SiO$_2$ (planar structure), 50 nm Pd/10 nm HfO$_x$/200 nm TiN/Si spacer /Si$_3$N$_4$/SiO$_2$(without raised structure, spacer-only structure) and 50 nm Pd/10 nm HfO$_x$/200 nm TiN/Si spacer /Si$_3$N$_4$/SiO$_2$ (with raised structure, proposed structure). The TEM image of proposed structure is shown in Fig. 2. The electrical field of Pd/10 nm HfO$_x$/200 nm TiN on Si spacer and raised structure (RS) are simulated by TCAD and the local electrical field in the HfOx film can be found, as shown in Figs. 3(a) and (b). Previous work [4] showed that the

conducting path is randomly formed in the HfOx film for the memory device without Pd NCs, but easily formed for Pd NCs device. This is because that the local electrical field appears around Pd NCs due to the stronger coupling effect of metal NCs in RRAM with Pd NCs. In this study, it is also found that the local electrical field in the HfOx film can be created by bottom electrode design. The bottom electrode deposited on Si spacer can create a local electric field in the HfOx film and improve the uniformity of set/reset voltage. Moreover, bottom electrode deposited on the raised structure (RS) also creates a local electric field in the HfOx film. Figs. 4(a) and(b) show I-V resistive switching characteristics of planar and proposed structures during continuous high/low resistance switching (HRS/LRS) by voltage sweeping. Results reveal that the memory window and switching uniformity is improved for the proposed structure. Compared with planar sample, the yield is higher for the proposed one. Under the same area, the proposed sample with spacer and RS design show a higher yield. For proposed samples, the yield gets better with the area of bottom electrode, as shown in Table. 1. Figs. 5(a) and(b) show I-V resistive switching characteristics of spacer-only and proposed structures during continuous high/low resistance switching (HRS/LRS) by voltage sweeping. It reveals that the sample with RS structure show a better characteristics due to an additional local electrical field in the HfOx film.

Conclusion

The memory device with locally electrical field in the HfOx film shows a better yield, set/reset uniformity and memory window. Experiments reveal that the sample with spacer and raised structures (RS) can improve the switching characteristics.

Acknowledgment

This work is supported by the NDL, R.O.C. and Ministry of Science and Technology under contract No. 103-2221-E-035 -082 .

References

[1] N. Xu, L. Liu, X. Sun, X. Liu, D. Han, Y. Wang, R. Han, J. Kang, and B. Yu, Appl. Phys. Lett. Vol. 92 (2008), p.232112.

[2] A. Chen, S. Haddad, Y. C. Wu, T. N. Fang, S. Kaza, and Z. Lan, Appl. Phys. Lett. Vol. 92 (2008), p. 013503.

[3] Y. T. Tsai, T. C. Chang, C. C. Lin, S. C. Chen, C. W. Chen, Simon M. Sze, F. S. Yeh , and T. Y. Tseng: Electrochemical and Solid-State Letters, Vol.14, (3), (2011) , p.H135-H138.

[4] T. K. Kang, C. K. Wang, Y. Y. Yang, Advances in Applied Materials and Electronics Engineering II, 3-6, 2013.

[5] T. K. Kang and Y. Y. Yang: IEEE Transactions on Electron Devices, Vol. 60, No. 7, July 2013, p.2415.

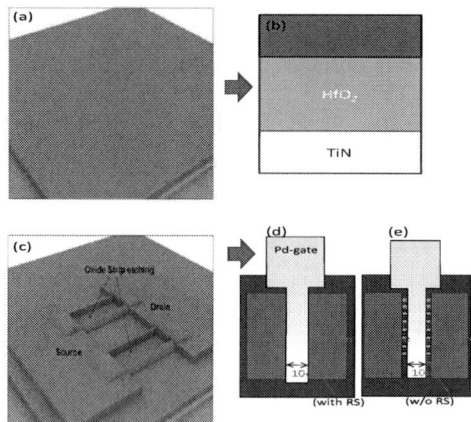

Fig.1. Schematic diagrams (a) 50 nm Si_3N_4/500 nm SiO_2 (b) 50 nm Pd/10 nm HfO_x/200 nm TiN on Si_3N_4/SiO_2 (c) Si spacer formation (d) Proposed sample: 50 nm Pd/10 nm HfO_2/200 nm TiN on Si spacer/ Si_3N_4/SiO_2 with RS structure (e) Spacer-only sample: 50 nm Pd/10 nm HfO_x/200 nm TiN on Si spacer/ Si_3N_4/SiO_2 without RS structure. The source and drain will be connected with ground.

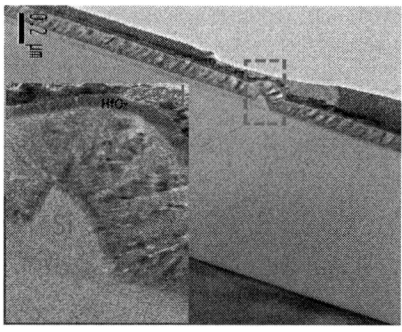

Fig. 2. The cross-sectional transmission electron microscope (TEM) images of 50 nm Pd/10 nm HfO_x/200 nm TiN on Si spacer/ Si_3N_4/SiO_2.

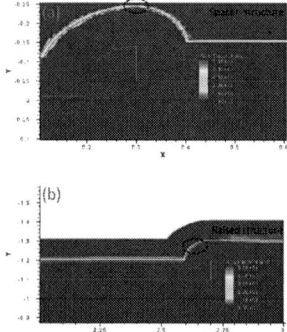

Fig. 3. Local electrical field of (a) Pd/10 nm HfO_x/200 nm TiN on Si spacer and (b) Raised structure (RS). The local electrical fields are indicated by two black circles in (a) and (b).

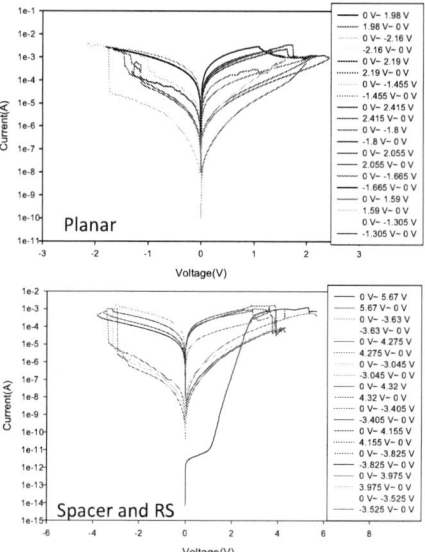

Fig. 4. I-V resistive switching characteristics of planar and proposed structure (spacer and RS). More uniform switching characteristics can be achieved by proposed structure design.

Table1. Yields of planar and proposed structures (spacer and RS). L ×W is the area of bottom electrode. Under the same area, The sample with spacer and RS design show a higher yield (at least a switching cycle). For proposed samples, the yield gets better with the area of bottom electrode.

	2 spacer (L=10 um, W=4um)	8 spacer (L=10um, W=22um)	16 spacer (L=10um, W=46um)	20 spacer (L=10um, W=58um)	Planar (L=10um, W=4um)
Yield	6/10	9/10	9/10	9/10	4/10

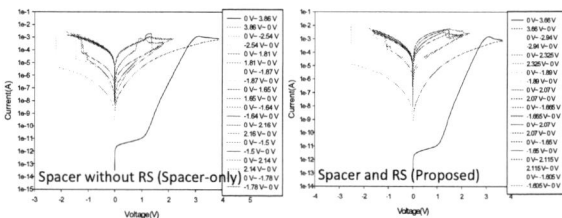

Fig. 5. I-V resistive switching characteristics of spacer-only and proposed structure.

38

Position and Number Control of Donor-QD Potential
by Pattern-doping in SOI-FET Channels

K. Tyszka[1,2], D. Moraru[3], T. Mizuno[1], R. Jablonski[2] and M. Tabe[1]

[1]Research Institute of Electronics, Shizuoka University, 3-5-1 Johoku Hamamatsu 432-8011, Japan

[2]Institute of Metrology and Biomedical Engineering, Warsaw University of Technology, Poland

[3]Faculty of Engineering, Shizuoka University, 3-5-1 Johoku Hamamatsu 432-8011, Japan

Email: tyszka@rie.shizuoka.ac.jp

Abstract — **We study the statistical way to control the position and number of lowest-energy donor-induced quantum dots (QDs) formed in the channels of SOI-FETs, even using the conventional random doping process. By Kelvin probe force microscopy (KPFM) and simulations, we find that selective-patterned-doping works effectively to form a unique QD nearby the central position of the pattern, associated with bell-shaped potential background due to positive donors. It should be noted, however, that the controllability of position and number is gradually lost with increasing screening by coverage of free electrons.**

I. INTRODUCTION

Donor atoms work as QDs in nanoscale-channel Si transistors [1-4] even under random donor distributions. Although a few papers have reported studies of donor-QDs with respect to "patterned channel" or "patterned doping" [4-6], controllability of the position and the number of lowest-energy QDs has not been directly studied. In this work, in order to clarify the controllability of naturally-formed QDs, we study the effects of pattern-doping in a wide concentration range by KPFM and corresponding simulations. We find that patterning the doping profile works effectively for the formation of a unique QD with a nearby central position within the pattern.

II. KPFM MEASUREMENTS AND SIMULATIONS

We fabricated SOI-FETs, as shown in **Fig. 1**(a) [6]. Doping with phosphorus (P) donors was done by thermal diffusion either uniformly or selectively to form a doped-slit across the channel. Two different doping concentrations were used: $N_D^H \approx 1 \times 10^{19}$ cm^{-3} (case H) and $N_D^L \approx 1 \times 10^{18}$ cm^{-3} (case L). The channel is depleted of electrons by applying negative substrate bias (V_{sub}=-4 V), while source and drain are all the time grounded [7,8]. In simulation, surface potential is obtained by self-consistently solving Poisson and Thomas-Fermi equations, as schematically shown in **Fig. 2**(a).

III. OBSERVATION OF DEEPEST-POTENTIAL QDS

KPFM images measured at room temperature for V_{sub}=-4 V [**Figs. 1**(b)-(e)] and corresponding simulation [**Figs. 2**(b)-(e)] are shown for N_D^H and N_D^L, and for channels with and without selective-doping patterns. Darker contrast indicates lower electronic potential, while circles mark several QDs with deep potentials (solid circle indicates the deepest QD and dotted circles are second deepest).

Case H. **Fig. 1**(b) shows the potential of a channel slit-doped with concentration N_D^H, whereas **Fig. 1**(c) shows that of uniform doping after removal of bell-shaped background. Dark spots (QDs) are ascribed mostly to clusters of P-donors. Although there is some statistical fluctuation, it can be seen that the patterned-doping is helpful in position- and number-control of the deepest QDs. In **Fig. 1**(c), several deep-potential QDs become now dispersed at scattered positions. Corresponding simulations [**Figs. 2**(b)-(c)] show consistent results.

Case L. **Fig. 1**(d) shows the potential for N_D^L. In this concentration regime, most of the QDs are likely to be formed by individual P-donors. Corresponding simulations are shown in **Figs. 2**(d)-(e). Again, for this lower concentration, the deepest QD is found to be closer to the center position, when slit-doping is used. **Fig. 2**(f) schematically shows the effect of the slit-doping pattern on donor-induced potentials.

IV. STATISTICAL ANALYSIS OF PATTERN EFFECT

In **Fig. 3**, we plot the average distance of deepest-potential QD from the center position as a function of N_D for 50 different donor-distributions. It is obvious that, for patterned-channels, the QDs tend to be located near the center, especially in high N_D range. **Fig. 4** shows the number of competing (comparable-energy) QDs in certain energy windows of 5 or 30 mV from bottom of the channel potential as a function of N_D for fully-depleted channels. The slit-doping is obviously advantageous for the formation of a unique (single) QD, in particular, for 30 mV window.

Once more positive V_{sub} is applied, free electrons are injected from source and drain into the channel. Then, the positively-charged donor potential is significantly screened, leading to flatter potential landscape. Therefore, the number of lowest QDs is also sensitive to electron screening (electron coverage fraction), as shown in **Fig. 5**. For high screening, an increasing number of competing QDs is expected, but a unique QD can be expected for full depletion (f_e=0%) in a wide range of N_D.

V. CONCLUSIONS

We showed, by direct KPFM observations and supporting simulations, that patterned-doping of FET channels is effective for the control of location of donor-induced QDs (near center) and for the formation of a single QD. Further tuning of channel doping pattern and electron screening can provide improved controllability of donor-QDs in location and number in a wide range of concentrations.

ACKNOWLEDGMENTS

This work was supported by MEXT Kakenhi (23226009, 25630144) & Coop. Res. Project (RIE, Shizuoka Univ.).

REFERENCES

[1] H. Sellier *et al.*, Phys. Rev. Lett. **97**, 206805 (2008).
[2] G.P. Lansbergen *et al.*, Nature Phys. **4**, 656 (2008).
[3] M. Pierre *et al.*, Nature Nanotechnol. **5**, 133 (2010).
[4] M. Tabe *et al.*, Phys. Rev. Lett. **105**, 016803 (2010).
[5] E. Hamid *et al.*, Phys. Rev. B **87**, 085420 (2013).
[6] D. Moraru *et al.*, Sci. Rep. **4**, 6219 (2014).
[7] M. Ligowski *et al.*, Appl. Phys. Lett. **93**, 142101 (2008).
[8] M. Anwar *et al.*, Appl. Phys. Lett. **99**, 213101 (2011).
[9] G. A. Thomas *et al.*, Phys. Rev. B **23** 5472 (1981).

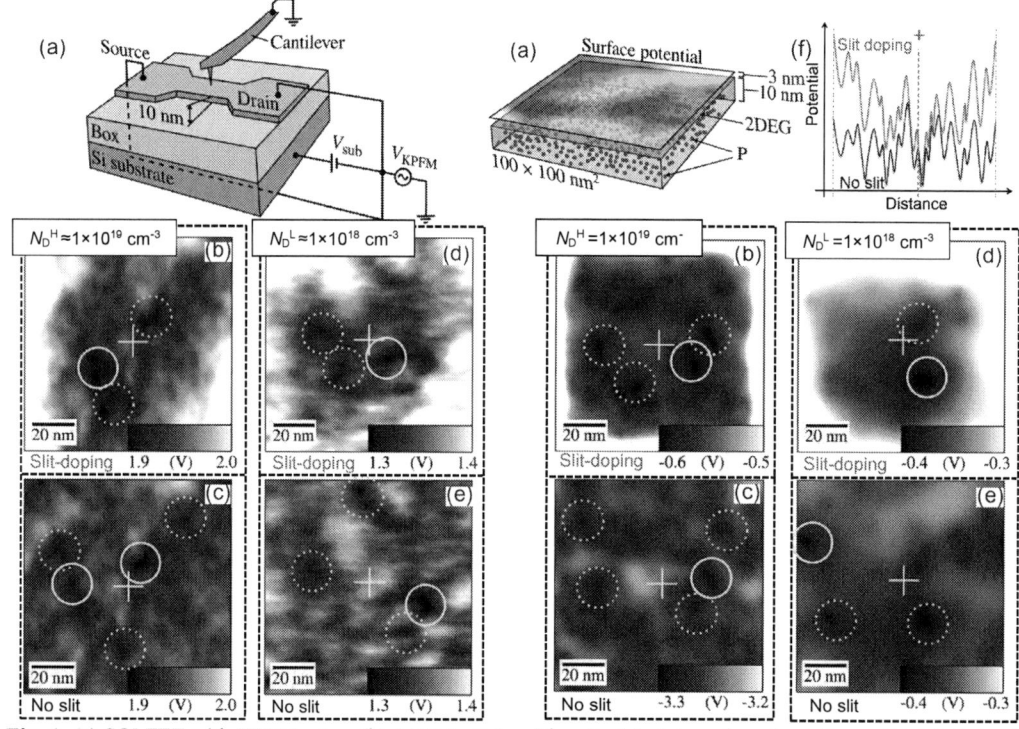

Fig. 1. (a) SOI-FET with KPFM setup. (b)-(e) Potential landscapes for $N_D^H \approx 1 \times 10^{19}$ cm^{-3} (left) and $N_D^L \approx 1 \times 10^{18}$ cm^{-3} (right). For all images, V_{sub}=-4 V. Top panels: slit-doping pattern. Bottom panels: uniform doping (no slit). Circles indicate deepest QDs.

Fig. 2. (a) Setup for simulations. (b)-(e) Potential landscapes simulated for $N_D^H \approx 1 \times 10^{19}$ cm^{-3} (left) and $N_D^L \approx 1 \times 10^{18}$ cm^{-3} (right). Top panels: slit-doping pattern. Bottom panels: uniform doping (no slit). (f) An illustration of the effect of slit-doping on potential.

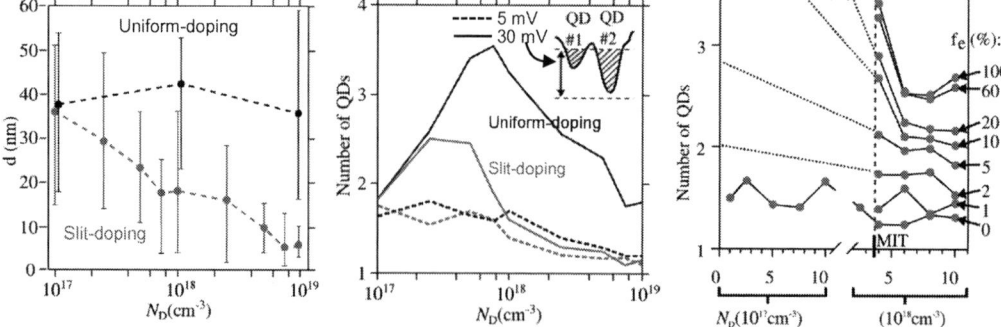

Fig. 3. Average distance of deepest QD from the center (50 samples) for slit-doped channels (bottom line) and uniformly-doped channels (upper line).

Fig. 4. Number of QDs for slit-doped channels (red) and uniformly-doped channels (black). Windows: 5 mV and 30 mV (inset).

Fig. 5. Number of QDs as a function of N_D (below and above metal-insulator transition) for different degrees of screening (f_e).

Nanodamascene metal-insulator-metal single electron transistor prepared by atomic layer deposition of tunnel barrier and subsequent reduction of metal surface oxide

Golnaz Karbasian, Alexei O. Orlov, and Gregory L. Snider

Electrical Engineering Department, University of Notre Dame, IN 46556

Golnaz.Karbasian.1@nd.edu

Abstract — We present an experimental demonstration of a metallic single electron transistor fabricated using plasma enhanced atomic layer deposition (PEALD) of tunnel barrier dielectric followed by reduction of the metal surface that was oxidized during ALD. We found that ALD deposition of a thin SiO_2 layer results in the formation of NiO on the surface of Ni that significantly increases the effective tunnel barrier thickness in $Ni-SiO_2-Ni$ tunnel junctions. We demonstrate that NiO can be fully reduced to Ni by annealing in a 95% Ar - 5% H_2 ambient at 400 °C.

I. INTRODUCTION

Single-electron transistors (SETs) are the most sensitive electrometers to date [1]. Fabrication of metallic SETs has always been challenging mainly due to lithographic constraints and difficulty in forming ultra-thin tunneling barriers with a low dielectric constant on metals. Atomic layer deposition (ALD) is a promising method to form different dielectrics on various types of substrates with monolayer precision. Utilizing plasma enhanced atomic layer deposition of SiO_2 on Ni, we successfully fabricated metallic SETs.

II. EXPERIMENT

A. Device preparation

Figure 1 illustrates the steps in the fabrication process. SET structures were fabricated on a thermally grown SiO_2 insulating layer on a silicon wafer (Fig. 1a). First, the pattern of an isolated vertical "island" is formed in poly methylglutarimide (PMGI) using a Vistec EBPG 5000 100 keV electron beam lithography (EBL) system. PMGI has been shown to have a higher contrast than poly methylmethacrylate (PMMA) and can be used as an etch mask due to its higher etch resistance [2]. The pattern of the island in then etched in the oxide using Ar, C_4F_8, CHF_3, and CF_4 based inductively coupled plasma (ICP) and Ni is deposited to fill the trench in SiO_2 (Fig. 1b and c). Chemical mechanical polishing (CMP) is used to remove the Ni overburden on the oxide, while leaving the island trench filled with Ni (Fig. 1d). Next, an SiO_2 PEALD process is performed to form a dielectric barrier covering the island, followed by the annealing in a 95% Ar-5% H_2 ambient at 400 °C (Fig. 1e). Finally, Ni source and drain are defined by a second EBL and

liftoff (Fig. 1f). An image of a finished device in a scanning electron microscope (SEM) is shown in Figure 2.

B. Result and discussion

Our experiments show that the NiO formed during 2 SiO_2 ALD cycles (~1.5 Å of SiO_2), increases the tunneling resistance to greater than 25 MΩ. Upon reducing the surface NiO back to metallic Ni, the resistance drops to 0.8 kΩ ($R_s \approx 20$ Ω/□), proving that 1.5 Å SiO_2 barrier is transparent to electron tunneling. We have found that after the anneal step, 13 cycles of ALD SiO_2 leads to resistance of 3 MΩ that is well above the quantum resistance ($R_Q = h/e^2 \approx 25.8$ kΩ) required for operation of SETs [3].

Figure 3 shows the charge stability diagram, "Coulomb diamonds", of a finished device at 5.4K. Junction capacitances, extracted from fitting simulations to experimental data, also in line with the geometry of the measured device, are approximately 40 aF. This amounts to the charging energy of 1.95 meV ($E_c = e^2/2C$, where C is the total capacitance of the island). Every step used in this process is completely compatible with CMOS fabrication, which enables the integration of SETs with standard integrated circuits (ICs).

ACKNOWLEDGEMENTS

This work was supported by National Science Foundation grants CHE-1124762 and DMR-1207394.

REFERENCES

[1] G. Zimmerli, *et al.*, "Noise in the Coulomb blockade electrometer," *Appl. Phys. Lett.*, vol. 61, pp. 237-239, 1992.

[2] G. Karbasian, *et al.*, "High aspect ratio features in poly(methylglutarimide) using electron beam lithography and solvent developers," *J. Vac. Sci. Technol., B,* vol. 30, p. 06FI01, 2012.

[3] M. H. Devoret and H. Grabert, "Introduction to Single Charge Tunneling," in *Single Charge Tunneling* vol. 294, H. Grabert and M. Devoret, Eds., Springer, 1992, pp. 1-19.

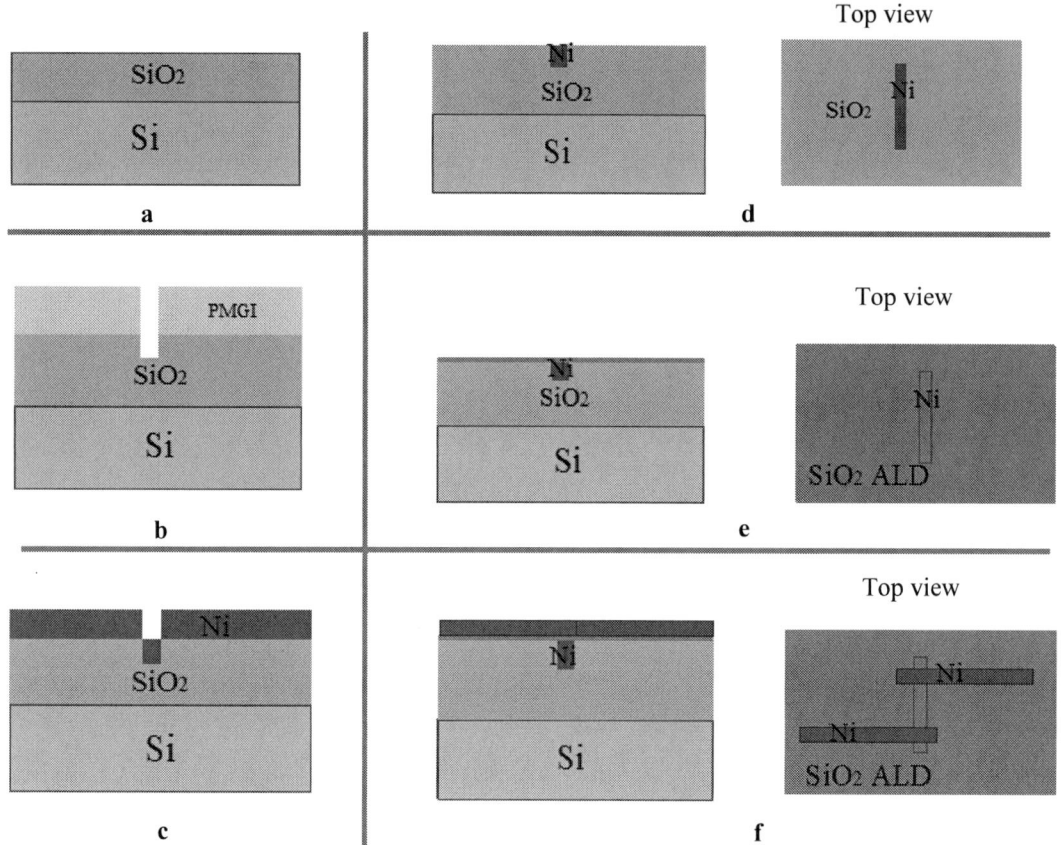

Figure1 Process steps in fabricating the single electron transistor.

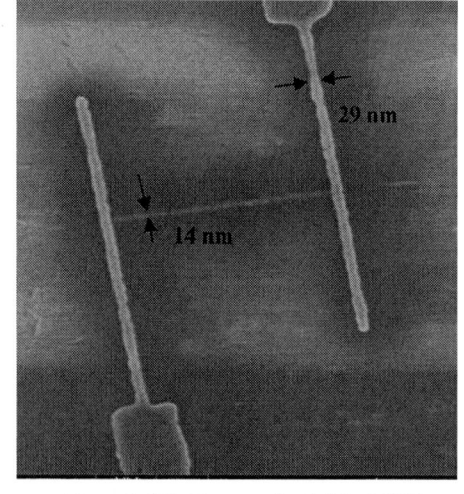

Figure2 SEM image of a finished SET.

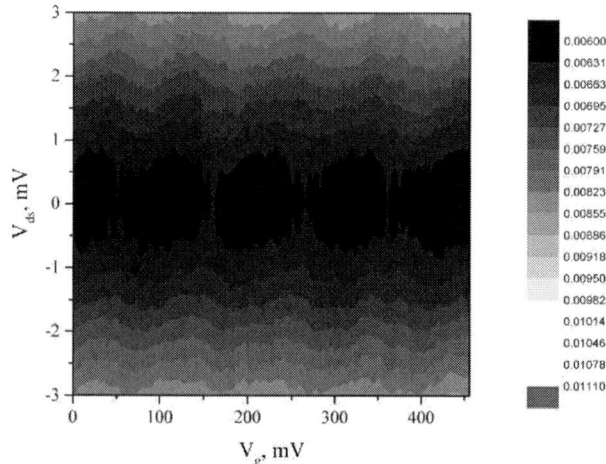

Figure3 Coulomb diamonds of a finished device at 5.4K.

Impacts of Channel Constriction Dimensions of Graphene Single-Carrier Transistors on the Coulomb Diamond Characteristics

Takuya Iwasaki[1], Manoharan Muruganathan[1], and Hiroshi Mizuta[1,2]

[1] School of Material Science, Japan Advanced Institute of Science and Technology (JAIST),
1-1 Asahidai, Nomi, Ishikawa 923-1292, Japan

[2]Nanoelectronics and Nanotechnologies Research Group, Electronics and Computer Science, Faculty of Physical Sciences
and Engineering, University of Southampton, Highfield, Southampton SO17 1BJ, United Kingdom

Phone: +81-761-51-1573, FAX: +81-761-51-1149, Email: s1440002@jaist.ac.jp

Abstract — We present a new engineering method to achieve high-performance graphene single-carrier transistors (GSCTs) by suppressing the formation of unwanted carrier (electrons/holes) puddles in GSCTs by comparing the three devices with different constriction dimensions. GSCTs with wider and longer constrictions exhibited the presence of multiple carrier puddles in the channel regions. The GSCTs with narrower constrictions showed the fully-lifted blockade regions in the Coulomb diamond characteristics, which indicate the absence of the multiple quantum dots in the channel. As the channel constrictions size is reduced to the graphene intrinsic carrier puddles dimension, formation of unwanted carrier puddles is avoided.

I. INTRODUCTION

Graphene-based quantum dots (GQDs) is expected to be a promising candidate for the spin quantum bit for the quantum information processing system [1]. Graphene single-carrier transistors (GSCTs) are fabricated by the geometrically defined GQD along with two channel constrictions [2]. However, defects, edge-disorder, and random strain in the etched graphene channel lead to the formation of unintentional carrier (electrons/holes) puddles, which act as QDs [3]. In this paper, we report an approach to avoid the uncontrolled formation of these carrier puddles in GSCTs by scaling down the channel constriction regions.

II. EXPERIMENT

Fig. 1(a) shows the GSCTs fabrication process flow. Initially we transferred the graphene flakes from the highly ordered pyrolytic graphite to the n-doped Si substrate with 300 nm SiO$_2$ by mechanical exfoliation. Raman spectroscopy was used to identify the number of graphene layers. After that, the contact electrodes were fabricated by deposition of Ti/Au and the lift-off processes. Then, electron beam lithography (EBL) resist, Polymethyl methacrylate or Hydrogen

silsesquioxane, was spun onto the samples. The EBL exposure and the reactive ion etching were then performed to define the GSCT structure containing a single GQD and two constrictions. Three GSCTs were fabricated with various sizes of the GQDs and constrictions as shown in Fig 1(b). GSCT with wider (75 nm) and longer (200 nm) constrictions is called as Device A, and 55 nm and 45 nm constriction width GSCTs are called as Device B and Device C, respectively. Electric measurements were carried out in the Helium cryostat and the cryogenic prober station as shown in Fig. 1(c).

III. RESULTS AND DISCUSSION

Figures 2(a)-(c) show the backgate modulation characteristics for Device A, B, and C, respectively. Device A shows the multiple peaks, which changed with V_d values. The background current increased with the increase in the V_d values. Device B exhibits Coulomb oscillation characteristic with a number of clear Coulomb oscillation peaks, which are stable over the whole range of V_d. In contrast to Device A and B, Device C shows the Coulomb oscillation characteristic with remarkably suppressed background current.

Coulomb diamond measurement results for the three devices are presented in Figs. 3(a)-(c). Device A shows no Coulomb blockade lifting, which is attributed to the presence of strongly-coupled multiple QDs defined by induced carrier puddles in the longer and wider constrictions. Device B Coulomb diamond characteristics exhibits the presence of few QDs. On the other hand, Device C (narrowest constrictions) shows the Coulomb diamond characteristic with fully-lifted blockade regions. These characteristics suggest that the formation of unwanted carrier puddles are suppressed with the decrease in the channel constriction dimensions. The intrinsic disorder length associated with the carrier puddles in graphene is reported to be 30 nm [4]. As Device C constrictions width is around 45 nm, the formation of unwanted carrier puddles are avoided.

These results clearly manifest the importance of controlling the channel constriction dimensions in order to realize GSCTs with well-controlled Coulomb blockade properties.

ACKNOWLEDGEMENT

This work is supported by Grant-in-Aid for Scientific Research No. 25220904 from Japan Society for the Promotion of Science and the Center of Innovation Program from Japan Science and Technology Agency.

REFERENCES

[1] B. Trauzrttel, D. V. Bulaev, D. Loss, and G. Burkard, "Spin qubit in graphene quantum dots", *Nature Phys.*, vol. 3, no. 3, pp. 192-196, Feb. 2007.

[2] T. Ihn, J. Güttinger, F. Molitor, S. Schnez, E. Schurtenberger, A. Jacobsen, S. Hellmüller, T. Frey, S. Dröscher, C. Stampfer, and K. Ensslin, "Graphene single-electron transistors", *Materials Today*, vol. 13, no. 3, pp. 44-50, Mar. 2010.

[3] P. Gallagher, K. Todd, and D. G. Gordon, "Disorder-induced gap behavior in graphene nanoribbons", *Phys. Rev. B*, vol. 81, no. 11, 115409, Mar. 2010.

[4] J. Martin, N. Akerman, G. Ulbricht, T. Lohmann, J. H. Smet, K. Von Klitzing, and A. Yacoby, "Observation of electron-hole puddles in graphene using a scanning single-electron transistor", *Nature Phys.*, vol. 4, no. 2, pp. 144-148, Nov. 2008.

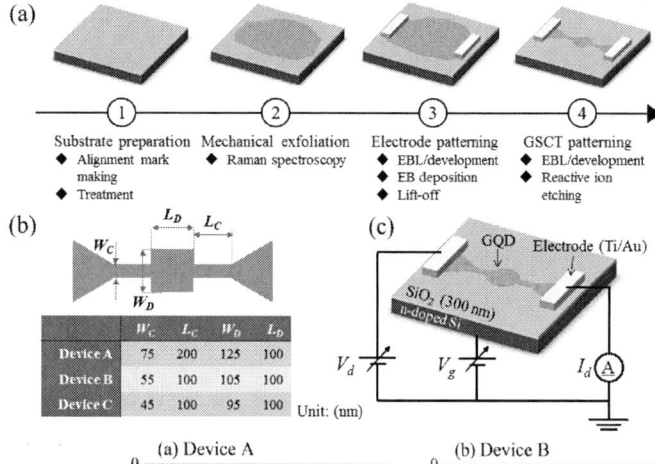

Fig. 1 . (a) Fabrication processes of the GSCTs. (b) Schematic and table for the design dimension of the dot and constriction for our three GSCTs. W_C, L_C, W_D, and L_D represent the constriction width, length, dot width, and length, respectively. (c) Electric measurement configuration for the GSCTs.

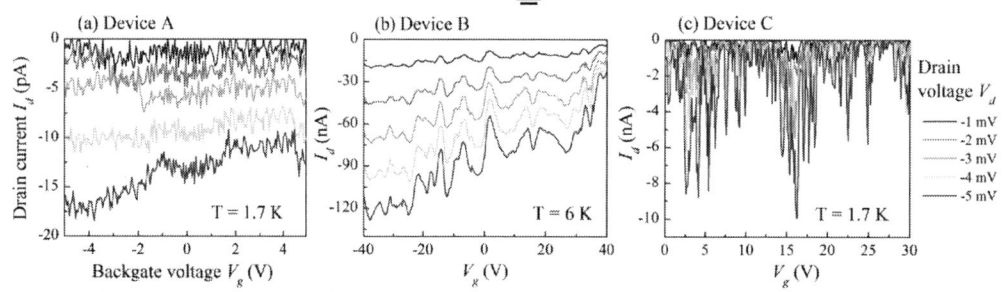

Fig. 2 . Drain current I_d plot as a function of backgate voltage V_g for (a) Device A, (b) B , and (c) C, respectively. The legend in (c) is common (a) and (b).

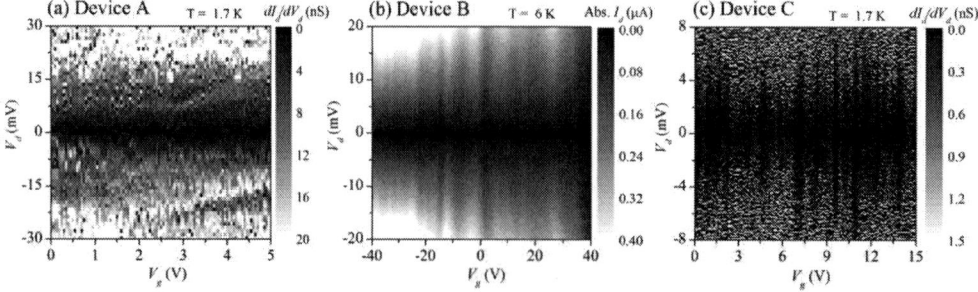

Fig. 3 . (a),(c) Differential conductance dI_d/dV_d, and (b) the absolute value of I_d as a function of V_d and V_g for (a) Device A, (b) B, and (c) C, respectively. Dark area corresponds to the low conductance regions.

Series-triple quantum dots fabricated under each control gate by the use of thermal oxidation

Takafumi Uchida[1], Hikaru Sato[1], Atsushi Tsurumaki-Fukuchi[1],
Masashi Arita[1], Akira Fujiwara[2], and Yasuo Takahashi[1]

[1] Graduate School of Information Science and Technology, Hokkaido University, Sapporo 060-0814, Japan

[2] NTT Basic Research Laboratories, NTT Corporation, 3-1 Morinosato Wakamiya, Atsugi, 243-0198, Japan

Email: takafumi-uchida@frontier.hokudai.ac.jp

Abstract —**Triple quantum dots (QDs) connected in series were successfully fabricated by the use of pattern-dependent oxidation (PADOX) [1] of a Si nanowire and additional oxidation through the gap of three fine gate electrodes attached on the Si nanowire. This method made a QD just under each gate. The fabricated devices were confirmed that they actually consisted of three dots by the electrical measurements, in which gate capacitances between the gates and dots were successfully evaluated. These results demonstrated the applicability of PADOX method to make many QDs connected in series by increasing the number of fine gate electrodes attaching on the nanowire.**

I. INTRODUCTION

Capacitively coupled quantum-dot-systems are usable as the tools for quantum information processing [2,3] and development of a current standard [4,5]. For improving the performances in these applications, it would be required to increase the number of coupled QDs. To reduce the total device size, we have to develop a new fabrication and compact design methods of QDs together with the gate electrodes attached to each dot. Here, we propose a method to make Si QDs connected in series by the use of thermal oxidation through the gap of poly-Si gate electrodes which formed on a Si nanowire. Series-triple QDs were actually fabricated in the Si nanowire, and the formation was confirmed by the electrical measurements.

II. EXPERIMENTAL DETAILS

The schematic cross section of our triple QDs device is shown in Fig. 1. The Si nanowire was formed by electron-beam lithography and dry etching of the top Si layer on a SOI wafer. The width and length of the nanowire were 40 nm and 220 nm, respectively. The tunnel barriers at both sides of the nanowire were formed by using PADOX of the nanowire at 1000°C in the dry oxygen ambient. Three fine gate electrodes (G1, G2, G3) of 40 nm width, made of n-type poly-silicon, were attached on the nanowire with an inter-electrode gap of 70 nm. The tunnel barriers between the QDs were formed by oxidation through gaps between the gate

electrodes at 700°C and 1000°C, and the series-triple QDs (QD1, QD2, QD3) were fabricated in the nanowire under the gate electrodes [6]. After depositing 50-nm-thick SiO_2 gate insulator, a top gate electrode was attached. Formation of the triple QDs was investigated by measuring electrical characteristics under application of gate voltages (V_1, V_2, and V_3) from the three different gates of G1, G2, and G3. The equivalent circuit with the triple QDs is shown in Fig. 2. In this experiments, all the gate capacitances were evaluated from three kinds of charge stability diagrams measured at about 8 K and the drain voltage of 5 mV. In the evaluation, the inter-dot coupling capacitances, C_{M1} and C_{M2}, were ignored.

III. RESULTS AND DISCUSSION

Fig. 3 shows the contour plots of the drain current (I_D) as a function of V_1 and V_2 measured at constant V_3. This corresponds to the charge stability diagram of the double QDs composed of QD1 and QD2 because the effect of QD3 was almost negligible due to small variation in V_2. The dotted and broken lines represent the boundaries of charge transition in QD1 and QD2, respectively. The gate capacitances evaluated from the period and slope of these lines are shown in the table of Fig. 3. Similarly, the charge stability diagram of QD2 and QD3 measured at fixed V_1 and the estimated gate capacitances are shown in Fig. 4. On the other hand, the charge stability diagram as a function of V_1 and V_3 measured at fixed V_2 was relatively complicated because of the effect of QD2. The charge transitions in the triple QDs were observed with three kinds of slopes shown in Fig. 5. The gate capacitances evaluated from Fig. 5 were almost consistent with those evaluated from Figs. 3 and 4. Small errors are thought to come from the fact that the gate capacitance of a small Si QD was not constant but depended on the number of electrons in QDs [6]. The averaged gate capacitances are summarized in Table I. From the results, each QD is thought to be formed under each gate electrode since the largest gate capacitance for each QD was obtained by the overlying gate for each QD. In other words, each QD can be mainly controlled by the gate on the QD. All the gate capacitances of the triple QDs can be evaluated using the method described above.

Table I. Averaged gate capacitances of triple quantum dots calculated from the tables in Fig. 3 - 5.

Gate	QD1 [aF]	QD2 [aF]	QD3 [aF]
G1	C_{11}: 3.25	C_{12}: 0.23	C_{13}: 0.06
G2	C_{21}: 2.29	C_{22}: 3.42	C_{23}: 1.10
G3	C_{31}: 0.03	C_{32}: 0.58	C_{33}: 1.41

IV. SUMMARY

Triple QDs were successfully fabricated by the use of PADOX of Si nanowire and an additional oxidation through the gap of three fine gates. All the gate capacitances were evaluated from the three kinds of charge stability diagrams. This results clearly show that each QD was formed under each gate electrode. This fact implies that the number of quantum dot in the Si nanowire can be increased when the number of gate electrodes attached on the nanowire is increased.

ACKNOWLEDGEMENT

This work was partly supported by the KAKENHI by MEXT and JSPS (24360128, 25420279, 26630141).

REFERENCES

[1] Y. Takahashi *et al.*, Electron. Lett., **31**, 136 (1995).
[2] A. M. Tyryshkin *et al.*, Nat. Mater., **11**, 143 (2012).
[3] M. Veldhorst *et al.*, Nat. Nanotechnol., **9**, 981 (2014).
[4] G. Yamahata *et al.*, Appl. Phys. Lett., **98**, 222104 (2011).
[5] A. Rossi *et al.*, Nano. Lett., **14**, 3405 (2014).
[6] T. Uchida *et al.*, J. Appl. Phys., **117**, 084316 (2015).

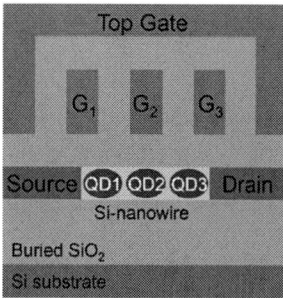

Fig. 1. Schematic cross section of triple quantum dots device.

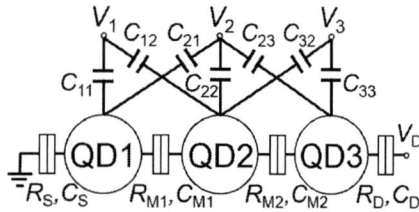

Fig. 2. Schematic equivalent circuit of triple quantum dots with three gate voltages. For simplicity, C_{13} and C_{31} are not shown.

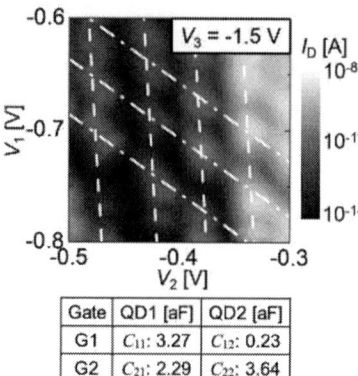

Gate	QD1 [aF]	QD2 [aF]
G1	C_{11}: 3.27	C_{12}: 0.23
G2	C_{21}: 2.29	C_{22}: 3.64

Fig. 3. Charge stability diagram of QD1 and QD2 as a function of V_1 and V_2. The dotted and broken lines represent the charge transition in QD1 and QD2, respectively. The table at the bottom shows the gate capacitances caluculated from the periods and slopes of these lines.

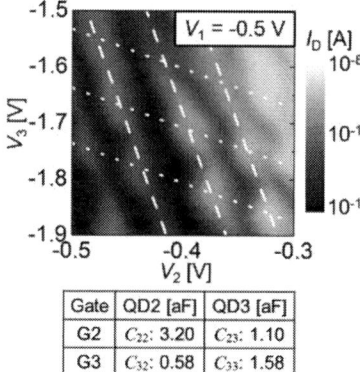

Gate	QD2 [aF]	QD3 [aF]
G2	C_{22}: 3.20	C_{23}: 1.10
G3	C_{32}: 0.58	C_{33}: 1.58

Fig. 4. Charge stability diagram of QD2 and QD3 as a function of V_2 and V_3. The broken and chain lines represent the charge transition in QD2 and QD3, respectively. The gate capacitances caluculated from the periods and slopes of these lines are shown at the bottom table.

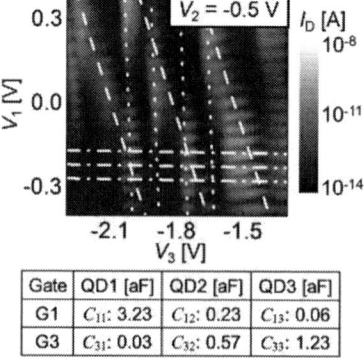

Gate	QD1 [aF]	QD2 [aF]	QD3 [aF]
G1	C_{11}: 3.23	C_{12}: 0.23	C_{13}: 0.06
G3	C_{31}: 0.03	C_{32}: 0.57	C_{33}: 1.23

Fig. 5. Charge stability diagram of QD1, QD2, and QD3 as a function of V_1 and V_3. The dotted, broken, and chain lines represent the charge transition in QD1, QD2, and QD3, respectively. The gate capacitances caluculated from the periods and slopes of these lines are shown at the bottom table.

Fabrication and characterization of physically-defined double quantum dots without unintentional localized states on highly-doped silicon substrate

Y. Yamaoka, T. Kodera, S. Oda

Department of Physical Electronics, Quantum Nanoelectronics Research Center, Tokyo Institute of Technology, Japan
Email: yamaoka.y.aa@m.titech.ac.jp

Abstract — **We fabricated double quantum dots (DQDs) and a single electron transistor (SET) as a charge sensor (CS) on highly-doped silicon-on-insulator (SOI) substrate. We observed Coulomb oscillations, and Coulomb diamonds in the SET, and estimated its charging energy to be ~15meV. We detected the change in the number of electrons of the DQDs with the CS, and observed a honeycomb-like charge stability diagram which is typical characteristics for DQDs. The regular shape of the charge stability diagram indicates that we succeeded in fabricating the DQDs without unintentional localized states.**

INTRODUCTION

Qubits based on electron spins in silicon double quantum dots (DQDs) have been well studied recently, since long coherence time is expected because of weak hyper fine interaction and spin-orbit interaction [1]. To implement the electron spin qubits, it is necessary to fabricate DQDs as designed without unintentional localized states [2]. On highly-doped silicon substrate, to obtain DQDs as designed has been considered to be difficult because of fluctuated dopant potentials.

Here, we improved the precision of electron beam lithography (EBL) and dry etching techniques, and then fabricated physically well-defined DQDs on highly-doped silicon-on-insulator (SOI) substrate. Measured transport properties show we obtained the DQDs as designed without unintentional localized states. This indicates that fluctuated dopant potentials may not be so problematic in our current device size scale (~30 nm in diameter), because there are still many dopants (several thousand) in our device. If the QD size becomes smaller, in other words, the number of dopants becomes smaller down to ~10 or less, then fluctuated dopant potentials would affect electron transport.

DEVICE FABRICATION

We implanted PH_3 at a dose of 1.0×10^{15} cm^{-2} into the 30-nm-thick SOI through 10-nm-thick oxide. This dose creates effective carrier concentration of 1.0×10^{19} cm^{-3} after annealing. We need this value in the leads for ideal Ohmic contacts with metal electrodes. Then we fabricated DQDs and a single electron transistor (SET) as a charge sensor (CS) on the highly-doped SOI substrate by EBL and SF_6 dry etching.

Figure 1 shows a scanning electron microscope (SEM) image of the device. It is composed of the DQDs with two side gates and the CS with one side gate. We can control the energy levels of the QDs by applying voltage to the side gates.

MEASUREMENT RESULTS AND DISCUSSION

The results of the measurement at 4.2 K are shown in Fig. 2, 3, and 4. Figure 2 shows the current I_{CS} through the CS as a function of the side gate voltage V_{sg} when drain voltage $V_{CS\_d} = 1$ mV and source electrode is grounded. We obtained clear Coulomb oscillation, which is typical for SET. Figure 3 shows the I_{CS}/V_{CS_ds} as a function of the side gate voltage V_{sg} and drain-source voltage V_{CS_ds}. We observed Coulomb diamond, from which we estimated charging energy E_C to be ~15 meV [3]. The charge stability diagram of the DQDs was obtained by measuring the transconductance dI_{CS}/dV_{sg1} of the CS (Figure 4 (a)) [2,4]. Abrupt change in I_{CS} corresponds to the charge transition in the DQDs. The charge transition lines of the DQDs are indicated in Figure 4 (b). It reveals that the change in the number of electrons in the DQDs is detected by the CS. The regular shape of the charge stability diagram indicates that we successfully obtained the DQDs without unintentional localized states.

CONCLUSIONS

In conclusion, we fabricated DQDs and CS on highly-doped SOI substrate without unintentional localized states. The CS detected the charge transitions in the DQDs. This work indicates that device using the DQDs on highly-doped SOI substrate has still potential for quantum information processing.

ACKNOWLEDGEMENT

Part of this work is financially supported by Kakenhi Grants-in-Aid (Nos. 26709023, 26630151, and 26249048), the Murata Science Foundation, and the Project for Developing Innovation Systems of the Ministry of Education, Culture, Sports, Science and Technology (MEXT) of Japan.

REFERENCES

[1] B. M. Maune, *et al.*, "Coherent single-triplet oscillations in a silicon-based double quantum dot," *Nature* 481, 344-347 (2012).

[2] K. Horibe, *et al.*, "Lithographycally defined few-electron silicon quantum dots based on a silicon-on-insulator substrate," Appl. Phys. Lett. 106, 083111 (2015).

[3] L P Kouwenhoven, *et al.*, "Few-electron quantum dots," *Rep. Prog. Phys.* 64, 701-736 (2001)

[4] W. G. van der Wiel *et al.*, "Electron transport through double quantum dots," Rev. Mod. Phys., Vol. 75, No. 1, January (2003).

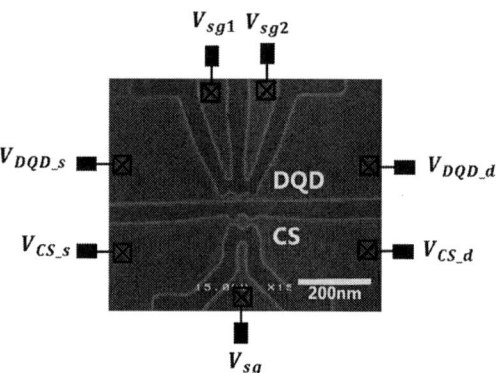

Figure 1 SEM image of the device. Dark region is BOX and bright region is SOI. The upper part is DQD and the lower part is CS.

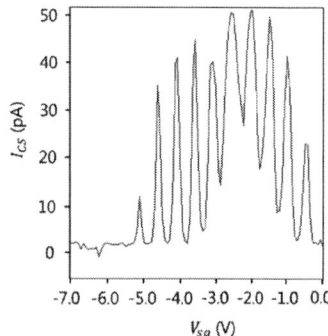

Figure 2 Coulomb oscillation of the CS when V_{CS_ds}=1mV.

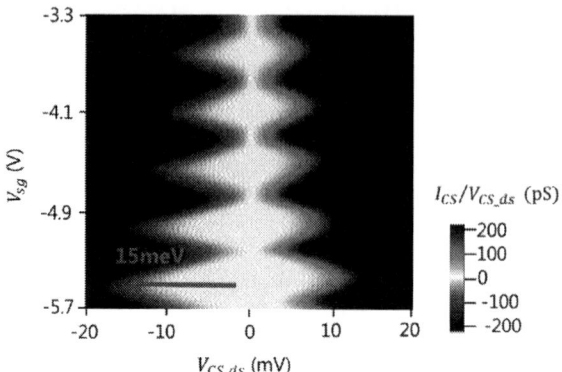

Figure 3 Coulomb diamond of the CS, from which the charging energy E_C is estimated to be 15 meV.

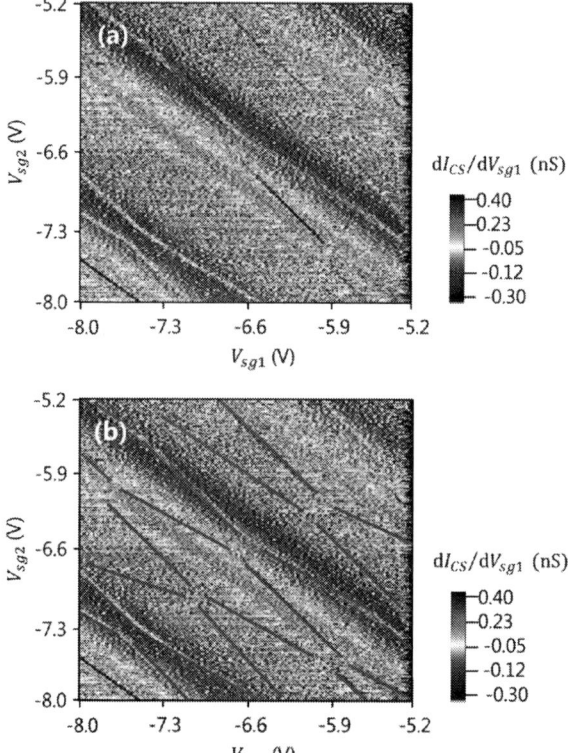

Figure 4 Charge sensing experiment. (a) Charge stability diagram of the DQDs obtained by measuring the transconductance dI_{CS}/dV_{sg1} of the CS. (b) Charge transition lines of the DQDs are indicated in the same figure as Fig. 4(a).

Characterization of carrier dynamics in Ge quantum dots through Ge quantum-dot MOSFETs using pulsed voltage technique

Ming-Hao Kuo,[1] Ho-Chane Chen,[1] Wei-Ting Lai,[1,2] and Pei-Wen Li[1,2]

[1]Department of Electrical Engineering, National Central University, Chung Li, Taiwan, 320, Republic of China
[2]Department of Electronic Engineering, National Chiao Tung University, Hsin Chu, Taiwan, 300, Republic of China
Email: pwli@nctu.edu.tw

Abstract — We demonstrated a novel CMOS approach for the fabrication of high-performance Ge quantum dot (QD) MOSFETs offering great promises as optical switches and transducers for Si-based optical interconnects. Measured photocurrent to dark current ratio (I_{ph}/I_{dark}) and photoresponsivities from the Ge QD PT were as high as 272,000 and 1.7 A/W, respectively, under incident power of 20 μW at 980 nm illumination. Carrier dynamics in Ge QDs were characterized by applying a depleting voltage pulse to the gate of MOSFET and by time-resolved photoluminescence on Ge QD array.

I. INTRODUCTION

Since their inception in the early 1980s semiconductor quantum dots (QDs) have found widespread application in advanced electronics, photonics, memories, precision metrology and biosensing [1-3]. Thanks to quantum confinement effects, electronic structures and optical properties of a semiconductor QD are no longer governed by its material composition alone but rather also depend on its size, shape, and embedding matrix. Contingent advantages of size-tunable transition energy, enhanced oscillator strength, enhanced electron-electron interactions, less electron-phonon scattering, and intermediate-band formation between QDs not only improve performance of QD-based photonic devices but also open up completely new features and functionality.

To obtain optimally desired properties of QD-based devices, fundamental understanding of steady-state carrier transport and carrier dynamics within specific QDs is crucial. This in turn requires well-controlled growth techniques, proper test structures/platforms, and efficient, accurate characterization methods. Electronic and optical properties of QDs are most commonly studied using current-voltage (*I-V*), capacitance-voltage (*C-V*) and photoluminescence (PL) techniques, which usually require a large ensemble of QDs with high uniformity in QD size/shape in a large spatial distribution.

In this letter, we reported that nanometer-scale Ge QDs with desired sizes, locations, and depths of penetration into the Si substrate are achievable by thermal oxidation of poly-SiGe nano-pillars over buffer Si_3N_4 layers on the Si substrate [4]. The designer Ge QDs were then incorporated into the gate stack of Si MOSFETs for enhancing light absorption and producing high-responsivity Ge-QD MOS phototransistors. Steady-state *I-V* characteristics measured on our Ge-QD phototransistor at 300 K showed an extremely low off-state leakage ($I_{off} \sim 9.25$ fA/μm^2) in the darkness, high responsivity of 1.7 A/W and fast temporal response time of 1.4ns under 850nm illumination. In this paper, we

further employed time-resolved PL (TRPL) and pulsed gate voltage techniques to study carrier dynamics including the generation lifetime for minority carriers and recombination lifetime for photocarriers within the Ge QDs by driving the Ge-QD MOSFETs from accumulation through deep depletion to the ultimate thermal equilibrium states in the darkness and under illumination at temperature 150–300K.

II. EXPERIMENTAL RESULTS AND DISCUSSION

Process flow and the corresponding TEM/EDX micrograph of Ge-QD phototransistors are summarized in Fig. 1. Designer Ge QDs with desired size (50-100 nm), location and depth of penetration into the Si substrate are generated using the control available through lithographical nano-patterning and selective oxidation of SiGe pillars grown over buffer Si_3N_4 layers on Si substrates. Typical *dc* I-V_g characteristics of Ge-QD MOSFETs ($W_g/L_g = 50$μm/3μm) with a gate oxide thickness of 15nm measured under variable-power 980nm illumination at 300 K are displayed in Fig. 2. MOSFETs with no QDs are also measured for comparison. It is clearly to see that 980nm light pumping induces significant current enhancement for Ge-QD MOSFETs as a result of QD-induced tremendous photocarrier generation, whereas no observable changes in current of MOSFETs with no QDs by illumination or not.

To gain insight on dynamics of thermal carrier and photocarrier within Ge QDs, depleting voltage pulses with various durations are applied to the gate of Ge-QD MOSFETs in the darkness and under 980nm illumination, respectively, to drive them from the initial accumulation into non-equilibrium deep depletion, and ultimately to thermal equilibrium inversion when the duration of voltage pulses is greater than the recovery time, t_f (Fig. 3−6). The recovery time is determined by (1) thermal electron-hole pair (ehp) generation properties of the Si substrate and (2) gate oxide/Si channel interface, (3) injection efficiency of minority carriers from n$^+$-source, (4) photogenerated ehp by illumination, and (5) ehp from Ge QDs (Fig. 1c). Time-resolved current of MOSFETs containing no QDs exhibits a typical sharp current rise followed by a very gentle current dip as the space-charge region (scr) under the gate varies with time. This is paralleled by a capacitance decrease in deep depletion. Once the inversion layer forms and the gate-induced depletion is pinned to its final value, the surface generation component diminishes and the current reaches to a *dc* steady-state value. Extracted thermal generation time, $t_{f,Si}$, for electrons in Si channel is 0.2ms. A dramatic different transient current behaviour is observed for Ge-QD MOSFET with a sharp current overshoot in the deep-depletion domain, indicating that tremendous ehps generation induced by Ge QDs predominates the Si surface generation and effectively shortens the generation lifetime by approximately one order in magnitude ($t_{f,QD} \sim 30\mu s$). The predominance of ephs induced by the Ge QDs over that in Si is further evidenced by a significant enhancement in the overshooting current peak by reducing temperature and by illumination. Another interesting observation is that under low-power illumination ($<1.2\mu W$), current overshoot increases and t_f reduces with increasing illumination power due to excess photocarrier generation induced by Ge-QD. Whereas high-power optical pumping ($>1.2\mu W$) reduces not only t_f but also declines overshoot current peak with a concurrent appearance of a non-diminishing current plateau when turning off the voltage pulse. Measured PL lifetime of the Ge QDs at 20K is approximately 1.7ns (Figure 7). According to the temperature dependency of PL intensities, we estimate that the recombination lifetime at 300K is shorter than 0.2ns. Both generation and recombination lifetimes of carriers in Ge QDs are obviously shorter than those in Si, enabling fast switching for optical interconnection.

III. CONCLUSION

High-performance Ge-QD PTs have been demonstrated in a CMOS compatible approach. The Ge-QD PT exhibits significant enhancement in photocurrent and high photoresponsivity thanks to a strong absorption of Ge QD based on both photoconductive and photovolatic effects, offering a great promise for future Si-based optical interconnection applications.

ACKNOWLEDGEMEN

This work was supported by the Ministry of Science and Technology of R. O. C. (MOST 102-2221-E-009-195-MY3).

REFERENCES

[1] L. Robledo et al., *Science*, vol. **320**, no.5877 (208).
[2] O. Astafiev et al., *Nature*, vol. **449**, no.7162 (2007).
[3] J. M. Yang et al., *ACS Nano*, vol. **5**, no.6 (2011).
[4] M. H. Kuo et al., *Appl. Phys. Lett.*, **101**, 223107 (2012)

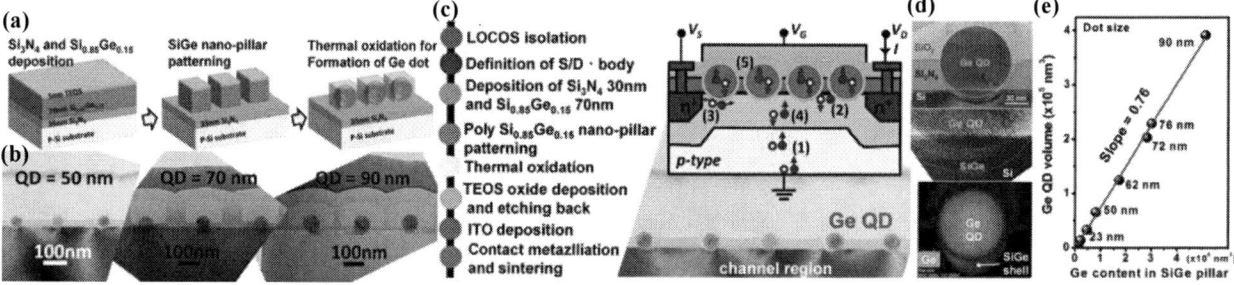

Fig. 1 Designer Ge dots with desired sizes and spatial locations fabricated by selective oxidation of SiGe nano-pillars. (a–c) experimental procedure for the production of Ge QDs and Ge-QD phototransistor and the corresponding CTEM images of the SiO$_2$/50–90nm Ge dots/Si$_3$N$_4$ heterostructures after thermal oxidation, (d) HRTEM and EDX of Ge QDs/SiO$_2$/SiGe heterostructure formed in a single one-step oxidation, and (e) size "tunability" relationship of the Ge dots is based on the Ge content of the un-oxidized Si$_{0.85}$Ge$_{0.15}$ nano-pillars.

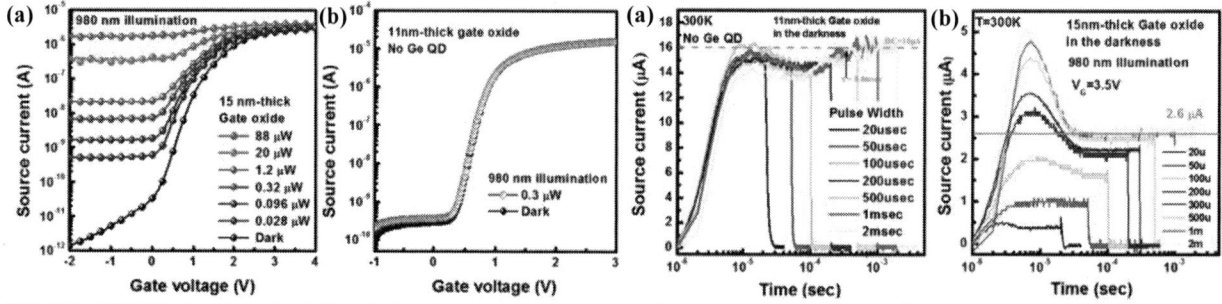

Fig. 2 DC steady-state I-V_g characteristics of MOSFETs with and without QDs measured at 300K in the darkness and under 980nm illumination.

Fig. 3 Transient current measured by applying voltage pulse with various pulse duration to the gate of MOSFETs with and without Ge QDs at 300K.

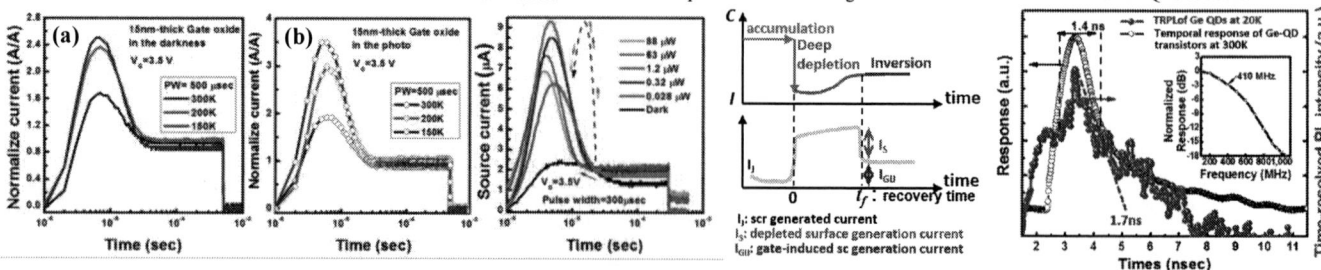

Fig. 4 Transient current of Ge-QD MOSFETs measured at 150–300K in the darkness and under illumination.

Fig. 5 Power-dependent transient current of Ge-QD MOSFETs at 150K.

Fig. 6 Deep-depletion MOSFETs characteristics: C-t and I-t transient.

Fig. 7 Temporal and frequency response of the Ge-QD MOSFETs under 0.6 mW, 850 nm illumination.

50

Simultaneous two gate reflectometric spectroscopy of Si coupled donor-dot system

Xavier Jehl[1], Alexei O. Orlov[2], Romain Maurand[1], Patrick Fay[2], Gregory L. Snider[2],
Sylvain Barraud[3], and Marc Sanquer[1]

[1]DSM-INAC, CEA-Grenoble, France
[2]Department of Electrical Engineering, University of Notre Dame, USA
[3]DRT-Leti, CEA-Grenoble, France
Email: aorlov@gmail.com

Abstract — Gate-coupled reflectometric spectrometry has recently emerged as a tool for studies of transport in nanostructures. Here we report reflectometric spectroscopy of a double-gate single electron device in which two coupled quantum dots are formed under the gates. The spectroscopy is performed by detection of charging processes in the system using a dual port reflectometer. The potential application of this scheme for detection of qubits is also discussed.

I. INTRODUCTION

Double quantum dots are the central building blocks for the realization of spin qubits in solid-state devices nowadays. Characterization of these systems is typically performed by means of external charge detection, wherein separate sensors (either single-electron transistors or quantum point contacts) are integrated with the device/system under test. In practice, these external sensors are often difficult to implement and limit the ability to scale up qubit-based systems. The use of reflectometric measurements of nanoscale charge-based devices has gained in popularity over the past few years, since this technique provides a unique combination of fast and sensitive charge sensing that does not require external detectors. Here we report the operation of a dual-channel, double-gate reflectometer in which two gate sensors are used to probe the charge state of the device.

II. EXPERIMENTAL RESULTS

A. Samples and experimental setup

A Si nanowire surrounded by two gates was fabricated using fully depleted SOI [1] (Fig. 1a); spacers above undoped regions form the tunnel barriers. Transport in these devices at temperatures <20 K is dominated by Coulomb blockade, and the population of the dots formed by the two top gates and the back gate is controlled by the applied voltages, V_{g1}, V_{g2}, and V_{bg}. To investigate charging processes in the system, we use reflectometric spectroscopy with two channels acquired simultaneously [2] (Fig. 1b). Note that the resonant networks at each gate (composed of surface mount inductors and parasitic components) provide the resonators needed for high charge sensitivity.

B. Experimental result and discussion

Figure 2 shows the results of gate reflectometry measurements obtained for a device of the type shown in Fig. 1a. We concentrate here on a regime that corresponds to a charge exchange between a centered 20nm×40nm dot accumulated near the buried oxide ($V_{bg} \gg 0V$, equally coupled to V_{g1} and V_{g2}) and a donor in the channel near gate 1 (Fig. 2). By employing a sensor at gate 2 we can detect charge population change in a donor state located next to this gate (Fig 2 b,d). Near $V_{g1}=0.05V$, however, there is a break in the line, beyond which there is no detectable signal. However, the signal from other senor at gate 1 provides the necessary information: one can see that the negatively sloped line in Fig. 2 a, c, which corresponds to charging of a central dot near $V_{g1}=0.05V$, $V_{g2}=-0.2V$, experiences a perturbation stemming from charge exchange with the donor state. This charge transition along the dashed line in Fig. 2 can be exploited for qubit operation. The demonstrated two-gate sensing scheme enables unambiguous identification of charging processes. Our study points the way toward a sensitive reflectometry scheme that would allow single-shot readout of silicon spin qubits and other nanoscaled logic devices such as quantum-dot cellular automata [3].

ACKNOWLEGEMENTS

R. Maurand, S. Barraud, X. Jehl, and M. Sanquer acknowledge financial support from the EU through FP7 initiative under Project TOLOP (318397) and SiAM (610637); A. Orlov and P. Fay acknowledge support of the NSF Foundation grant DMR-1207394.

REFERENCES

[1] S. Barraud *et al.*, *Elec. Dev. Lett.*, 33, 1526 (2012).
[2] A.O.Orlov *et al.*, *IEEE nanotechnology*, (2015).
[3] Lent C. S. *et al.*, *Nanotechnology* (1993).

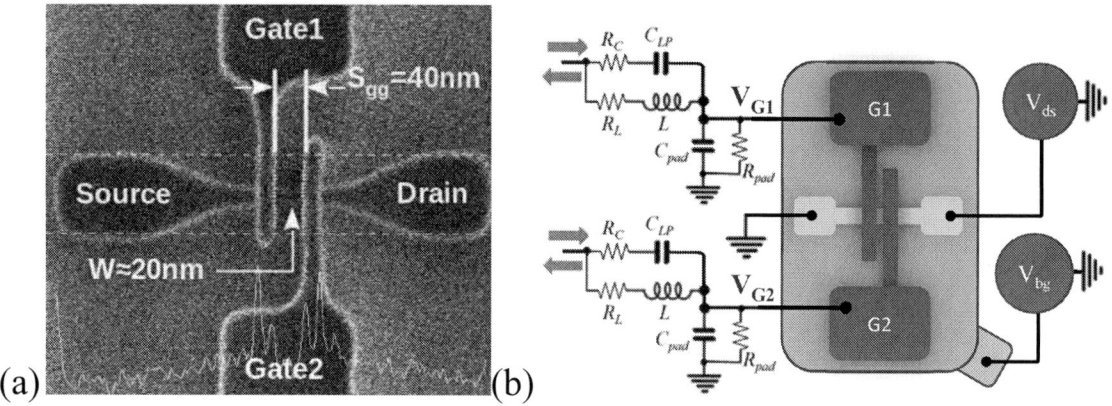

Figure 1. (a) SEM micrograph of a device nominally identical to the one studied. W=20 nm, separation between the gates is about 40 nm, the nanowire thickness is 11 nm; (b) sketch of the dual channel reflectometry experiment. Realistic model networks sensitive to high load impedances are shown. Blue arrows indicate RF incoming and reflected signals at two different frequencies, f_{g1} = 332 MHz and f_{g2} =414 MHz.

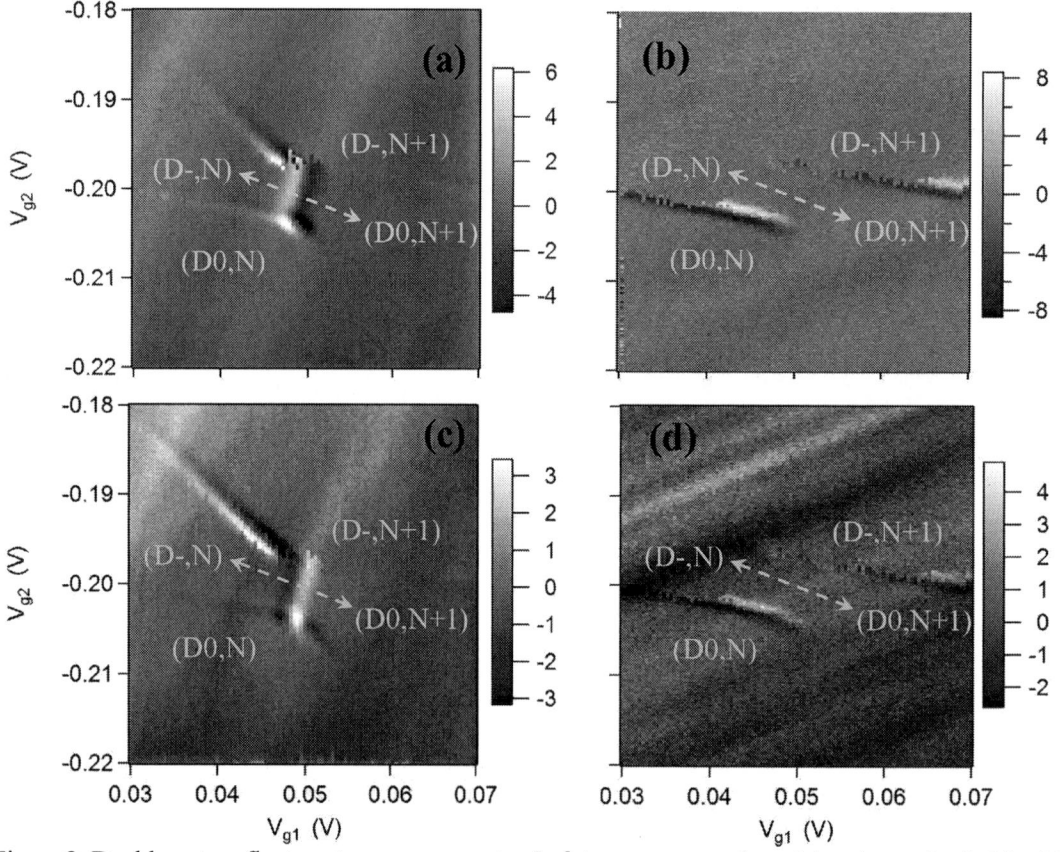

Figure 2. Double-gate reflectometry measurements. Left two maps are phase (a) and magnitude (c) of the signal reflected at gate 1; the two maps on the right are phase (b) and magnitude (d) of the signal reflected at gate 2. Gray scale represents the strength of the corresponding signals. The experiment is performed at a mixing chamber temperature of 100 mK and a back-gate voltage of V_{bg}=21.9V. Indices: D0 – ionized donor; D- – donor filled with an electron; N, N+1 – respective number of electrons populating central dot.

52

Fabrication of a highly controllable Si-MOS quantum dot device

Takumu Honda[1], Jun Yoneda[2,3], Kenta Takeda[2,3], Tetsuo Kodera[1,4], Seigo Tarucha[2,3,4], Shunri Oda[1]

[1]Department of Physical Electronics, Quantum Nanoelectronics Research Center, Tokyo Institute of Technology, Japan
[2]Department of Applied Physics, The University of Tokyo, Japan
[3]Center for Emergent Matter Science (CEMS), RIKEN, Japan
[4]Institute for Nano Quantum Information Electronics, The University of Tokyo, Japan
Email: honda.t.ag@m.titech.ac.jp

Abstract — Integrating a micro-magnet (MM) to a quantum dot (QD) is a promising route to realize highly coherent control of single electron spins on a chip [1]. In order to apply this spin-manipulation technique to Si QDs, miniaturizing the QD while maintaining the controllability is essential to lift the valley degeneracy [2]. Here we fabricated a MOS-QD under a thin gate oxide integrated with Al gates to accumulate and define a QD. At 4.2 K, we observed Coulomb diamonds indicating the formation of a QD with a charging energy E_C of 15 meV. This relatively large E_C for gate defined QDs encouragingly suggests that this device structure enables to reduce the size of fully-tunable QDs. We also claim that this structure with thinly oxidized Al gates is compatible with electron-spin-resonance (ESR) experiments with MMs.

I. INTRODUCTION

Electron spins in semiconductor QDs are a strong candidate for solid-state quantum bits and find many applications in spintronics. Si QDs are especially attractive since confined electron spins have long coherence times [3], owing to weak spin-orbit couplings and hyperfine interactions with nuclear spins. However, the size of Si QDs needs to be made small, due to its relatively heavy effective mass and small valley splitting [2]. This makes it challenging to achieve the full device controllability with multiple biasing gates; to operate spins electrically, one would need to tune precisely and independently the QD levels and the couplings. Here we report on the fabrication and measurement results of a Si-MOS QD as a promising device structure to meet these conditions. We note that this structure with oxidized Al gates will help to put a MM close to the QD, to facilitate high-speed ESR.

II. FABRICATION

We fabricate a Si-MOS QD device with Al gates on an undoped Si substrate (see Fig. 1(a)). The two-dimensional electron gas (2DEG) is formed at the interface of Si/SiO2, beneath the 10-nm-thick gate oxide. This shallow 2DEG will help to independently control the local confinement potentials by biasing surface gates. Two layers of Al gates are electrically isolated by alumina formed at 150 deg. C in air with a typical breakdown voltage above 5 V. This thin alumina will help to put a MM very close to QDs in future devices to mediate a fast electrically driven ESR under large local MM fields. Precise alignment between two layers of Al gates is achieved by using proximal Pt marks. This multiple exposure enables the miniaturization of QDs. Outside the device area is covered with a 200-nm-thick field oxide to avoid carrier accumulation. Ohmic contacts are formed by ion implantation, followed by the drive-in process at 1000 deg. C. Figure 1(b) shows a scanning electron microscope (SEM) image of the device. The top gate voltage V_{top} is applied to accumulate carriers in the 2DEG, while V_{pl} is to finely tune the electrochemical potentials in the QD. V_R and V_L are biased to control tunnel barriers between the QD and reservoirs. A 250-nm-thick Co is deposited using an e-beam evaporator as a MM.

III. RESULTS & DISCUSSION

To characterize the fabricated Si-MOS QD, we perform transport measurements on devices without MMs at 4.2 K. Figure 2 shows the current I_{sd} through the QD region with sweeping V_{top}. At the onset of current, it shows the Coulomb oscillation, a unique feature to QD devices. When V_{top} becomes more positive, it gradually shows FET characteristics. This shows that the topgate modulates the electron density and QD energies. Figure 3 plots I_{sd} as a function of V_{sd} and V_{pl}, when V_{top} is fixed at 0.78 V, and V_L and V_R are properly tuned. From the observed Coulomb diamonds, we extracted a large E_C of 15 meV as an indication of a decrease in the QD size. Assuming a flat-disk QD, the QD radius is calculated to be as small as 12 nm. This value is fairly small compared with other typical gate-defined QD systems [4]. The result shows that we have succeeded in fabricating a small QD system applicable to spintronics applications.

ACKNOWLEDGEMENT

Part of this work is financially supported by the SCOPE Program from the MIC, Japan; Kakenhi Grants-in-Aid (Nos. 26709023, 26630151, 26249048, and 26220710); the Murata Science Foundation; the Project for Developing Innovation Systems of the MEXT of Japan; the ImPACT Program from Cabinet Office, Government of Japan.

REFERENCES

[1] J. Yoneda et al., "Fast Electrical Control of Single Electron Spins in Quantum Dots with Vanishing Influence from Nuclear Spins," *Phys. Rev. Lett.*, vol. 113, pp. 267601, December 2014.

[2] X. Hao, et al., "Electron spin resonance and spin–valley physics in a silicon double quantum dot," *Nature communications*, vol. 5, pp. 3860, May 2014.

[3] E. Kawakami et al., "Electrical control of a long-lived spin qubit in a Si/SiGe quantum dot," *Nature nanotechnology*, vol. 9, pp. 666-670, September 2014.

[4] F. A. Zwanenburg, et al. "Silicon quantum electronics," *Rev. Mod. Phys.* vol. 85, pp. 961, July 2013.

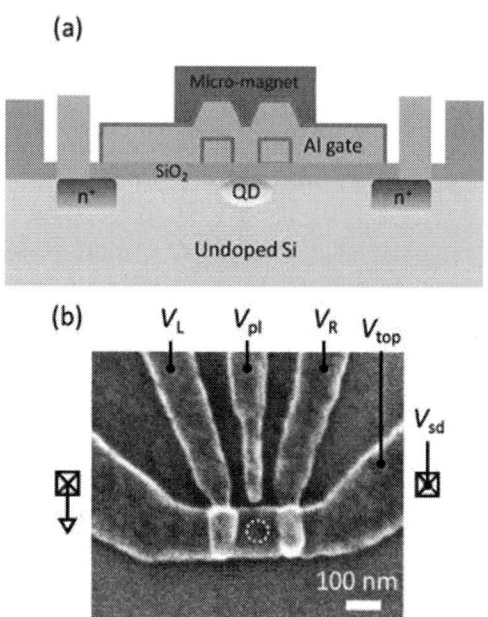

Fig 1. (a) Device structure. (b) SEM image of the QD device. The QD is formed in the yellow dotted circle.

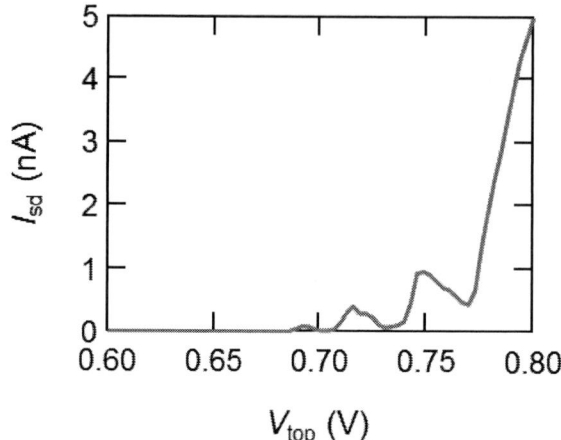

Fig. 2. QD current I_{sd} as a function of topgate bias V_{top}. Coulomb oscillations are observed with gradually overlapping FET characteristics.

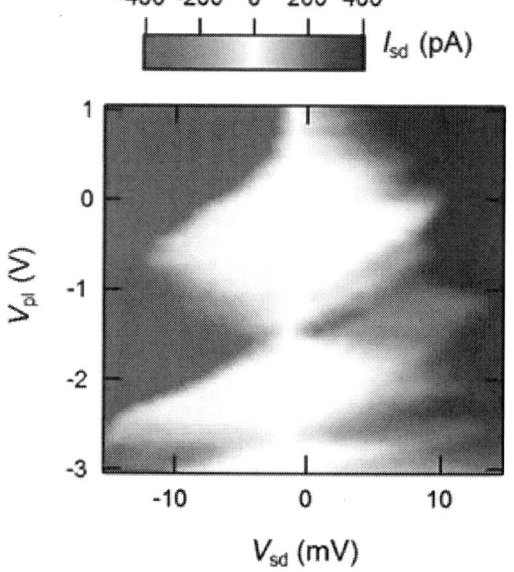

Fig. 3. Coulomb diamond characteristics monitored by the QD current as a function of V_{sd} and V_{pl}.

Tunneling Field-Effect Transistor with a Grown Si Epitaxial Layer for Boosting ON Current

Junil Lee[1], Jang Hyun Kim[1], Dae Woong Kwon[1], Euyhwan Park[1], Tae Hyung Park[1] and Byung-Gook Park[1,*]

[1]Inter-university Semiconductor Research Center (ISRC) and Department of Electrical and Computer Engineering,
Seoul National University, Gwanak-ro, Gwanak-gu, Seoul 151-742, Republic of Korea
Tel.: +82-2-880-7282, Fax: +82-2-882-4658, *Email: bgpark@snu.ac.kr

Abstract — **In this paper, we propose an advanced tunneling field-effect transistor (TFET) structure. The fabrication method of proposed TFET is designed. The characteristics of the proposed TFET are investigated by device simulation.**

I. INTRODUCTION

Reduction of supply voltage (V_{DD}) is one of the most important requirements for the Complementary Metal-Oxide-Semiconductor (CMOS) design. The dynamic and static power can be reduced by scaling V_{DD}. In order to decrease V_{DD} aggressively, tunnel field-effect transistors (TFETs) have been proposed [1]. Although the TFETs operate at very low V_{DD}, the TFETs have critical problem of low current drivability. In order to boost ON current, researchers have used an epitaxially grown layer as the channel or the source [2, 3]. In this work, we propose a TFET structure with an epitaxially grown layer as the channel and source and confirm the performance by TCAD simulation.

II. SIMULATION AND OPTIMIZATION

The proposed TFET structure is shown in Fig. 1(a) and the control TFET structure is shown in Fig. 1(b). TFET uses band-to-band tunnelling as a conduction mechanism as shown in Fig 1.(c). Key fabrication processes are shown in Fig. 2. The SiGe and Si layers were grown on the SOI substrate. The SiGe layer can be removed in highly selective isotropic etching process because the presence of germanium offers the possibility to distinguish it from Si [4]. The source can be refilled with highly doped silicon by selective epitaxial growth (SEG) after dry and wet etching process. The tunnelling rate of the proposed device is shown in Fig. 3. Electron generation by band-to-band tunnelling increases, as the gate bias voltage increases. Figure 4 shows the transfer characteristics of proposed TFET and planar TFET at the drain bias of 0.1 V and 1 V. The output characteristics are shown in Fig. 5.

To analyse the proposed TFET devices, simulations are conducted with various parameters in Fig. 1(a), T_{Si}, T_{SiGe}, and D_{SiGe}. First, the channel thickness, T_{Si}, is changed from 2 to 8 nm. Fig. 6 shows the effect of channel thickness. The subthreshold swing (SS) is reduced from 64 to 5 mV/dec. as T_{Si} is decreased. But, the threshold voltage (V_{Th}) does not change much. In a similar way (Fig. 7), T_{SiGe} is changed from 10 to 30 nm. There is little change of SS with different T_{SiGe}. So, T_{SiGe} is not critical for device performance. Lastly, simulation was done with various depth of etched SiGe (D_{SiGe}). The longer D_{SiGe} means that the tunnelling area is widened. Figure 8 shows the effect of increasing D_{SiGe}. SS and V_{Th} decreases with increasing D_{SiGe}. But, in case of D_{SiGe} = 45 nm, the channel potential is pinned at the drain bias.

III. CONCLUSION

In this paper, we designed an advanced TFET structure and investigated its properties in terms of various device dimensions such as T_{Si}, T_{SiGe} and D_{SiGe}. In addition, we introduced the fabrication method of the proposed TFET.

ACKNOWLEDGMENT

This work was supported by the Future Semiconductor Device Technology Development Program (10044842) funded By MOTIE (Ministry of Trade, Industry & Energy) and KSRC (Korea Semiconductor Research Consortium).

REFERENCES

[1] W. Y. Choi, B.-G. Park, J. D. Lee, and T.-J. King Liu, *Electron Devices Letters, IEEE*, No.8, Vol. 28 **(2007)**.

[2] R. Asra, K. V. R. M. Murali, and V. R. Rao, *Japanese Journal of Applied Physics*, 49, **(2010)**.

[3] A. Mallik, A. Chattopadhyay, S. Guin, and A. Karmakar, *Transaction on Electron Devices IEEE*, No. 3, Vol. 60, **(2013)**.

[4] S. Park, *Ph. D. Dissertation*, Dept. Electrical and Computer Science, Seoul National University, 2014, pp.71-73.

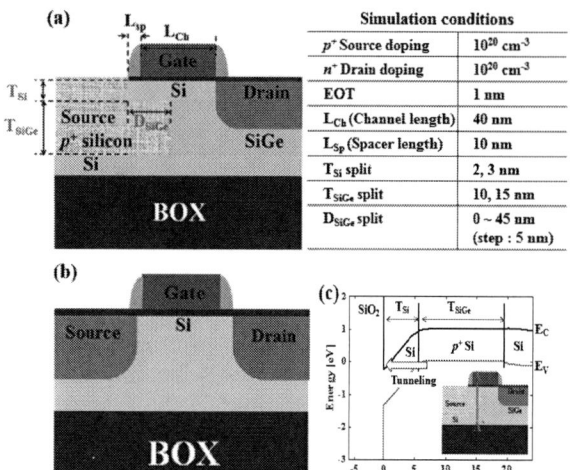

Fig. 1. (a) Schematic views of proposed TFET with selective epitaxial growth source and (b) conventional planar TFET with ion implanted source. Physical length of (a) and (b) are the same. (c) TCAD simulation of energy band diagram across AA'.

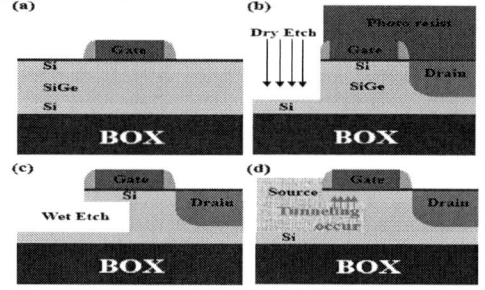

Fig. 2. Key processes of fabrication. (a) Gate etch and formation of sidewall spacer, (b) Half side dry etching after drain implant, (c) Selective wet etching of SiGe, (d) Source refilling with highly doped Si SEG.

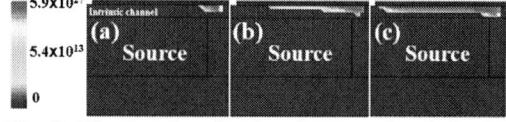

Fig. 3. Electron band to band generation it the intrinsic channel at V_{GS} (a) 0.7 V, (b) 0.9 V, (c) 1.1 V

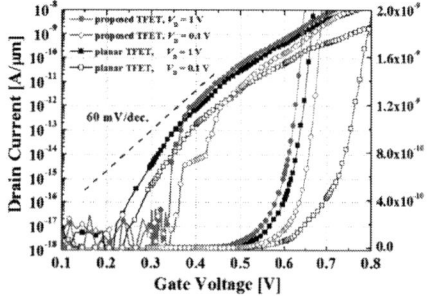

Fig. 4. Simulated transfer curves of proposed TFET in Fig. 1(a) and conventional planar TFET in Fig. 1(b).

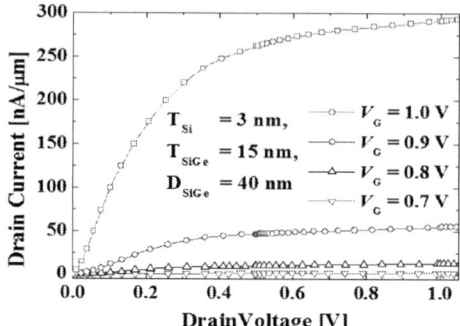

Fig. 5. Simulated output curves at different values of the gate voltage.

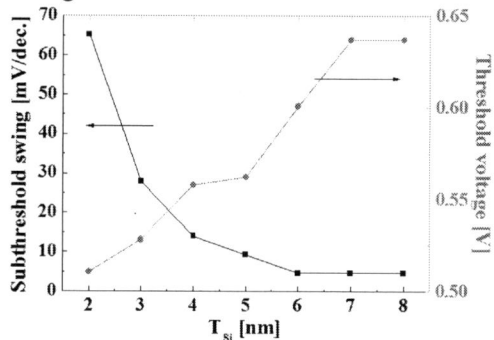

Fig. 6. Impact of T_{Si} in terms of *SS* and V_{Th}.

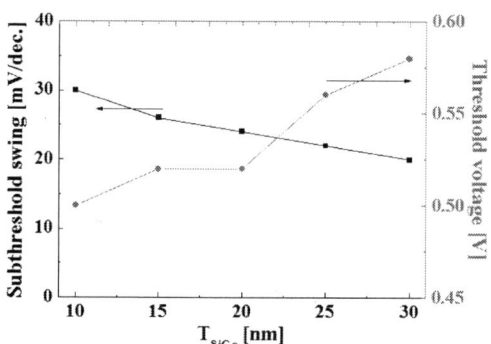

Fig. 7. Impact of T_{SiGe} in terms of *SS* and V_{Th}.

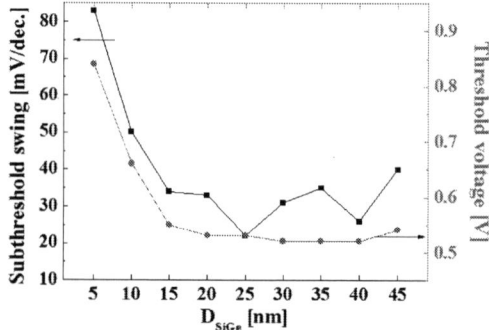

Fig. 8. Impact of D_{SiGe} in terms of *SS* and V_{Th}.

Short-Drain Effect of 5 nm Tunnel Field-Effect Transistors

Yu-Hsuan Chen[1], Nguyen Dang Chien[2], Jr-Jie Tsai[1], Yan-Xiang Luo[3], and Chun-Hsing Shih[1*]

[1]Department of Electrical Engineering, National Chi Nan University, Nantou 54561, Taiwan

[2]Faculty of Physics, University of Da Lat, Lam Dong 671463, Vietnam

[3]Institute of Electronics Engineering, National Tsing Hua University, Hsinchu 30013, Taiwan

[*]E-mail: shihch@ncnu.edu.tw

Abstract- **Unique short-drain effect was reported in extremely scaled tunnel field-effect transistors. This study numerically elucidated the short-drain effect and the difference between the short-channel and short-drain effects. For the short-drain effect, the off-state barriers reduce because of decreased drain lengths, generating ambipolar off-state current and degraded short-channel effect. The shorter channel is, the more pronounced is the short-drain effect.**

I. INTRODUCTION

Tunnel field-effect transistors (TFETs) have demonstrated to serve as an attractive candidate for energy-efficient applications [1]. Scaling the TFETs is essential to follow the pace of CMOS technologies. However, scaling TFETs down into sub-10 nm regimes is a considerable challenge [2], [3]. Different to the short-channel effect, this work reported the unique short-drain effect in extremely scaled TFETs. Two-dimensional simulations [4] were performed to elucidate the short-drain effect in sub-10 nm TFETs.

II. STRUCTURE AND PARAMETER

Fig. 1(a) shows the schematic structure of double-gate TFETs used in studies. This work employed a p+ source of 10^{20} cm^{-3} and a p-body of 10^{17} cm^{-3}. A gate-workfunction of 4.4 eV and 2 nm HfO$_2$ gate dielectric were used with 5 nm silicon body. An n-drain of 10^{18} cm^{-3} was utilized with a drain length (L_d) of 20 nm, unless otherwise specified. The 5 nm channel length (L_g) TFET was typified as an extremely scaled device.

III. RESULTS AND DISCUSSION

A. Short-Channel Effect

Fig. 1(b) shows the current-voltage curves of TFETs from L_g = 30 to 5 nm. The long-channel 30 or 20 nm TFETs exhibit favorable on-off switching. Notably, the short-channel effect produced degraded on-off transition and worse subthreshold swing in sub-10 nm TFETs. Fig. 2 presents the off- and on-state energy-band diagrams with various TFET lengths. The on-state tunnel barriers are almost unchanged, whereas the off-state tunnel barriers decrease because of scaling

L_g lengths. The shorter the channel length is, the more severe is the short-channel effect and thus the swing.

B. Short-Drain Effect

Both the drain and channel lengths are the key parameters in scaled TFETs to determine off-state tunnel barriers. Fig. 3 shows the (a) current-voltage curves and (b) off-state energy-band diagrams of 20 nm TFETs with various drain lengths. Different to the short-channel effect, the ambipolar off-state currents associate with the drain lengths, whereas the values of subthreshold swing are nearly unchanged. The off-state barriers reduce because of the decreased drain lengths. Fig. 4 and 5 sketch the drain current and energy-band diagrams of 10 and 5 nm TFETs, respectively. For sub-10nm TFETs, the short-drain effect couples with the short-channel effect, affecting both the subthreshold swing and the ambipolar off-state current. Notably, the effective off-state barrier relies highly on the drain length in sub-10 nm TFETs. Fig. 6 shows the minimum subthreshold swing of scaled TFETs with various drain lengths. For long-channel TFETs, variations of drain lengths produce negligible influence on the swing. However, the short-drain effect causes a substantial degradation in sub-10 nm TFETs. The shorter channel is, the more pronounced is the short-drain effect.

IV. SUMMARY

Unique short-drain effect was reported in scaled TFETs. Because the short-drain effect deteriorated in sub-10nm TFETs, the drain must be concurrently designed with the channel in extremely scaled TFETs.

REFERENCE

[1] Dmitri E. Nikonov and Ian A. Young, "Overview of beyond-CMOS devices and a uniform methodology for their benchmarking," *Proc. of the IEEE*, p. 2498, 2013.

[2] L. Liu, D. Mohata, and S. Datta, "Scaling length theory of double-gate interband tunnel field-effect transistors," *IEEE Trans. Electron Devices*, p. 902, 2012.

[3] C.-H. Shih and N. V. Kien, "Sub-10-nm asymmetric junctionless tunnel field-effect transistors," *IEEE Journal of the Electron Devices Society*, p. 128, 2014.

[4] Synopsys Sentaurus Device User Guide, Synopsys Inc., Mountain View, CA, USA, 2010.

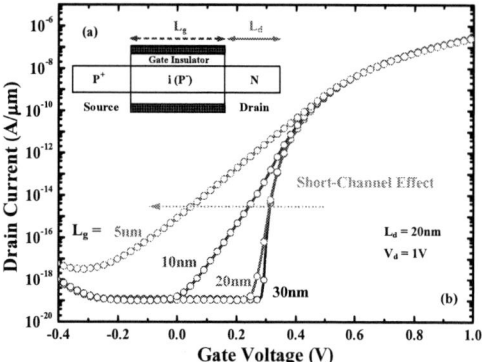

Fig. 1. (a) Schematic TFET structure. (b) Current-voltage curves of scaled TFETs. Short-channel effect caused degraded on-off switching in sub-10 nm TFETs.

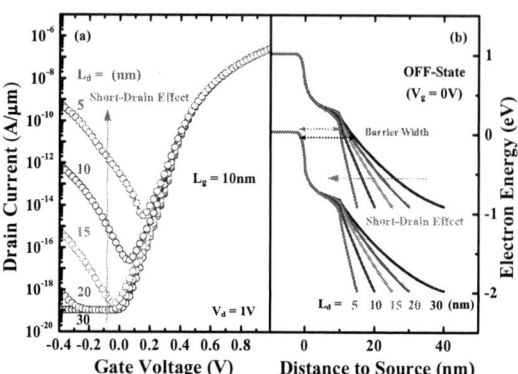

Fig. 4. (a) Current-voltage curves, and (b) Off-state energy-band diagrams, of 10 nm TFETs. The short-drain effect caused severe off current and worse swing.

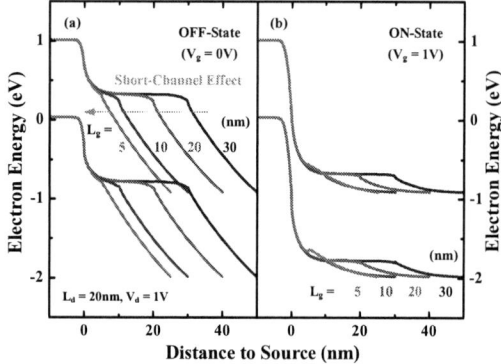

Fig. 2. (a) Off-state, and (b) On-state, energy-band diagrams of scaled TFETs with various channel lengths L_g from 30 nm to 5 nm.

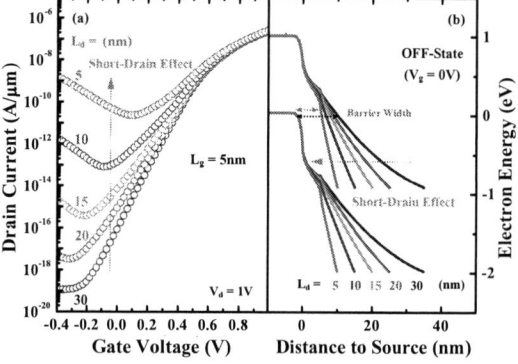

Fig. 5. (a) Current-voltage curves, and (b) Off-state energy-band diagrams, of 5 nm TFETs. The short-drain effect severely coupled with the short-channel effect.

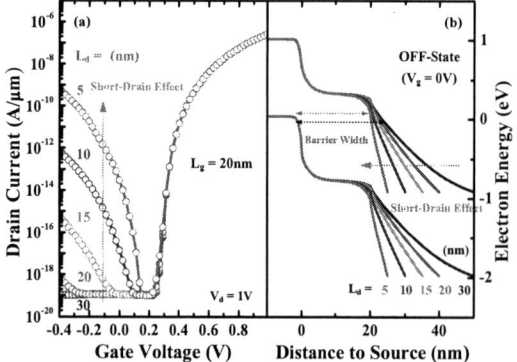

Fig. 3. (a) Current-voltage curves, and (b) Off-state energy-band diagrams, of 20 nm TFETs. The short-drain effect caused ambipolar off-state current.

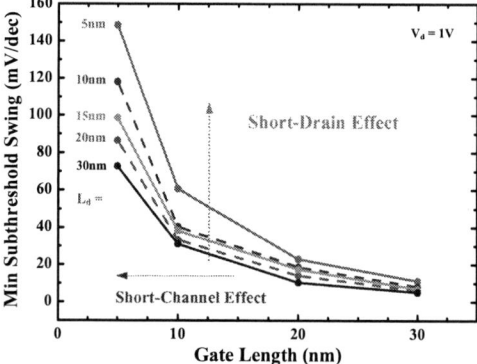

Fig. 6. Minimum subthreshold swing of scaled TFETs with various drain lengths. The short-drain effect was deteriorated as scaling down channel lengths.

Fabrication and Characterization of Silicon Nanowire Ultra-thin Channel Poly-Si Junctionless Field Effect Transistors with a Trench Structure

Ko-Wei Lin, Mu-Shih Yeh, Min-Hsin Wu and Yung-Chun Wu*

Department of Engineering and System Science, National Tsing Hua University, Hsinchu, Taiwan

Tel: +886-3-5715131 ext. 34287; Fax: +886-3-5720724; Email: cdstar9033@gmail.com

Abstract–This work demonstrates trench junctionless poly-Si thin-film transistor (JL-FET) [1] with ultra-thin body is obtained through dry etching process. JL-FET $L_G = 0.6\mu m$ shows excellent performance in a low drain-induced barrier lowering (DIBL), high I_{ON}/I_{OFF} (>10^8), excellent gate control and reduced sensitivity to temperature in terms of V_{TH} and SS.

I. Introduction

Scaling down of the IC device and 3D stacked is a recent studies. Nowadays the CMOS junction process is regarded as a critical issue. Therefore Junctionless-FETs (JL-FETs) seem the way to surmount the problem [2]-[5]. For the JL-FETs, the channel thickness of junctionless device is the most important parameter, which needs to be thin enough to deplete the heavily doped channel. In this work, we define different structures to discussion; the whole trench covered by gate (G_{out}) (Fig. 1a) and the gate inside the trench only (G_{in}) (Fig. 1b). L_G of G_{out} is the length of trench structure and L_G of G_{in} is defined by e-beam lithography [1].

II. Device Fabrication

Fig. 1(c) is the process flow of fabrication in the trench JL-FET. First, this trench JL-FET device's channel was a solid-phase-recrystallized (SPC) poly-Si film. The SPC layer was implanted BF_2 ions at a dose of 2×10^{14} cm^{-2}. The active NWs of the device were defined by e-beam lithography and reactive-ion etch (RIE). The trench structure was patterned by e-beam lithography and time-controlled RIE. Subsequently, an 8-nm-thick dry oxide was deposited as the gate oxide layer, consuming around 8-nm-thick poly-Si to form 6-nm-thick channel. Furthermore, 150-nm-thick in-situ doped n$^+$ poly-silicon deposition as a gate electrode, and pattern by e-beam lithography and RIE. A 200-nm-thick TEOS passivation layer was deposited. Finally, 300-nm-thick Al-Si-Cu metallization was performed and sintered.

III. Results and Discussion

Fig. 2 presents the trench JL-FET, including its Scanning Electron Microscope (SEM) images of NWs and Atomic Force Microscope (AFM) images of NWs. Fig. 3 displays a transmission electron microscopic (TEM) images, (a)-(b) show the interface of gate SiO_2/Si shows roughness, (c) presents the image along perpendicular to the gate direction; the NWs are surrounded by the gate electrode to form a trench structure with an effective width of 78

nm×10, $L_G = 0.6\mu m$ and $T_{CH} = 5.7$ nm.

Fig. 4 shows the I_D-V_G curve of trench JL-FET of G_{out} and G_{in} with $L_G = 0.6\mu m$. The G_{out} device has a better SS (115mV/dec.) and a lower DIBL (5 mV/V) than of the G_{in} device. But both of them have an apparent threshold voltage (V_{TH}) (-1.28/-1.71 V). V_{TH} refers to the gate voltage at $I_D = 10^{-9}$A. However, G_{in} has a lower OFF current; because it has smaller leakage region.

Fig. 5 shows the I_D-V_G and transconductance (g_m) characteristics of G_{out} and G_{in} trench JL-FET with $L_G = 0.6\mu m$. The G_{out} trench JL-FET has a better SS than G_{in} device. The g_m curves exhibit the G_{out} has a high maxima g_m. However, G_{out} has a high mobility, because g_m is directly proportional to mobility.

Fig. 6 shows the total resistance (R_{total}) of G_{out} and G_{in} trench JL-FET as a function of gate voltage at $V_d= -0.1$V. The relationship between the S/D series resistance (R_{SD}) and the R_{total} are described in the inset of Fig. 5. And R_{total} is inversely proportional to mobility. So we can confirm the truth that G_{out} has a high mobility again.

Fig. 7 compares the I_D-V_D output characteristics of the G_{out} and G_{in} NWs trench JL-FET for $L_G = 0.5$ μm, it shows that G_{out} device has a higher saturation current than G_{in} device.

Fig. 8 (a)-(b) show temperature dependence on I_D-V_G curves of G_{out} and G_{in} trench JL-FET. It reveals that the I_{ON} and V_{TH} exhibit a positive and negative shift with increasing temperature in both devices, owing to the poly-Si channel phenomenon. At high temperature, the carrier mobility is increased by emission out of the acceptor-liketrap level of grain boundaries.

IV. Conclusion

This poster demonstrated the fabrication and excellent performance of trench JL-FET. Fabrication process of the trench structure by dry etching is simple and easily integrated into the JL-FET device. The trench JL-FET has improved device characteristics such as a low OFF currents, a low SS (111 mV/dec.), a negligible DIBL (<5mV/V) and a high I_{ON}/I_{OFF} current ratio of 10^8. In addition to these improvements, our process is also compatible with the existing CMOS technologies. Such a trench JL-FET with NWs is promising for use in 3-D stacked IC applications.

V. Reference

[1] M.-S.Yeh et al., IEDM, p. 26-6, 2014.

[2] J.-P. Coling et al., Nature Nanotechnol., vol. 5, p. 225–229, 2010.

[3] S. Migita et al.,IEDM, p. 8.6.1–8.6.4., 2012.

[4] H.-B. Chen et al.,VLSI Technol. Circuits, pp. T232–T233., 2013.

[5] H.-C. Lin et al.,IEEE Trans. ED., vol. 60,p. 1142–1148, 2013.

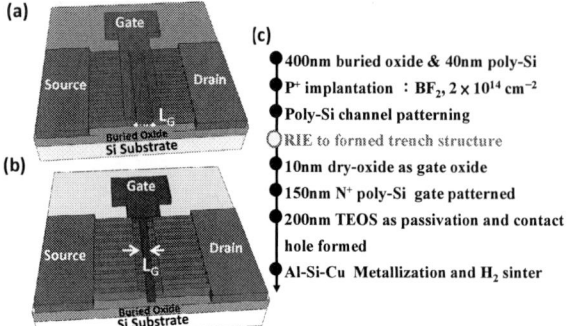

Fig. 1. (a) G_{out} structure trench JL-FET device. (b) G_{in} structure trench JL-FET device. (c) Details of the process flow chart of the fabricated devices.

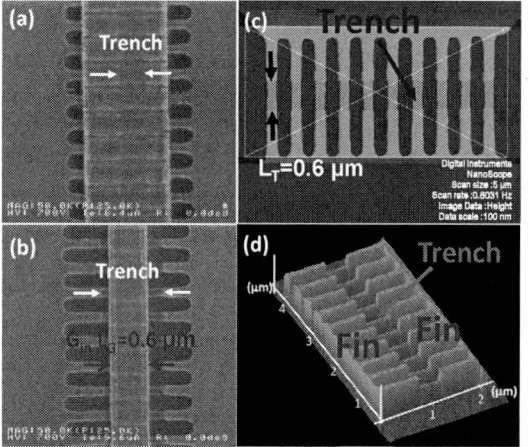

Fig. 2. (a)-(b) Top-view SEM image of trench JL-FET of G_{out} and G_{in} with L_G=0.6μm. (c)-(d) AFM image of channelregion of trench JL-FET with tenNWs.

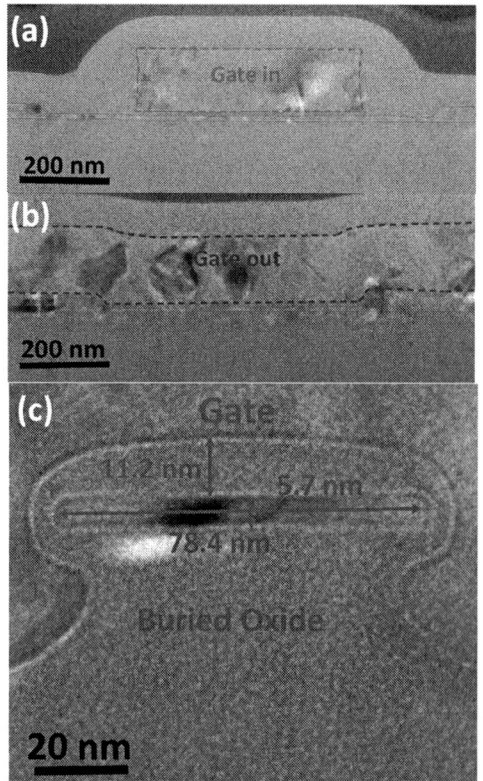

Fig. 3. (a)-(b) TEM image of trench JL-FET of G_{out} and G_{in} with L_G=0.6μm. (c) TEM image of channel region of trench JL-FET with NW.

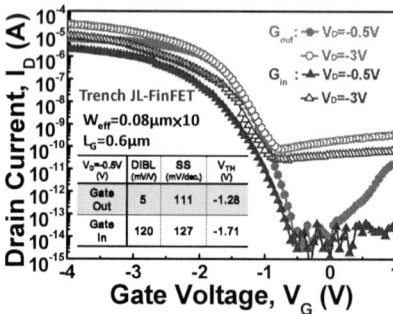

Fig.4. Comparison of the I_D-V_G curves of G_{out} and G_{in} NWs trench JL-FET with L_G=0.6μm at V_d= -0.5V and -3V.

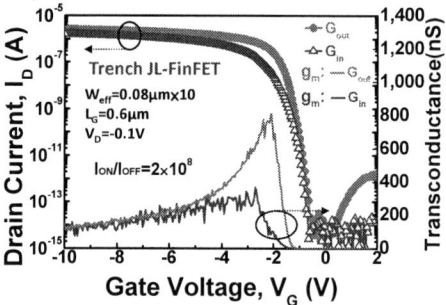

Fig.5. Comparison of the I_D-V_G curves of G_{out} and G_{in} trench JL-FET with L_G=0.6μm. G_{out} device has a better transconductance (g_m) than G_{in} device.

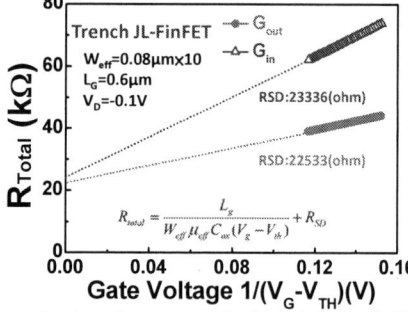

Fig.6. R_{total} as a function of gate voltage in G_{out} and G_{in} trench JL-FET with L_G=0.6μm at V_d= -0.1V.

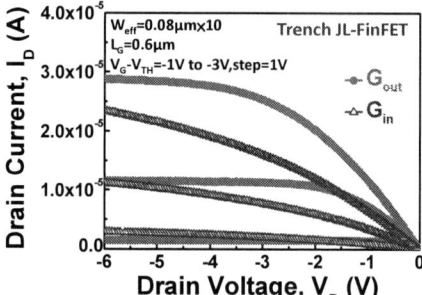

Fig. 7. I_D-V_D curves of G_{out} and G_{in} NWs trench JL-FET with L_G=0.6μm.

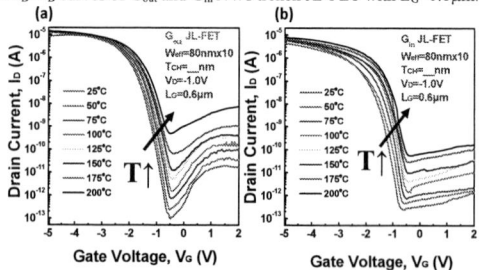

Fig. 8. Temperature dependence (25°Cto 200°C) on I_D-V_G characteristics of G_{out} and G_{in}. The I_{on} and V_{TH} exhibit positive and negative shift with increasing temperature, respectively.

Hybrid Channel Poly-Si Junctionless Field-Effect Transistors with Trench Structure Formed by Dry Etching Process

Cheng-Ping Wang, Yi-Ruei Jhan, Jun-Ji Su and Yung-Chun Wu*

Department of Engineering and System Science, National Tsing Hua University, Hsinchu, Taiwan

Tel: +886-3-5715131 ext. 34287;Fax: +886-3-5720724;Email: remotecpw@gmail.com

Abstract–This work demonstrates p-type hybrid poly-Si fin channel junctionless field-effect transistor (JL-FET) with trench structure by dry etching process. This JL-FET shows superior performance in a low drain-induced barrier lowering (<10mV/V) and high I_{ON}/I_{OFF} (>10^8) for L_{eff} = 1μm, excellent gate control.

I. Introduction

Nowadays, scaling down device faces more challenge, like short channel effect (SCE), and shallow junction etc. fabrication problem. JL-FET might be the way to solve the challenge, not only the simple fabrication and lower thermal budget but also suppress the short channel effect [1]. The JL-FET is heavily doped silicon nanowires (NWs) with Pi-gate electrode. The channel depleted region adjusts by the gate-to-NW work function different. However, the channel thickness of JL-FET should be thinner enough to deplete the channel. In this work, trench structure by dry etching and hybrid layer had been demonstrated [2]-[4]. Therefore, this novel JL-FET was compared with the conventional hybrid JL-FET.

II. Device Fabrication

Fig. 1(a) shows the fabrication process flows of the hybrid P/N fin channel JL-FET with trench structure. Initially grew a 400nm thermal SiO_2 layer on 6 inch silicon wafers. And deposited 45 nm undoped amorphous silicon (a-Si) by low-pressure chemical vapor deposition (LPCVD) on SiO_2. Then, the a-Si layer was solid-phase recrystallized (SPC) and formed large grain size. The SPC layer was implanted P ions at a dose of 2×10^{14} cm^{-2} serving as n-type substrate layer, followed by rapid thermal annealing (RTA). Subsequently, 45 nm channel layer was deposited on n-type substrate by the aforesaid SPC method. Then, this layer was implanted BF_2 ions at a dose of 2×10^{14} cm^{-2} and annealed to activate acceptor, forming p-type channel. The active layers which composed of p-type channel and n-type substrate were defined by e-beam lithography and reactive-ion etching (RIE). Next, trench structure was defined by e-beam lithography and RIE. An 8-nm-thick dry oxide grew as the gate oxide layer, consuming around 8-nm-thick poly-Si to form 6-nm-thick channels. Finally, gate formation, passivation and metallization were performed.

III. Results and Discussion

Fig. 1(b) and (c) schematically present the proposed devicestructure of the hybrid p-channel JL-FET with trench structure and the cross-sectional view along gate direction.

Fig. 2(a) shows scanning electron microscope (SEM) image of the ten NWs and active region for the hybrid JL-TFTs with trench structure. Fig. 2(b) shows conventional hybrid JL-FET L_G = 1μm. Fig.2(c) and (d) show atom force microscope (AFM) image of the ten NWs and the trench length (L_T) is 1μm. L_T means the effect gate length (L_{eff}) [2].

Fig. 3(a) and 3(b) show the cross-sectional transmission electron microscopic (TEM) images of a single NW along gate and channel direction. 3(c) shows the single NW enlarged view. The gate oxide and channel layer are 8nm and 6nm respectively.

Fig. 4 compares the I_D-V_G curve of hybrid JL-FET with trench structure and conventional hybrid JL-FET. When the V_D = -0.5V, the device with trench has steeper SS (109 mV/dec.), lower DIBL (9 mV/V), and higher ratio (>10^8) than the device without trench. However, the device with trench has lower leakage current, and it means the trench structure suppress the leakage.

Fig. 5 compares the I_D-V_D characteristics of hybrid JL-FET with trench structure and conventional hybrid JL-FET. The device with trench has a higher saturation current than the device without trench.

Fig. 6 shows the I_D-V_G and transconductance (g_m) characteristics of hybrid JL-FET with trench structure and conventional hybrid JL-FET. The g_m curves present the device with trench has a higher maxima g_m than conventional hybrid JL-FET. Trench structure has a higher mobility than conventional hybrid JL-FET because g_m is proportional to mobility.

Fig. 7 shows the total resistance (R_{total}) of hybrid JL-FET with trench structure and conventional hybrid JL-FET as a function of gate voltage at V_D = -0.1V. The JL-FET with trench structure has lower slope than conventional device. Because the slope of curve is reciprocal to the mobility, the device with trench structure has the lager mobility than conventional hybrid JL-FET.It also confirms the conclusion from fig. 6.

IV. Conclusion

In this study, the hybrid JL-FET with trench structure are success fully fabricated and characterized. Since the channel/substrate junction produces an additional depletion region and trench structure, the effective channel thicknessis reduced, the hybrid P/N JL-FETs provide a favorable SCE control, which results in a better electrical performance (such as SS, I_{on}/I_{off}, DIBL). Hence, this hybrid channel JL-FET with trench structure is highly promising for use in advanced system-on-chip and 3D stacked ICs applications.

V. Reference

[1]J.-P. Coling et al., Nature Nanotechnol., vol. 5, p. 225–229, 2010.

[2] Mu-Shih Yeh et al., IEDM, p. 26-6, 2014.

[3]Ya-Chi Cheng et al., IEDM, p. 26-7, 2014

[4] Ming-Hung Han et al., EDL, VOL. 34, NO. 2,Feb 2013,.

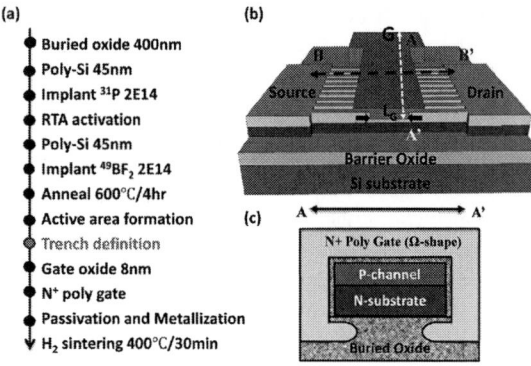

Fig.1 (a) Fabrication process flows of the device. (b) Schematic diagram of the hybrid p-channel JL-FET with trench structure. (c) The cross-sectional view along gate direction.

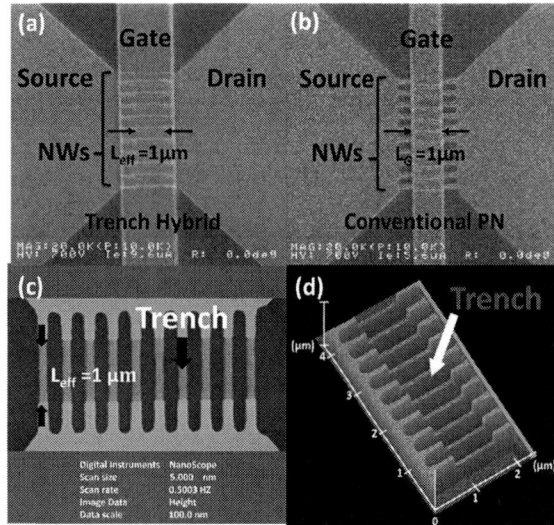

Fig. 2(a)-(b) Top-view SEM image of hybrid JL-FET with trench structure L_{eff} = 1µm and conventional hybrid JL-FET with L_G =1µm. (c)-(d) AFM image about channel region of trench hybrid JL-FET with ten NWs.

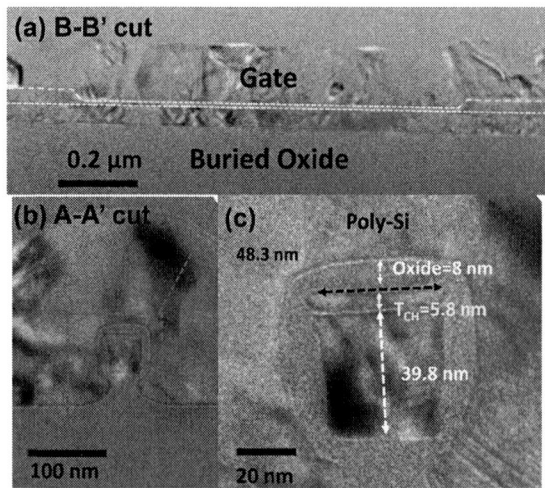

Fig. 3(a) and 3(b) show the cross-sectional TEM images of a single NW along gate and channel direction. 3(c) shows the single NW enlarged view. The gate oxide and channel layer are 8nm and 6nm respectively.

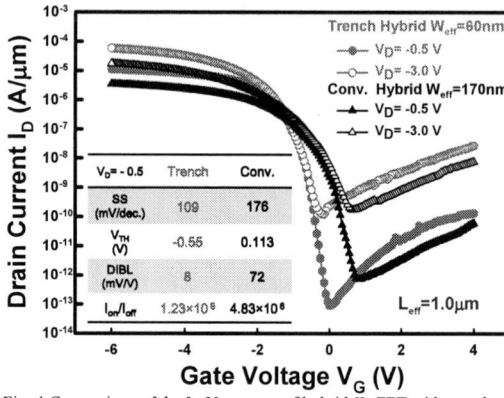

Fig. 4 Comparison of the I_D-V_G curves of hybrid JL-FET with trench structure L_{eff} = 1 µm and conventional hybrid JL-FET L_G = 1 µm at V_D = -0.5, -3 V.

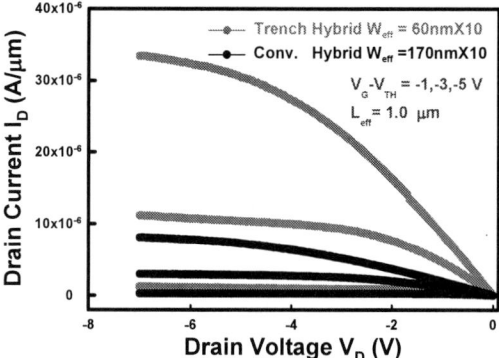

Fig. 5 I_D-V_D curves of hybrid JL-FET with trench structure L_{eff} = 1 µm and conventional hybrid JL-FET L_G = 1 µm.

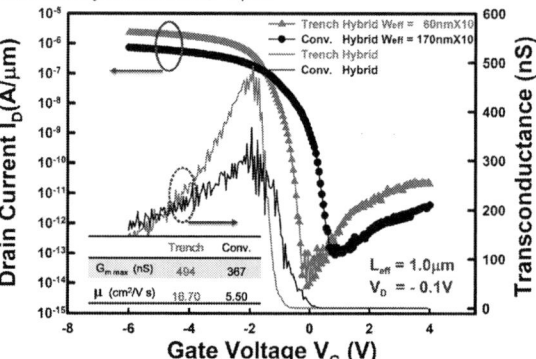

Fig. 6 Comparison of the I_D-V_G curves of hybrid JL-FET with trench structure L_{eff} = 1 µm and conventional hybrid JL-FET L_G = 1 µm. Hybrid JL-FET with trench structure has a better transconductance (g_m) than conventional.

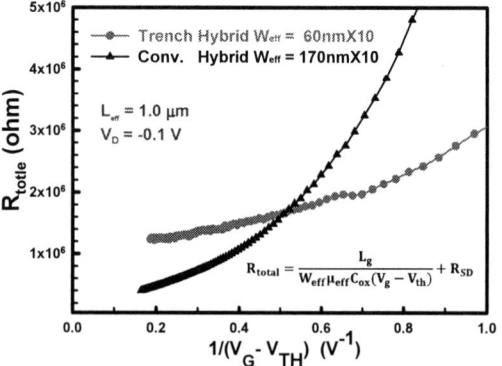

Fig. 7 R_{total} as a function of hybrid JL-FET with trench structure L_{eff}= 1 µm and conventional hybrid JL-FET L_G = 1 µm at V_d = -0.1V.

Built-in Effective Body-Bias Effect in UTBB Hetero-Channel MOSFETs and Its Suppression

Chang-Hung Yu and Pin Su

Department of Electronics Engineering & Institute of Electronics, National Chiao Tung University, Taiwan
E-mail: pinsu@faculty.nctu.edu.tw

Abstract

This work presents a built-in effective body-bias effect ($V_{BS,eff}$) in ultra-thin-body and BOX (UTBB) hetero-channel MOSFETs. This effect stems from the discrepancies in the electron affinity, the effective density-of-states, and the band-gap between the high-mobility channel and conventional Si channel. Physical $V_{BS,eff}$ models quantifying this effect are presented for nFET and pFET, respectively. Our study indicates that the DIBL of various hetero-channel devices can be worse than what permittivity predicts because of the built-in forward body-bias effect. Moreover, we have shown that this detrimental effect can be suppressed by the quantum-confinement effect. This effect has to be considered when designing or benchmarking various UTBB hetero-channel MOSFETs.

Introduction

Heterogeneous integration of high-mobility channel materials (such as III-V and Ge) on Si substrates is an important trend for VLSI [1]. However, one intrinsic drawback of the hetero-channel devices is the higher permittivity and worse device electrostatic integrity (EI). Ultra-thin-body and BOX (UTBB) structure has been suggested to mitigate the EI problem [2-3]. In addition, the UTBB structure also enables more efficient threshold-voltage (V_T) modulation and power/performance optimization through body bias (V_{BS}) [2-4]. Besides the permittivity, whether there is any other intrinsic mechanism determining the EI of UTBB hetero-channel devices is an important question. In this work, through analytical modeling corroborated by TCAD simulation [5], we present a built-in effective body-bias effect [6] in UTBB hetero-channel MOSFETs. Its impacts on the drain induced barrier lowering (DIBL) of various hetero-channel devices are investigated. Its suppression is also discussed.

Built-in Effective Body-Bias Effect

In Fig. 1(a) the $In_{0.53}Ga_{0.47}As$-OI nFET exhibits larger DIBL than the GeOI counterpart even though it has smaller dielectric constant (ε_r) than Ge. Fig. 1(b) further shows that, although the DIBL of $In_{0.53}Ga_{0.47}As$-OI can be improved by reducing the ε_r to that of Si, it is still significantly larger than that of SOI for a given V_{BS}. These anomalous DIBL characteristics can be explained by Fig. 2, in which the slope of E_C (and thus the vertical field) in the channel region for the InGaAs-OI device is smaller than the SOI counterpart. This result infers that there exists an effective built-in forward body-bias in the $In_{0.53}Ga_{0.47}As$-OI device with respect to the Si device. This built-in effective body-bias ($V_{BS,eff}$) is intrinsic to hetero-channel MOSFETs with Si-substrate. For nFETs:

$$V_{BS,eff} = \frac{1}{q}\left(\chi_{ch} - 4.07eV\right) - \frac{kT}{q}\ln\left(\frac{2.86\times10^{19}\,cm^{-3}}{N_{C,ch}}\right) \tag{1}$$

where χ_{ch} and $N_{C,ch}$ are the electron affinity and the effective density-of-states of conduction band for the channel material, respectively. Since $\chi_{ch} = 4.5eV$ and $N_{C,ch} = 2.08\times10^{17}cm^{-3}$ for $In_{0.53}Ga_{0.47}As$, there exists a 0.3V built-in forward body bias in the $In_{0.53}Ga_{0.47}As$-OI device at room temperature based on (1).

Fig. 3 compares the electron density distribution along the vertical direction for the $In_{0.53}Ga_{0.47}As$-OI and SOI devices. It is seen that the electron centroid of the InGaAs channel is closer to the back interface, and the impact of dielectric constant is modest. However, if the built-in $V_{BS,eff} = 0.3V$ is further compensated by external body biasing, the electron profile of the InGaAs-OI channel shows a fairly good agreement with that of SOI. This validates the accuracy of (1) and demonstrates that, in addition to permittivity, it is the built-in $V_{BS,eff}$ that determines the electron profile and thus the EI of the $In_{0.53}Ga_{0.47}As$-OI device.

For hetero-channel pFETs, it can be shown that

$$V_{BS,eff} = \frac{1}{q}\left(\chi_{ch} - 4.07eV\right) + \frac{kT}{q}\ln\left(\frac{3.1\times10^{19}\,cm^{-3}}{N_{V,ch}}\right) - \frac{1}{q}\left(1.12eV - E_{g,ch}\right) \tag{2}$$

with $N_{V,ch}$ the effective density-of-states of valence band and $E_{g,ch}$ the band-gap for channel. The difference in (1) and (2) explains why in Fig. 4(a) the GeOI pFET exhibits worse DIBL than that of GeOI nFET. The band-gap term in (2) results in a significant built-in forward $V_{BS,eff}$ for Ge pFET and smaller vertical channel field as shown in Fig. 4(b).

Fig. 5 investigates and compares the impacts of the built-in $V_{BS,eff}$ (calculated by (1) and (2)) on the DIBL of various high-mobility hetero-channel devices. It can be seen that while the $V_{BS,eff} = 0.3V$ for $In_{0.53}Ga_{0.47}As$-OI nFET, the GeOI nFET possesses a reverse $V_{BS,eff} = -0.1V$. This explains Fig. 1(a). Another anomalous DIBL characteristic in Fig. 5(a) is that the InSb nFET exhibits smaller DIBL than the InAs counterpart in spite of its larger permittivity. This can also be explained by the difference of their built-in $V_{BS,eff}$. In addition, a reverse $V_{BS,eff} = -0.14V$ for GaSb nFET explains its much smaller DIBL than InAs with similar permittivity. Fig. 5(b) shows that both the Ge and GaSb pFETs possess significant forward $V_{BS,eff}$, opposite to their nFET counterparts.

Suppression of the Impact of $V_{BS,eff}$ on DIBL

Since the built-in effective body-bias effect can have significant impact on the DIBL of hetero-channel devices, it needs to be considered in device design. Fig. 8 shows that, for a given L, the impact of built-in $V_{BS,eff}$ on DIBL depends on the BOX thickness (T_{BOX}). Moreover, for a given channel material, the impact of built-in $V_{BS,eff}$ on DIBL also depends on the substrate doping type and BOX material (Fig. 9). It is seen that the Al_2O_3 BOX increases the impact of built-in $V_{BS,eff}$ due to its higher dielectric constant.

Fig. 10 compares the DIBL characteristics of various hetero-channel devices with a more advanced technology node, which possesses an ultra-scaled channel thickness ($T_{ch}=4nm$). It indicates that, as the quantum-mechanical (QM) effect becomes important, the DIBL of the hetero-channel devices are remarkably reduced. This is because the electron centroid is forced to stay in the middle of the channel by the QM effect even though the high permittivity and the forward built-in $V_{BS,eff}$ try to push it toward the back interface (Fig. 11). In other words, the hetero-channel device should be made very thin and the quantum confinement can be used to mitigate the adverse impacts of high permittivity and built-in forward $V_{BS,eff}$ on DIBL.

Acknowledgement

This work was supported by MOST 102-2221-E-009-136-MY2, MOST 104-2911-I-009-301 (I-RiCE), and Ministry of Education in Taiwan (ATU).

References

[1] http://www.itrs.net/
[2] F. Andrieu *et al.*, *Symp. VLSI Tech.*, p. 57, 2010.
[3] C. Fenouillet-Beranger *et al.*, *Symp. VLSI Tech.*, p. 65, 2010.
[4] L. Grenouillet *et al.*, *IEDM Tech. Dig.*, p. 64, 2012.
[5] Sentaurus TCAD, G2012-06-SP2.
[6] C.-H. Yu and P. Su, *IEEE EDL*, vol. 35, no. 8, p. 823, 2014.

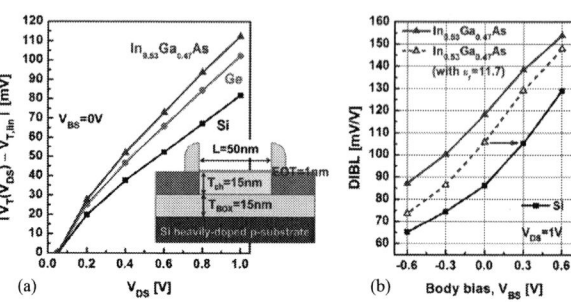

Fig. 1. (a) Simulated DIBL characteristics for SOI (ε_r=11.7), GeOI (ε_r=15.8) and In$_{0.53}$Ga$_{0.47}$As-OI (ε_r=13.9) nFETs showing worse DIBL for InGaAs-OI than GeOI. (b) Merely considering the impact of permittivity cannot explain the discrepancy in the DIBL vs. V_{BS} curve between InGaAs-OI and SOI devices.

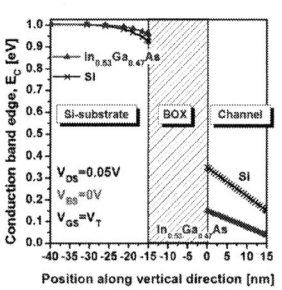

Fig. 2. Conduction-band edge (E_C) profiles along the T_{ch} direction showing the difference in vertical channel field (slope of E_C) between long-channel InGaAs-OI and SOI nFETs. The two devices have identical threshold voltage (V_T).

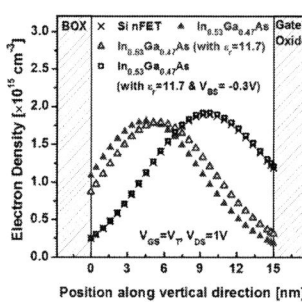

Fig. 3. After considering the impacts of permittivity and built-in $V_{BS,eff}$, the channel electron profile for the InGaAs-OI device coincides with that of SOI. The electron density is evaluated at the location where the minimum potential occurs along the L direction.

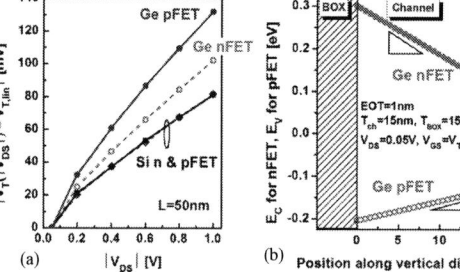

Fig. 4. (a) Comparison of simulated DIBL characteristics for SOI, GeOI n- and p-FETs. (b) Comparison of the vertical electric field for long-channel GeOI n- and p-FETs.

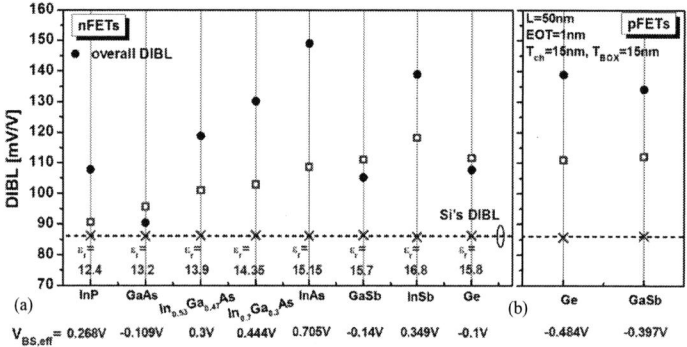

$V_{BS,eff}$ = 0.268V -0.109V 0.3V 0.444V 0.705V -0.14V 0.349V -0.1V   -0.484V -0.397V

Fig. 5. Dissection of DIBL for various hetero-channel (a) nFETs, and (b) pFETs. The gap between solid circle and open square indicates the impact of built-in $V_{BS,eff}$, while the gap between open square and blue cross indicates the impact of permittivity.

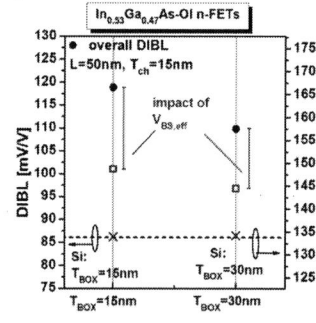

Fig. 8. The impact of built-in $V_{BS,eff}$ on DIBL depends on T_{BOX}. The gap between open square and blue cross indicates the impact of permittivity.

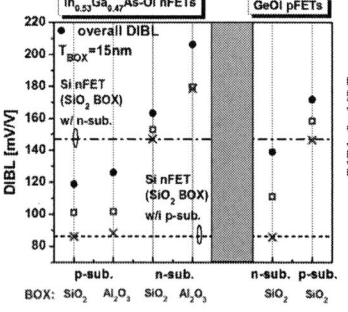

Fig. 9. The impact of built-in $V_{BS,eff}$ on DIBL depends on the substrate doping type and BOX material. The gap between solid circle and open square indicates the impact of $V_{BS,eff}$, while the gap between open square and blue cross indicates the impact of permittivity.

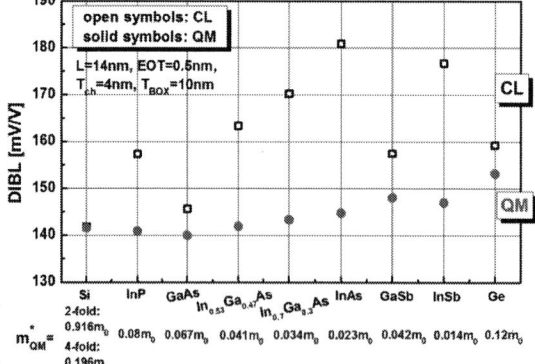

Fig. 10. Comparison of the DIBL characteristics for various hetero-channel n-MOSFETs with L=14nm and T_{ch}=4nm. The quantum confinement effect can be used to improve the DIBL as T_{ch} is ultimately scaled. m_{QM}^* denotes the quantization effective mass. CL denotes the classical condition, while QM denotes the quantum-mechanical condition.

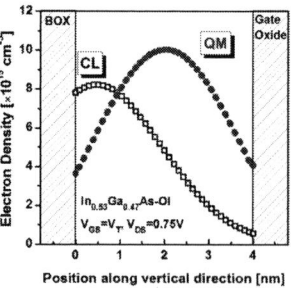

Fig. 11. Comparison of the channel electron density profile of the In$_{0.53}$Ga$_{0.47}$As-OI n-FET with L=14nm and T_{ch}=4nm between CL and QM conditions.

64

Boolean logic circuit implementaion using multi-input floating-body MOSFET

Min-Woo Kwon, Hyungjin Kim, Jungjin Park, and Byung-Gook Park*

Inter-university Semiconductor Research Center (ISRC) and
Department of Electrical and Computer Engineering, Seoul National University,
1 Gwanak-ro, Gwanak-gu, Seoul 151-742, Republic of Korea.
E-mail address: kmw1224@naver.com

Abstract — **In this paper, we describe the implementation of simple Boolean logic circuits using a multi-gate floating-body MOSFET, which was originally developed for neuromorphic circuits such as integrate-and-fire neuron circuits. By changing the channel doping level alone, the same circuit with a multi-gate floating-body MOSFET can represent different logic functions such as NAND or NOR. In addition, the circuit can operate as a multi-input logic gate without additional devices.**

Ⅰ. Introduction

Recently, the interest in biological system is increased and many researchers attempt to emulate neural networks that are characterized by parallel processing and low power consumption. We have proposed an I&F neuron circuit with a multi-gate floating-body MOSFET [1], and, in this paper, we have designed a Boolean-logic circuit with the floating body device.

The McCulloch and Pitts model have been regarded as the representative artificial neural network to emulate biological neural network [2]. Fig. 1 shows the schematic diagram of the neural network model. The input signals from different pre-neurons are integrated. If this integrated signal exceeds the threshold point, the output signal is generated. Depending on the threshold point, this network can act as an OR gate or an AND gate. Inspired by this McCulloch and Pitts model, we implement logic gates using a multi-gate floating-body MOSFET and one additional transistor. The simulation is performed by SIVACO ATLAS.

Ⅱ. Multi-input floating-body MOSFET

To emulate biological neuron properties, we have used the floating body MOSFET. Fig. 2 shows the 2-input floating body MOSFET structure. When one gate receives an input pulse, a channel is formed under the gate and excess holes are made by impact ionization and the holes are accumulated in the floating body. If two gates received input pulse, the channel is formed below each gate and holes are accumulated in the floating body. The number of holes made by each gate is integrated in the floating body. The holes make floating body potential lower

and the current increases. Because of the positive feedback between the current and the number of holes, they are increased very fast at the threshold point. This is the latch-up characteristic induced by the floating body effect. Fig. 3 shows the simulation result. As the drain and gate voltage is applied, the drain current and the number of holes are increased rapidly and the latch-up occurs. If the drain voltage is 0 V, the holes are removed. Fig. 4 shows the transient characteristic of the drain current depending on the floating body doping concentration.

Ⅲ. Logic circuit configuration

The basic description of a floating body MOS logic circuit is presented in Fig 5. The circuit receives binary signals X_1 and X_2 and generates output of V_{OUT}. The NOR and the NAND gate have the same circuit configuration. The only difference is doping concentration in the floating body. The NOR gate device has low doping ($4 \times 10^{16}/cm^3$), and the NAND gate device has high doping concentration ($1 \times 10^{17}/cm^3$). So the latch-up voltage of the NAND gate device is higher than that of the NOR gate device. In the NAND gate, the floating-body MOSFET latches up only when both X_1 and X_2 are 1, and V_{OUT} becomes 0. In the NOR gate, just one high input is required for the latch-up condition. Fig. 6 shows the simulation result of the drain current according to X_1 and X_2. If the logic circuit uses multi-gate floating body MOSFET, 3 or 4 input gate, the circuit operates as a multi-input logic gate without additional transistors.

Ⅳ. Conclusion

We have proposed logic circuits using multi-input floating-body MOSFETs. In the McCulloch and Pitts artificial neuron network model, the neuron acts like OR gate or AND gate depending on the threshold point. As in the case of the neural model, we change the latch-up voltage of the floating body MOSFET by varying the doping concentration. The logic circuit operation is obtained by using just two devices.

Acknowledgement

This work was supported by the Center for Integrated Smart Sensors funded by the Ministry of Science, ICT & Future Planning as Global Frontier Project" (CISS-2012M3A6A6054186).

References

[1] M.-W. Kwon et al., *Silicon Nanoelectronics Workshop*, 2013, pp. 113-114.

[2] W. A. McCulloch and W. Pitts, Math. Biophys. 5, 115-133, 1943.

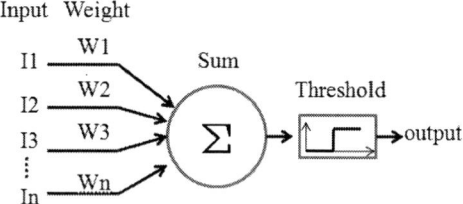

Fig.1. McCulloch and Pitts artificial neuron network model [2].

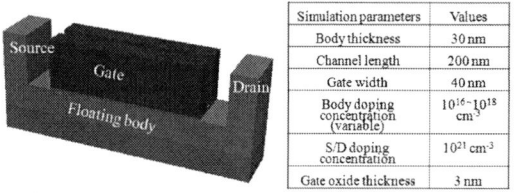

Fig.2. 2-input floating body MOSFET structure and simulation parameter.

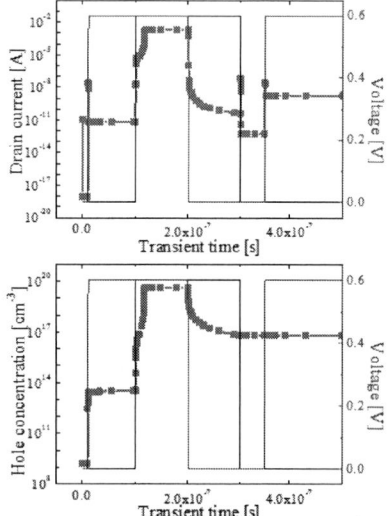

Fig.3. Simulation results of the drain current and the hole concentration the floating body MOSFET. Drain current increases very rapidly because of the accumulated holes .

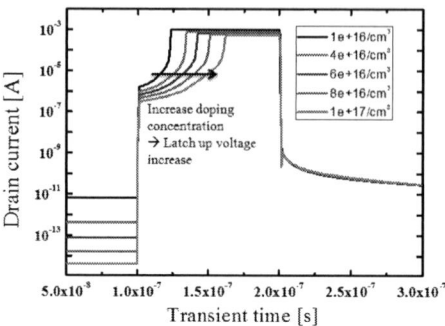

Fig.4. Simulation results of the drain current according to the floating body doping concentration. Latch up voltage is increase as increasing doping concentration.

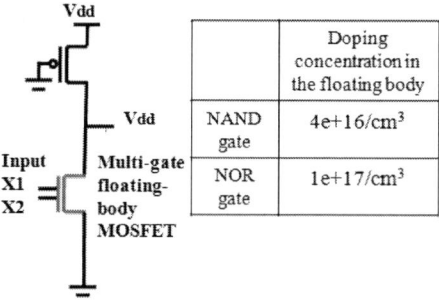

Fig.5. Simulation results of the drain current according to the floating body doping concentration. Latch-up voltage is increased as the doping concentration increases.

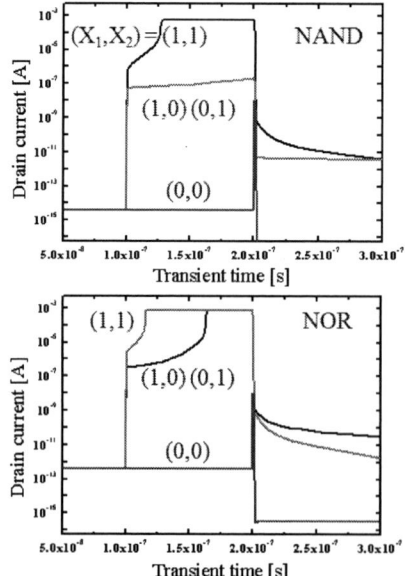

Fig.6. Simulation of the NAND and NOR logic operation. The floating body MOSFET ON/OFF is determined by latching-up or not.

Comparison of Electrical Characteristics of N-type Silicon Junctionless Transistors with and without Film Profile Engineering by TCAD Simulation

Jung-Ruey Tsai[1*], Horng-Chih Lin[2,3], Hsiu-Fu Chang[1], Bo-Shiuan Shie[2], Ting-Ting Wen[3], and Tiao-Yuan Huang[2]

[1]Department of Photonics and Communication Engineering, Asia University,
500 Lioufeng Rd., Wufeng, Taichung 41354, Taiwan, R.O.C.
[2]Department of Electronics Engineering and Institute of Electronics, National Chiao Tung University,
[3]Nano Facility Center, National Chiao Tung University,
No.1001, Ta-Hsueh Rd., Hsinchu 300, Taiwan, R.O.C.
*Email: jrtsai@asia.edu.tw

INTRODUCTION

Field-effect transistors (FETs) with junctionless (JL) channels have recently attracted much attention for various applications, such as metal-oxide semiconductor thin-film transistors (TFTs) [1], memory devices [2] and Si nanowire TFTs [3, 4]. The Si junctionless (JL) transistors employing high dopant concentration ($\geq 10^{19}$ cm^{-3}) in the source, drain, and nano-scaled channel have been demonstrated to provide excellent electrical characteristics. More recently, film profile engineering (FPE) concept for fabricating downscaled ZnO and IGZO TFTs [5, 6] have been proposed to obtain high-on/off current ratio and great subthreshold swing. Nevertheless, it emphasizes a significant issue of source/drain (S/D) series resistance on the downscaled device performance that needs to be further verified. In this work, electrical performance of downscaled N-type Si JL TFTs with FPE channel and conventional ones will be compared with each other by Sentaurus technology computer aided design (TCAD) simulation [7].

DEVICE STRUCTURE

Figure 1 shows the schematic of the n-type Si JL device with a thin FPE channel layer studied in this work. The thickness at the center of FPE channel is varied from 4 to 8 nm, which is also taken as the channel thickness of conventional JL device without FPE for comparison. The doping and channel length of n-type Si are set to be 1×10^{19} cm^{-3} and 400 nm, respectively. The thickness of gate oxide is given by 3 nm. The structures of FPE and conventional Si JL devices are performed by Sentaurus structure editor. The electrical characteristics of proposed devices are obtained based on the physical models used in the Sentaurus device simulation include basic mobility models, such as doping dependence, normal/high electric field and velocity saturation models. In addition, the interface fixed charge concentration and work function of gate electrode are reasonably given by 1×10^{19} cm^{-3} and 5.1 eV, respectively.

RESULTS AND DISCUSSION

Figure 2(a) plots the simulated transfer characteristics of Si JL transistors with the 8-nm-thickness channel in FPE and conventional configurations. It clearly suggests that the threshold voltage of FPE transistor is smaller than that of conventional one caused by the more electron concentration exist in the concave Si JL channel needed to be depleted at more negative gate bias. Additionally, both of the leakage current and on-state current are higher in the FPE device than that in the conventional one while the subthreshold swing is the same. As decreasing the thickness of the concave and conventional Si channel, not only the leakage current significantly decreases but the on-state current decreases slightly, as shown in Figs. 2(b) and 2(c). The threshold voltage shifts toward the positive value which indicates the improvement of gate control in the channel ability. With shrinking device size, the on-state resistance includes the source/drain (S/D) contact resistance and series/spreading resistance is becoming a key issue for device performance. Thus, the characteristics of square root of drain current vs gate voltage under the drain voltage of 1 V is shown in Fig. 3. It clearly shows that the on-state resistance of the conventional device is higher than that of FPE device. Based on the ignored S/D contact resistance and same series resistance under the S/D regions, the difference of the on-state resistance between FPE and conventional devices is reasonable mainly caused by the spreading/series resistance existed in the Si JL channel. Further transfer characteristics of devices with channel length of 200 nm are plotted for comparison, as shown in Fig. 4. With decreasing the channel length, the leakage current is depressed for thin channel thickness while more on-state current is appeared in FPE devices than in conventional ones. Phenomenon of threshold voltage shifts as decreasing the channel length suggests that the ability of gate control in the channel is worse in FPE devices than in conventional ones.

CONCLUSION

Comparison of Electrical performance of downscaled N-type Si JL TFTs with and without FPE channels have illustrated by TCAD simulation. There is a trade-off between the threshold voltage shift and on-state resistance should be addressed to obtain the high electrical performance of Si JL devices with FPE.

ACKNOWLEDGEMENT

This work is supported by the National Science Council, Taiwan under No. NSC 103-2221-E-468 -032 and Asia University, Taiwan under No. 102-asia-52.

REFERENCES

[1] T. Kamiya, K. Nomura, and H. Hosono, *Sci. Technol. Adv. Mater.*, vol. **10**, 2012, p 044305.

[2] X. Duan, Y. Huang, and C. M. Lieber, Nano Lett., vol. **2**, 2002, p. 487.

[3] R. Trvisoli, and M. Pavanello, *IEEE Trans. Electron Devices*, vol. **59**, 2012, p. 3510.

[4] H.-C. Lin, C.-I Lin, and T.-Y. Huang, *IEEE Electron Dev. Lett.*, vol. **33**, 2012, p. 53.

[5] R.-J. Lyu, H.-C. Lin, and T.-Y. Huang, *IEEE Trans. Electron Devices*, vol. **61**, 2014, p. 1417.

[6] H.-C. Lin, B.-S. Shie, and T.-Y. Huang, *IEEE Trans. Electron Devices*, vol. **61**, 2014, p. 2224.

[7] Synopsis, Sentaurus Device H-2013.03 (March 2013).

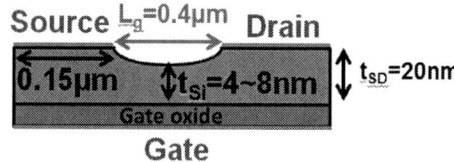

FIGURE 1. Schematic of the Si NW JL device with a thin FPE channel.

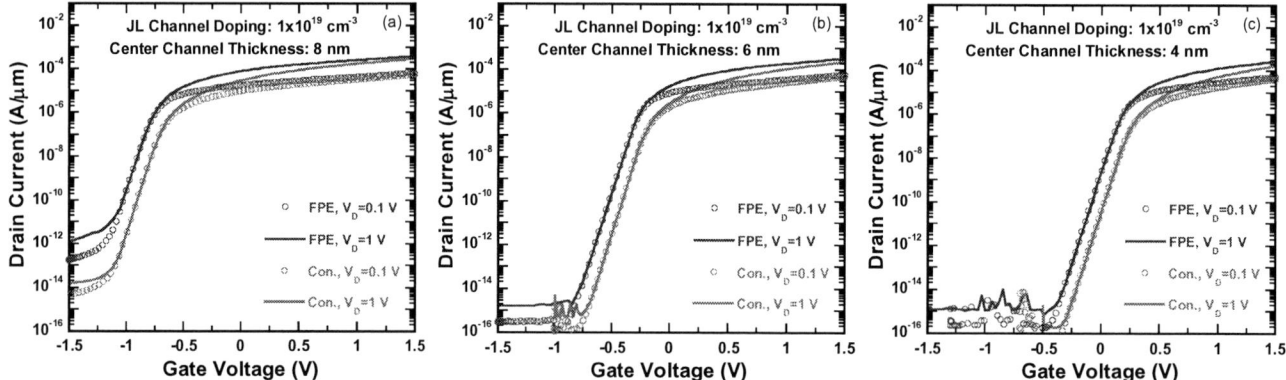

FIGURE 2. Comparison of the transfer characteristics of conventional and FPE Si JL transistors with the channel thickness of (a) 8 nm; (b) 6 nm and (c) 4 nm.

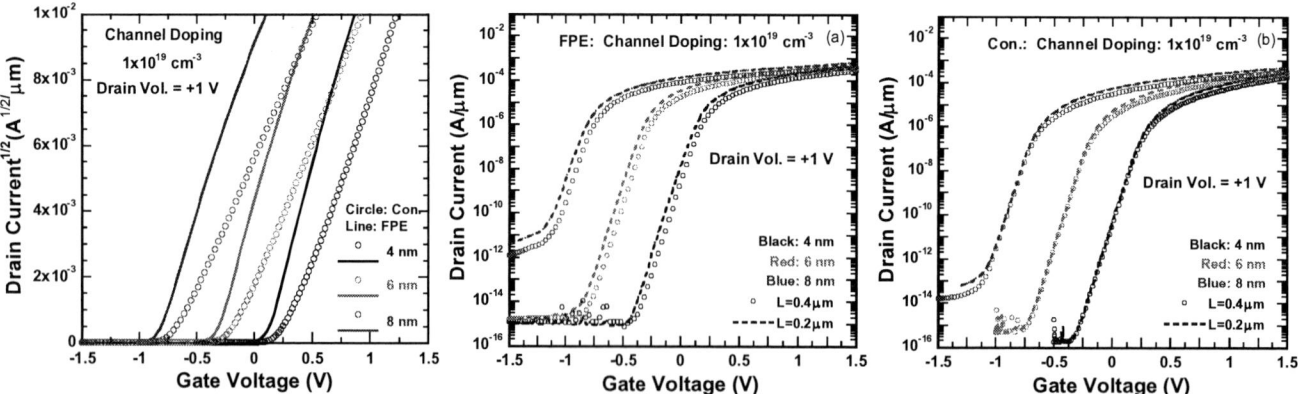

FIGURE 3. The output characteristics are extracted at drain voltage of 1 V for both conventional and FPE devices.

FIGURE 4. Comparison of the transfer characteristics of Si JL transistors with the various channel thickness ranging from 24 to 8 nm and channel length of 0.2 and 0.4 μm for the (a)FPE and (b)conventional device configuration.

Thermodynamic stability of high phosphorus concentration in silicon nanostructures

Michele Perego[1], Gabriele Seguini[1], Elisa Arduca[1,5], Jacopo Frascaroli[1], Davide De Salvador[2,4,*], Massimo Mastromatteo[2,4], Alberto Carnera[2,4], Giuseppe Nicotra[3], Mario Scuderi[3], Corrado Spinella[3], Giuliana Impellizzeri[4], Cristina Lenardi[5], and Enrico Napolitani[2,4,*]

[1] Laboratorio MDM, IMM-CNR, Via Olivetti 2, I-20864 Agrate Brianza, Italy
[2] Dipartimento di Fisica e Astronomia, Università degli Studi di Padova, Via Marzolo 8, I-35131 Padova, Italy
[3] IMM-CNR, Z. I. VIII Strada 5, I-95121 Catania, Italy
[4] MATIS IMM-CNR, Via S. Sofia 64, I-95132 Catania, Italy
[5] Dipartimento di Fisica, Università degli Studi di Milano, Via Celoria 23, I-20100 Milano, Italy

Email: Michele.perego@cnr.it

Abstract — **In this work we focus on P doping of Si nanocrystals (NCs) embedded in a SiO_2 matrix. We prove that, at equilibrium, high P concentrations within the Si NCs are thermodynamically favoured. We experimentally estimate the energy barriers for P diffusion in SiO_2 and trapping/de-trapping at the SiO_2/Si NCs interface, obtaining a complete picture of the system at equilibrium.**

I. INTRODUCTION

Deterministic doping of nanostructures is a key challenge for the fabrication of future advanced nanoelectronic devices.[1,2]. Unfortunately, a clear understanding of nanoscale doping is not available yet, as the physical mechanisms involved in this process are significantly different from those of bulk materials due to additional constrains related to the presence of the interfaces and of the surrounding oxide matrix. In this regard, Si NCs embedded in a SiO_2 matrix represent a paradigmatic system for the physical understanding of dopant incorporation in Si-based nanostructures.[3-4]

II. RESULTS

We investigate the thermodynamic stability of P dopant impurities in Si nanocrystals at thermodynamic equilibrium. To achieve this goal we fabricated samples with controlled diffusion sources that are spatially separated from the nanocrystals and we delivered a controlled amount of dopant atoms from the dopant source to the nanocrystals by diffusing the dopants trough the SiO_2 matrix (Fig. 1).
Energy Filtered Transmission Electron Microscopy (EF-TEM) was used to measure the size of the Si NCs (4.0 nm) and to demonstrate their stability during the P diffusion process (Fig. 2). P diffusion trough the SiO_2 matrix and incorporation in the Si NCs was monitored by a calibrated time of flight secondary ion mass spectrometry (ToF-SIMS) (Fig. 3) and X-ray photoelectron spectroscopy (XPS). (Fig. 4) The combination of the spatial and chemical information, obtained from ToF-SIMS and XPS analysis respectively, allowed determining the amount of P atoms effectively trapped (6 %) within the Si NCs and to demonstrate the possibility to largely exceed the P solid solubility in bulk Si. A simple model based on diffusion Fick's law in one dimension is used to determine the energy barrier (1 eV) that prevents P release from the Si NCs.

III. CONCLUSIONS

We demonstrate that high levels of impurities can be introduced in the Si NCs in a stable configuration by a simple solid-state diffusion process. The dopant content can be finely tuned by properly adjusting the annealing conditions. This methodology provides a simple way to introduce dopants in Si nanostructures and it is fully compatible with the stringent technological constrains of semiconductor industry

REFERENCES

[1] J. C. Ho, R. Yerushalmi, Z. Jacobson, Z. Fan, R.L. Alley, A. Javey, _Nature materials_, vol. 7, pp. 62–67, 2008.
[2] International Technology Roadmap for Semiconductors 2013 (ITRS) http://www.itrs.net.
[3] Seino, K.; Bechstedt, F.; Kroll, P. _Physical Review B_, vol. 82, 085320, 2010.
[4] Seguini, G.; Castro, C.; Schamm-Chardon, S.; BenAssayag, G.; Pellegrino, P.; Perego, M. _Applied Physics Letters_, vol. 103, pp. 023103, 2013.

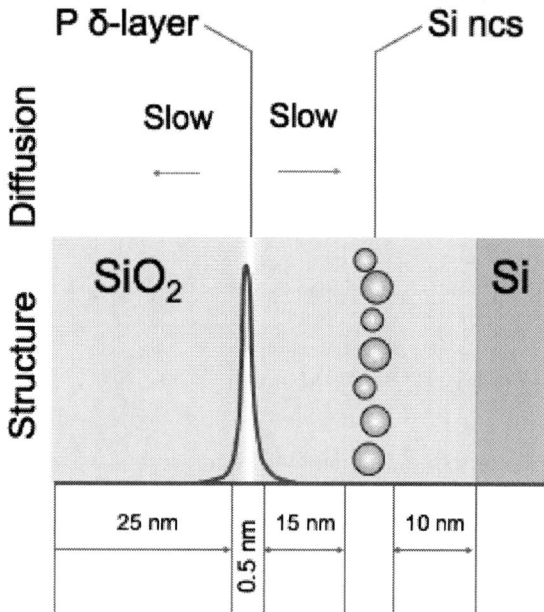

Figure 1: Schematic picture describing the experimental approach used to study P incorposration withi n the Si NCs

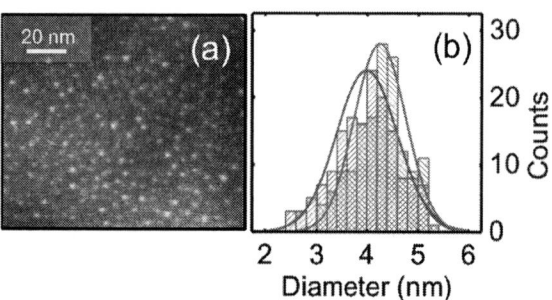

Figure 2: (a) EFTEM plan view image of the as deposited sample (b) size distribution of the NCs in the as deposited (green) and annealed (red) samples.

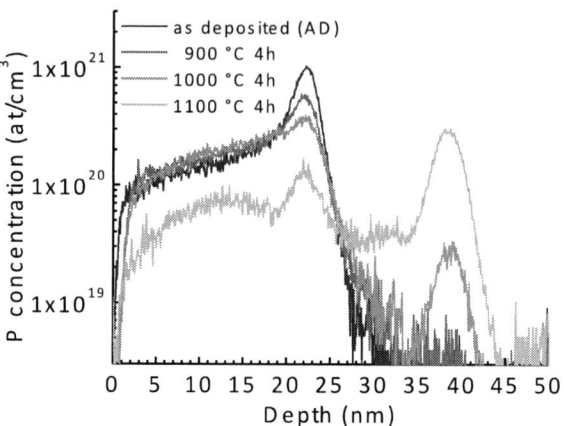

Figure 3: ToF-SIMS profiles of the as deposited and annealed samples showing progressive accumulation of P atoms in the Si NC region.

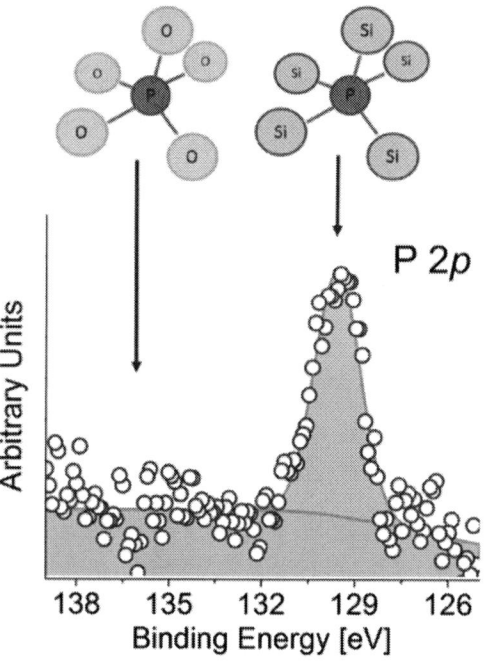

Figure 4: High-resolution XPS spectrum of the P 2p core level signal a sample annealed at 1100 °C for 4 hours.

3D-TCAD Simulation Study of the Novel T-FinFET Structure for Sub-14nm Metal-Oxide-Semiconductor Field-Effect Transistor

Chen-Han Chou[1], Chung-Chun Hsu[1], Steve S. Chung[1] and Chao-Hsin Chien[1*]

[1]*Department of Electronics Engineering, National Chiao Tung University, Hsinchu, Taiwan*

*E-mail: chchien@faculty.nctu.edu.tw, Tel: +886-3-571-2121 ext. 54252

Abstract — **We propose a novel device structure, namely T-FinFET, for sub-14nm MOSFET with using lighter anti punch through (APT) implant. According to 3D TCAD simulation, the T-FinFET is found to posses many advantages over the normal FinFET, such as better short channel effect (SCE) and drain induced barrier lowering (DIBL), having smaller S/D capacitance and junction leakage and fewer masks. Compared to gate-all-around (GAA) structure, the T-FinFET also has compatible electrical performance. All these features are obtained by depositing a self-aligned (SA) oxide after recessing the Si fin in the S/D region. It can be applied to Ge and III-V MOSFETs for suppressing the SCEs and S/D leakage, arising from higher permittivity and lower band gap than Si.**

I. INTRODUCTION

Intel has announced mass production of 14 nm technology node body-tied FinFET with 42 nm fin height and 6 nm fin width. The FinFET has been proven possessing strong gate control ability; however, the body-tied FinFETs are prone to suffer from punch through along the sub-channel region [1]. To tackle this problem, increasing sub-channel doping by anti punch through (APT) methods were developed, such as APT implant [2] and APT layer [3]. GAA device had been researched because of its aggressive gate control ability [4], but it still has few undesirable problems like self-heating issue [5], complicate process flow, etc. In this paper, we propose a novel body-tied T-FinFET aiming for the sub-14 nm application. According to the 3D simulation, the T-FinFET depicts better immunity against SCEs than the normal FinFET even with lighter APT concentration. Compared to the GAA FInFET, the T-FinFET still exhibits comparable electrical performance. As a result, we think the T-FinFET is a novel, high performance, cost-effective and easy-to-fabricate structure for the future application.

II. DEVICE PREPARATION

Illustrative structures and their cross sections along the channel of three kinds of FinFETs, including normal FinFET, T-FinFET and GAA-FinFET, are shown in Fig. 1. The process flow for the T-FinFET is shown in Fig. 2. **The key is Step 7 in which anisotropic oxide is deposited by HDPCVD or collimated Sputter. The oxide on the side wall of fin can be removed by HF gas before selective lateral S/D growth.** In this paper, we establish the 3D FinFET structure by SPROCESS simulator and calculate the electrical results by SDevice with the basis physical models, such as Density Gradient Quantization Model for the confined carrier distribution. The 3D simulation structure of T-FinFET is shown in Fig.3 with various fin widths and channel lengths; it is named by the indicated inverse T shape. The channel doping for all devices is set to be $1E15$ cm^{-3}. The sub-channel doping ($1E16/1E17/1E18/1E19$ cm^{-3}) is completed by APT technique to suppress punch through leakage current; **while in theT-FinFET channel it is kept to be 1E15** cm^{-3}.

III. RESULTS AND DISCUSSION

A. Electrical Characteristic between Varied Structures

Fig. 4 shows V_{TH} and V_{TH} roll-off and Fig. 5 DIBL as a function of L_G. We can see the T-FinFET depicts the minimal SCE and DIBL. Fig. 6 shows the current flows of the normal FinFETs with various APT doping concentrations when $I_{DS}=1E-7$ $\mu A/\mu m$ at which current level V_{TH} was defined. Apparently, serious punch through current beneath the channel region is seen for the $1E16$ cm^{-3} case. Fig. 7 shows the comparison of current flows of the T-FinFET and normal FinFET. We observe that the T-FinFET exhibits compatible punch through immunity with the normal FinFET with a doping level of $1E18$ cm^{-3} in the sub-channel. Fig. 8 shows the comparison between the normal FinFET and T-FinFET again. With a doping of $1E17cm^{-3}$ in the sub-channel, the FinFET suffers worse leakage problem as V_{DS} is increased; while the T-FinFET does not. Fig. 9 shows V_{TH} roll-off versus gate length with various fin widths. The T-FinFET has better SCE immunity than the FinFET as the fin width becomes broader. Fig.10 shows the V_{TH} roll-off and DIBL of three kinds of FinFETs as a function of L_G. Remarkably, the T-FinFET depicts compatible channel control ability with the GAA FinFET. This can be well understood with the result in Fig. 8; the punch through leakage path is effectively blocked by the SA oxide instead of highly doped impurities.

B. Influence of Process Variation on T-FinFET

In real device fabrication, process variation is a critical issue. We, thus, investigate the influence of SA oxide thickness on the electrical properties of the T-FinFET, as shown in Fig. 11 and Fig. 12. In Fig. 12, we can see that different SA oxide thicknesses only result in slight variation in V_{TH}. Even so, we need to keep in mind that thicker SA oxide will sacrifice the effective fin height, implying the effective channel width is decreased. Conceptually, the SA oxide is like a lock to restrict the current flow, as shown in the inset of Fig. 11.

IV. CONCLUSION

Table I summarizes the comparison of multi-gate FinFETs studied in this paper. Our T-FinFET can provide very good immunity against SCEs, less self-heating issue, lower junction leakage and capacitance. Compared to those devices with using ATP techniques and GAA structure, it has the simpler process flow and lower expense. Finally, process tolerance is good since the thickness variation of SA oxide seems not significantly impact the device performance.

REFERENCES

[1] C.R. Manoj, et.al., IWPSD 2007,pp. 134-137.
[2] Yiming Li, et.al., JJAP 2003,pp. 2152-2155.
[3] Shamarao, P., et.al., IEEE T-ED 1996,pp. 1942-1949.
[4] Singh, N., et.al.,IEEE EDL 2006,pp. 383-386.
[5] Bangsaruntip, S., et.al., VLSI 2010,pp. 21-22

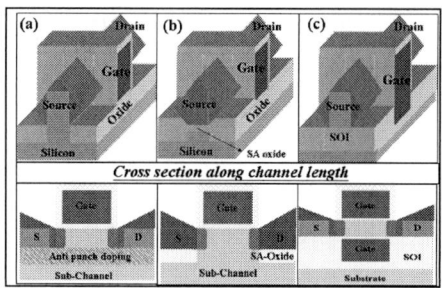

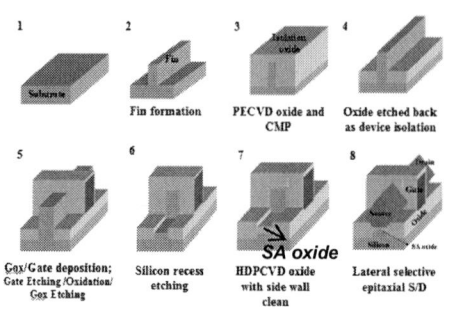

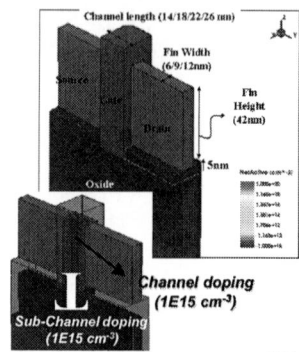

Fig. 1 Schematic structures of (a) normal FinFET (b) T-FinFET (c) gate-all-around (GAA) FinFET and the corresponding cross sections of devices along channel direction.

Fig. 2 Process flow for T-FinFET. The key is Step 7 in which anisotropic oxide is deposited by HDPCVD or Sputter and the oxide on the side wall of fin shall be removed before selective lateral S/D growth.

Fig. 3 Simulation structure of T-FinFET. Doping concentration in the channel and sub-channel region is 1E15 cm^{-3}.

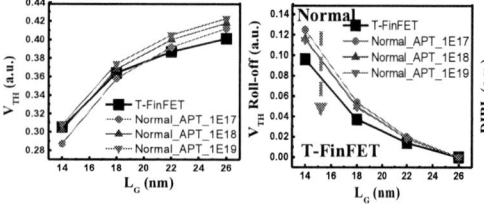

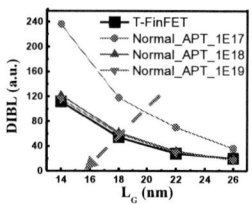

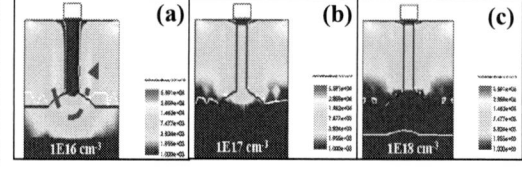

Fig. 4 V_{TH} and V_{TH} roll-off v.s. gate length for T-FinFET and normal FinFETs with various sub-channel doping concentrations. **EOT= 1nm for all simulated devices in this article.**

Fig. 5 DIBL v.s. gate length for T-FinFET and normal FinFETs.

Fig. 6 Current flow of normal FinFETs with various APT doping concentrations when I_D=1E-7 μA/μm. (a) 1E16 (b) 1E17 (c) 1E18 cm^{-3}. Serious punch through current below the channel region is seen for lightest doping case.

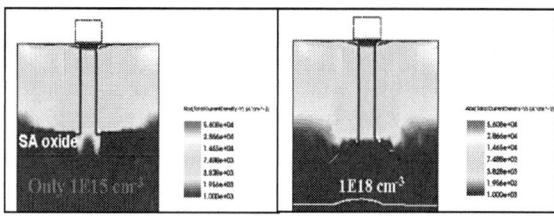

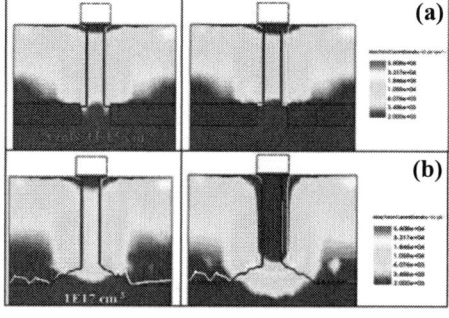

Fig. 7 Current flow of T-FinFET (left) and normal FinFET (right). SA oxide supplies strong punch through immunity even though the doping concentration of sub-channel is only 1E15 cm^{-3}. For normal FinFET, doping is 1E18 cm^{-3}. V_{DS}=0.5V. L_G=14 nm.

Fig. 8 Current flows of T-FinFET (a) and normal FinFET (b) with different V_{DS} as I_D=1E-7 μA/μm. T-FinFET with SA oxide does not need to add sub-channel doping. L_G=14 nm.

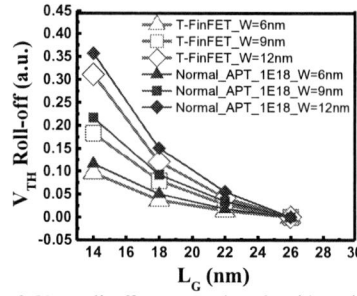

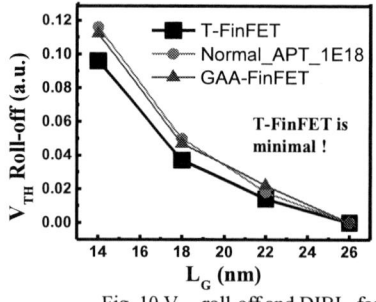

Fig. 9 V_{TH} roll-off v.s. gate length with various fin widths. T-FinFET has better SCE immunity than normal FinFET.

Fig. 10 V_{TH} roll-off and DIBL for three kinds of devices, i.e., T-FinFET, normal FinFET and GAA. T-FinFET depicts minimal V_{TH} roll-off and similar DIBL behavior with the others. W_{FIN}= 6nm.

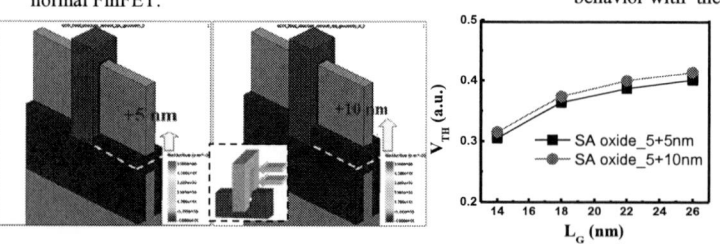

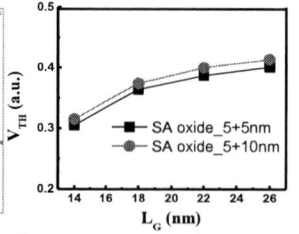

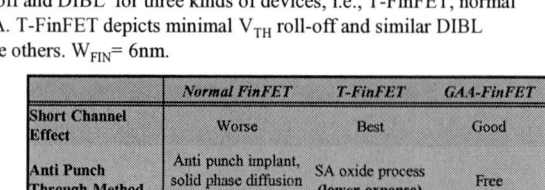

Fig. 11 Schematic structures of T-FinFET to investigate the effect of SA oxide thickness on the electrical properties. White dash line labels the STI surface. Total thickness of SA oxide are 10 & 15 nm.

Fig. 12 V_{TH} v.s. gate length. +5 & +10 nm indicate the thickness of SA oxide portion over the STI surface.

	Normal FinFET	T-FinFET	GAA-FinFET
Short Channel Effect	Worse	Best	Good
Anti Punch Through Method	Anti punch implant, solid phase diffusion (higher expense)	SA oxide process (lower expense)	Free
DIBL_L_G=14nm	Worse	Good	Good
Device Issue	Wider fin needs higher APT concentration	Lighter APT implant and easy process control	Complex process flow and self-heating issue

Table 1. Comparison table of various types of multi-gate FETs. T-FinFET has advantages of more superior SCEs, lighter APT & easy processing.

72

Characteristics of Inversion, Accumulation and Junctionless mode Silicon N-Type and P-Type Bulk FinFETs with optimized 3-nm nano-fin structure

Vasanthan Thirunavukkarasu[1,2], Yi-Ruei Jhan[1], Yan-Bo Liu[1] and Yung-Chun Wu[1,*]

[1]Department of Engineering and System Science, National Tsing Hua University, Hsinchu, Taiwan

[2]Nano Science and Technology Program, Taiwan International Graduate Program, Academia Sinica, Taipei, Taiwan
and National Tsing Hua University, Hsinchu, Taiwan

*Phone: +886-3-5715131 ext: 34287, Fax: +886-3-5720724, E-mail: nanodhasan@gmail.com

Abstract

We for the first time evaluate the 3-nm gate Length (L_G=3nm) inversion (IM) mode, accumulation (AC) mode and junctionless (JL) mode Silicon bulk FinFET performance with optimized nano-fin structure (F_W=F_H=3nm) using 3-D quantum transport device simulation. The excellent electrical characteristics of L_G=3nm Si bulk FinFET are reported. The sub threshold slope values (SS~65mV/dec.) and drain-induced barrier lowering (DIBL<17mV/V) are analyzed in all three IM, AC and JL modes bulk FinFET with $|V_{TH}|$ ~0.31 V. This research reveals that Moore's law can be scaled down to 3-nm nodes.

Keywords: Si, IM, AC, JL, bulk FinFET, 3nm, L_G

Introduction

For sub 10-nm node technologies, different device models have been proposed and intensively researched to overcome the several critical challenges that arise due to the relentless scaling to ever small dimensions. Various approaches have been proposed to attenuate the impact of short channel effects (SCE) on threshold voltage, drain-induced barrier lowering (DIBL), sub threshold swing [1-4]. Continuous scaling of transistors hinders the development of high quality junctions. Development of high performance sub 10-nm node bulk FinFET becomes more challenging [5]. In this work, we investigate the performance of the L_G=3nm and Si Bulk FinFET in all three IM, AC and JL modes of operation with optimized nano-fin (F_W=F_H=3nm). The transfer and output characteristics of L_G=3nm Ge & Si bulk FinFETs are discussed. In addition, the 3-D electron density distribution in off- state and on-state are also explained.

Model Descriptions

Fig. 1 presents the structure of the simulated devices with the important parameters. The doping concentration used are also tabulated **(Table 1)**. The devices have an EOT of 0.3nm and L_G=F_W=F_H=3nm. The work function (WF) can be tuned appropriately to get the desired V_{TH}(~0.31V). To obtain accurate results the device is simulated by solving 3-D quantum transport equations using the commercial Synopsys Sentaurus Device Tool [6]. The physical models considered includes, density gradient model, quantum confinement effect, band gap narrowing model, Shockley-Read-Hall recombination, and doping-dependent model. The mobility model used in device simulation is according to the Mathiessen's rule.

Results and Discussion

Fig. 2 shows the transfer curves of the L_G=3nm Si Bulk FinFETs. The saturation current I_{SAT} of Si IM, AC and JL NFET (at V_G= 0.7V, V_D = 1V) are 2.52×10^{-4} A/µm, 2.54×10^{-4} A/µm and 2.32×10^{-4} A/µm respectively; The I_{SAT} of Si IM, AC and JL PFET are 2.24×10^{-4} A/µm, 2.25×10^{-4} A/µm and 2.26×10^{-4} A/µm respectively.. **Fig. 2(a)**, clearly establishes that IM mode have similar characteristics like AC mode. The

sub threshold slope (SS) for n-type IM, AC and JL modes are respectively, 78.74mV/dec, 78.79mV/dec and 77.37mV/dec. The sub threshold slope (SS) for p-type IM, AC and JL modes are respectively, 67.63mV/dec, 67.65mV/dec and 62.28mV/dec. The drain induced barrier lowering (DIBL), defined as the difference in V_{TH} between $V_{D(low)}$ = 0.05V and V_D = 0.7V, for n-type IM, AC and JL modes equals only 16.04mV/V, 16.17mV/V and 26.80mV/V respectively. The similar performances are also achieved in p-type IM, AC and JL bulk FinFET (29.80mV/V, 31.89mV/V and 40.20mV/V). **Fig. 2(c)** shows the I_D-V_G of Si JL NFET which achieves almost ideal SS value of 77.37 mV/dec. The DIBL value of Si JL PFET is 40.20mV/V; which proves that simulated Si device has excellent device characteristics. It is noteworthy, the off-state current are all low in IM, AC & JL modes of Bulk FinFET owing to extensively scaled nano-fin. **Fig. 3(a) & Fig.3(b)** shows the I_D-V_D curves of IM & AC Si PFET. The result shows a excellent output characteristic behavior. **Fig. 4(a) & Fig. 4(b) and Fig. 5(a) & Fig. 5(b)**, compares on-state (V_{GS}=1V) & off-state (V_{GS}=1mV) electron density distribution at the 3D cross-sections of the 3nm nano-fin n-type Si-IM and Si-AC Bulk FinFETs, respectively. **Fig. 6(a) and Fig. 6(b)**, compares on-state (V_{GS}=1V) and off-state (V_{GS}=1mV) electron density distribution at the 3D cross-sections of the 3nm nano-fin n-type Si-JL Bulk FinFET. The conduction path is located at the middle of the nano-fin as expected. Notably, the on-state and off- state results of 3-D eDensity distribution from quantum transport simulation demonstrate that the device can be scaled down to a physical limit of 3nm. And the current conduction in all three (JL, AC & IM) modes are almost similar because the carriers fully occupy 3nm nano-fin cross-section.

Conclusions

We analyzed, the 3-nm gate length bulk Silicon FinFET operating in inversion (IM) mode, accumulation (AC) mode and junctionless (JL) mode. The observed transfer characteristics, output characteristics and electron density distribution results of the 3-D quantum transport device simulation reveal the fact that all the three IM, AC and JL mode of operation are perfectly feasible even at 3-nm gate length. Thus, enables the bulk FinFET devices to be scaled down to its least possible physical limits obeying Moore's scaling law for all future applications.

References

[1] B. Duriez, *et al., IEDM*, (2013), pp.522.
[2] W.Guo *et al., IEDM*, (2014), pp.168.
[3] K.Usuda *et al., IEDM*, (2014), pp.422 .
[4] S. Jin *et al., IEDM*, 2014, pp.184.
[5] J.P. Colinge *et al., Nature Nanotech.*, vol. 5, 2010, pp.225
[6] Synopsys TCAD, Version G-2012.06.

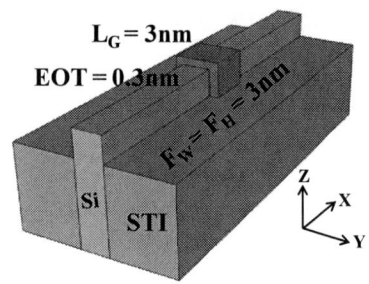

L_G = 3nm
EOT = 0.3nm
$F_W = F_H = 3nm$

Si STI

Fig.1. Device structure and important parameters of simulated 3-nm Gate Length (L_G) IM, AC & JL Silicon Bulk FinFET.

Table 1. Doping Concentration of simulated 3-nm Gate Length IM, AC & JL Si Bulk FinFET; Arsenic and Boron are used for N-type and P-type doping respectively.

Device Mode	Junctionless	Accumulation	Inversion
Source & Drain Doping Concentration	N : 1x10^{20} cm^{-3} P : 1x10^{20} cm^{-3}	N :1x10^{20} cm^{-3} P :1x10^{20} cm^{-3}	N :1x10^{20} cm^{-3} P :1x10^{20} cm^{-3}
Channel Doping Concentration		N : 1x10^{18} cm^{-3},N-Type P : 1x10^{18} cm^{-3},P-Type	N : 1x10^{18} cm^{-3},P-Type P : 1x10^{18} cm^{-3},N-Type
Substrate Doping Concentration	N : 5x10^{18} cm^{-3},P-Type P : 5x10^{18} cm^{-3},N-Type	N : 5x10^{18} cm^{-3},P-Type P : 5x10^{18} cm^{-3},N-Type	N : 5x10^{18} cm^{-3},P-Type P : 5x10^{18} cm^{-3},N-Type

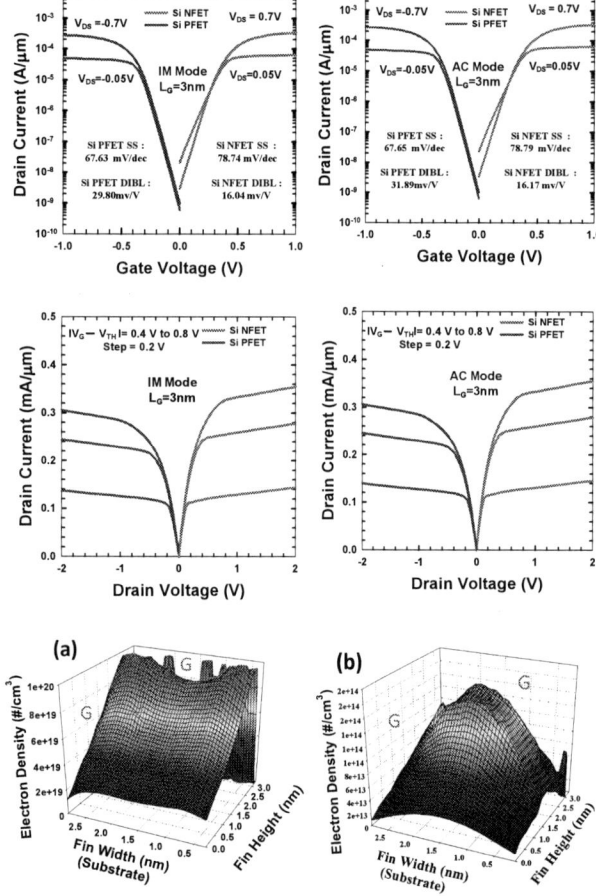

Fig. 2. I_D-V_G curves of the 3nm Gate length (L_G) with N-type and P-type Si Bulk FinFET operating in (a) IM, (b) AC & (c) JL mode with SS and DIBL values shown inset.

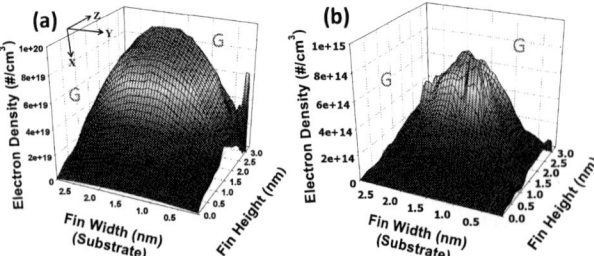

Fig. 3. I_D-V_D curves of the 3nm Gate length (L_G) with N-type and P-type Si Bulk FinFET operating in (a) IM, (b) AC & (c) JL modes with threshold voltage of (V_{TH}) ~ 0.3V and different over-drive voltages (V_{OV}).

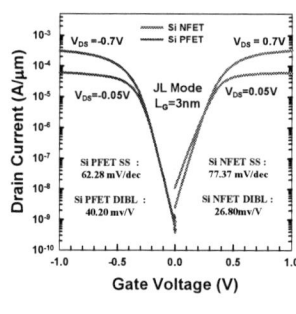

Fig. 4. 3D Mesh Plot for Electron Density distributions in the 3nm gate length (L_G) n-type IM mode Si Bulk FinFET in (a) on-state(V_{GS}=1V) & (b) off-state(V_{GS}=0.001V).

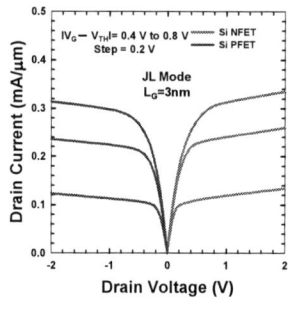

Fig. 5. 3D Mesh Plot for Electron Density distributions in the 3nm Gate length (L_G) n-type AC mode Si Bulk FinFET in (a) on-state(V_{GS}=1V) & (b) off-state(V_{GS}=0.001V).

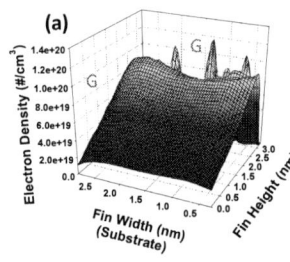

Fig. 6. 3D Mesh Plot for Electron Density distributions in the 3nm Gate length (L_G) n-type JL Si Bulk FinFET in (a) on-state(V_{GS}=1V) & (b) off-state(V_{GS}=0.001V).

Table 2. Summary of important numerical values of simulated 3-nm Gate Length Si IM, AC & JL Si Bulk FinFETs.

Device Mode	Junctionless	Accumulation	Inversion
Work Function	N : 4.55 eV P : 4.60 eV	N : 4.40 eV P : 4.80 eV	N : 4.40 eV P : 4.80 eV
$V_{TH (\sim 0.31V)}$	N : 0.3057 V P : -0.2969 V	N : 0.2987 V P : -0.2987 V	N : 0.3000 V P : -0.3019 V
SS	N :77.37 mV/dec P : 62.28 mV/dec	N : 78.79 mV/dec P : 67.65 mV/dec	N : 78.74 mV/dec P : 67.63 mV/dec
DIBL	N : 26.80 mV/V P : 40.20 mV/V	N : 16.17mV/V P : 31.89mv/V	N : 16.04mv/V P : 29.80mv/V
I_{ON}	N : 2.32x10^{-4} A/μm P : 2.26x10^{-4} A/μm	N : 2.54x10^{-4} A/μm P : 2.24 x 10^{-4} A/μm	N : 2.522x10^{-4} A/μm P : 2.240x10^{-4} A/μm

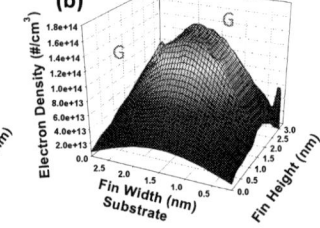

Bringing Physics to Device Design – a Fast and Predictive Device Simulation Framework

M. Karner, Z. Stanojević, F. Mitterbauer, C. Kernstock, H. Demel

Global TCAD Solutions GmbH., Landhausgasse 4/1a, 1010 Vienna, Austria

Email: {z.stanojevic|m.karner|f.mitterbauer|c.kernstock|h.demel}@globaltcad.com

Abstract—We present a physically grounded modeling, simulation, and parameter-extraction framework that targets design and engineering of ultra-scaled devices and next-generation channel materials. The framework consists of a fast and accurate Schrödinger-Poisson solver/mobility extractor coupled to a device simulator. The framework brings physical modeling of semiconductor channels to device design and engineering which until now has been the domain of TCAD tools based on purely empirical models.

I. METHODOLOGY

Our method for simulating ultra-scaled devices involves physically grounded models for electronic structure and transport [1], implemented in a fast numerical Schrödinger-Poisson solver and mobility calculator (VSP [2]), and coupled to a device simulator (Minimos NT), both being part of GTS Framework [3].

II. ELECTRONIC STRUCTURE AND TRANSPORT

In this work, we also present a novel model for the electronic structure near the valence band edge - the *Six-Valley-Model*. The model is derived from well-known the 3-band $\mathbf{k \cdot p}$ model for valence band in diamond and zincblende crystals [4]. The six valley in question are projections of the $\mathbf{k \cdot p}$-Hamiltonian onto the states $|1,1,0\rangle/\sqrt{2}$, $|1,\bar{1},0\rangle/\sqrt{2}$, $|1,0,1\rangle/\sqrt{2}$, $|1,0,\bar{1}\rangle/\sqrt{2}$, $|0,1,1\rangle/\sqrt{2}$, and $|0,1,\bar{1}\rangle/\sqrt{2}$, respectively. The valleys are Visualized in Fig. 1.

The six valleys correctly reproduce the orientation-dependent confinement behavior of the valence bands. Stress, while altering the coupling energies between bands in the $\mathbf{k \cdot p}$ model, translates to mere energy shifts of the six valleys relative to each other. Fig. 2 shows stress-dependent mobility of a 5 nm thin silicon channel. The table in Fig. 1 gives an overview of the model parameters.

Transport in the channel is analyzed using linearized Boltzmann transport equation for small fields. The available scattering processes in our framework [5] include acoustic and nonpolar optical phonons, polar optical phonons (for III/V materials), ionized impurities (local and remote), surface roughness, and alloy disorder.

For comparison, a full $\mathbf{k \cdot p}$-based electronic structure calculation, scattering model and mobility extraction were also implemented [6, 7]. However, parabolic models such as the single-band model for Δ-valley electrons and the Six-Valley-Model for holes allow a semi-analytical treatment of occupancy and transport, which makes their execution thousands of times faster. Thus, with proper calibration, they are ideally suited for a TCAD framework.

III. COUPLING TO DEVICE SIMULATION

The coupling between Electronic structure is done by decomposing the simulation domain along the channel direction into two-dimensional slices, as illustrated in Fig. 3. On each iteration step, Minimos NT invokes an instance of VSP on each slice to calculate the confined carrier densities and mobilities. The principle of operation is shown in Fig. 5; Minimos NT passes electrostatic potential and quasi-Fermi energis for electrons and hole, from which VSP can determine the subband structure and its occupancy. The data on each slice is then extruded and interpolated back onto the three-dimensional simulation domain before starting the next iteration step.

Rather than passing the carrier densities directly to Minimos NT, VSP instead calculates a correction potential. The correction potential is such that adding it to the input electrostatic potential would result in Minimos NT reproducing exactly the same carrier densities on the slice as VSP does. This *self-consistent quantum correction* approach ensures a stable and fast convergence of the three-dimensional problem. We observe that, upon convergence, carrier densities from Minimos NT and VSP become identical.

IV. RESULTS AND CONCLUSION

To demonstrate the capabilities of our approach we conducted a simulation study of the nanowire gate-all-around transistors fabricated and characterized by Bangsaruntip et al. [8, 9]. The geometric model of the device can be seen in Fig. 3. The ensemble of simulated devices consists of n-type and p-type nanowire transistors, both with a circular cross-section, 12.8 nm in diameter, with an EOT of 1.5 nm. The gate length was set to 25 nm. Transfer characteristics were extracted for the linear and saturated regimes, shown in Fig. 6.

The gate length was then varied from 25 nm to 150 nm to observe DIBL behavior and threshold voltage roll-off, as shown in Fig. 7. Finally, the variation of the transfer characteristic under tensile stress along the nanowire axis was investigated; the results shown in Fig. 8 are consistent with the experimental findings from [9].

All these analyses could be done in a fairly short amount of time, thanks to the advanced models and fast algorithms used in this work, highlighting the main novelty of the work: combining the efficiency of TCAD simulation with physical models.

REFERENCES

[1] Z. Stanojević *et al.*, Solid-State Electronics (2015).
[2] O. Baumgartner *et al.*, J. Comput. Electron. (2013).
[3] GTS Framework, http://www.globaltcad.com/en/products/.
[4] G. Dresselhaus *et al.*, Phys. Rev. **98**, 368 (1955).
[5] Z. Stanojevic *et al.*, *IEDM* (2013), pp. 332–335.
[6] Z. Stanojevic *et al.*, *SISPAD* (2014), pp. 181–184.
[7] Z. Stanojevic *et al.*, *2014 IEEE Silicon Nanoelectronics Workshop* (2014), pp. 83–84.
[8] S. Bangsaruntip *et al.*, *IEDM* (2013), pp. 20.2.1–20.2.4.
[9] S. Bangsaruntip *et al.*, *VLSIT* (2010), pp. 21 –22.
[10] T. Manku *et al.*, J. Appl. Phys. **73**, 1205 (1993).

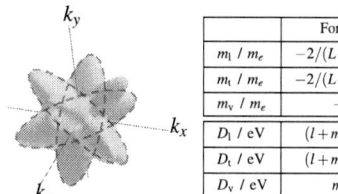

	Formula	Analyt.	Fitted
$m_{\mathrm{l}}\,/\,m_e$	$-2/(L+M-N)$	2.353	2.723
$m_{\mathrm{t}}\,/\,m_e$	$-2/(L+M+N)$	0.114	0.127
$m_{\mathrm{v}}\,/\,m_e$	$-M$	0.275	0.269
$D_{\mathrm{l}}\,/\,\mathrm{eV}$	$(l+m-n)/2$	-3.985	-4.376
$D_{\mathrm{t}}\,/\,\mathrm{eV}$	$(l+m+n)/2$	2.815	3.254
$D_{\mathrm{v}}\,/\,\mathrm{eV}$	$m/2$	-3.99	-4.586

Fig. 1. An energy-iso-surface of the valence dispersion relation in Si; the surface exhibits a warped structure with "fingers" pointing towards each of the equivalent $\langle110\rangle$-directions. Three of the six valleys of the Six-Valley-Model are highlighted with dashed lines. Parameters of the model can be derived from coefficients of the **k·p**-Hamiltonian, L, M, and N and the strain parameters l, m, and n. The analytical values are based on the Hamiltonian from [10].

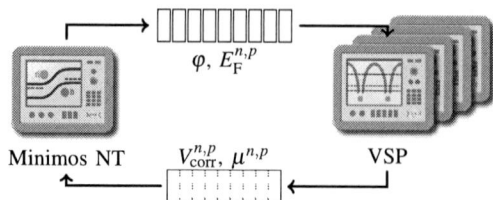

Fig. 5. Schematic overview of the coupling between Minimos NT and VSP; The simulation domain is decomposed into slices on each of which Minimos NT invokes a VSP instance (in parallel) and passes the electrostatic potential along with the quasi-Fermi-energies. VSP returns the quantum correction potential and the mobility for the slice which are interpolated back onto the original 3D simulation domain.

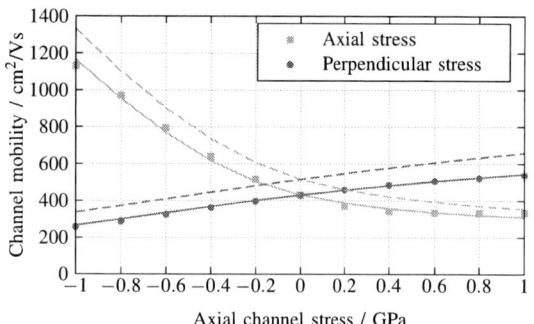

Fig. 2. Stress dependent hole mobility of a 5 nm thin silicon channel with a (001)-surface; stress is applied in transport direction ([110]) and perpendicular to it ([1$\bar{1}$0]). Symbols were obtained from a **k·p**-Hamiltonian, solid lines from the Six-Valley-Models with parameters fitted to the **k·p**-result, and dashed lines with analytical six-valley parameters.

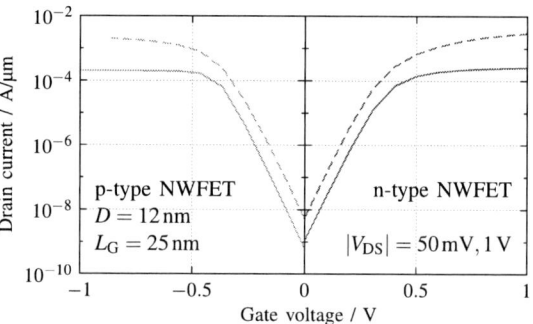

Fig. 6. Transfer characteristic in linear and saturated mode of both n-type and p-type nanowire transistors; the characteristic was normalizes, so that the off-state $I_{\mathrm{D}}^{\mathrm{sat}}$ for n- and p-type device coincide.

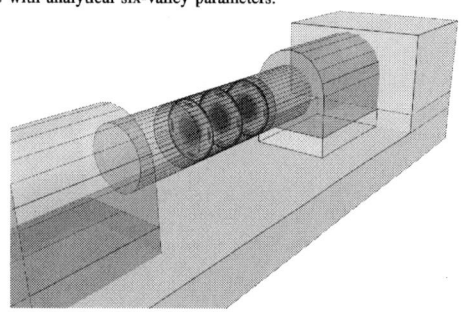

Fig. 3. Visualized slices of nanowire transistor channel displaying the electron density

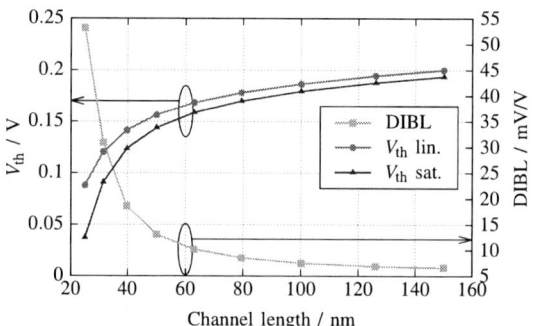

Fig. 7. Short-channel properties of the n-type nanowire transistor; the threshold voltage roll-off and drain-induced barrier lowering (DIBL) are shown.

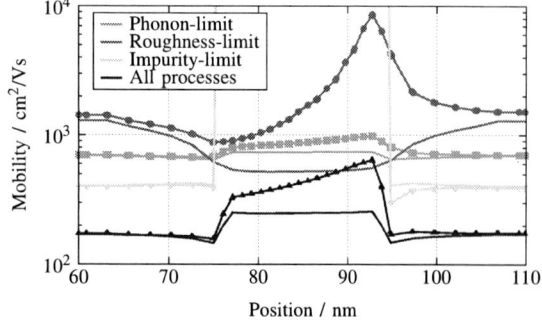

Fig. 4. Electron mobility along channel showing the scattering processes that limit the channel current at $V_{\mathrm{DS}} = 50\,\mathrm{mv}$ (solid lines) and 1 V (symbols)

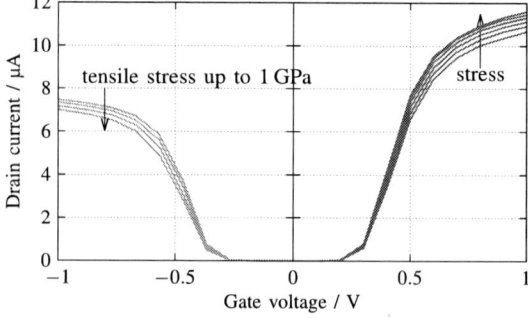

Fig. 8. Variation in linear transfer characteristic with respect to tensile stress along the axis of the nanowire transistors

A Capacitance-Voltage model for DG-TFET

Arnab Biswas, Adrian M. Ionescu

Ecole Polytechnique Fédérale de Lausanne, Station 11, CH-1015 Lausanne, Switzerland
Phone: +41 21 693 69 72, Fax: +41 21 693 3640, Email: arnab.biswas@epfl.ch

Abstract: In this work we develop a simplified capacitance model for Double Gate TFETs. Capacitance-voltage measurements were done on all-Silicon SOI TFETs at different biasing schemes to support the model development. TCAD simulations [1] of DG-TFETs were used to validate the model.

Measurement Setup and Principle: Capacitance-Voltage (CV) measurements were performed on all-Silicon SOI TFETs [2]. The HP 4284A precision LCR meter was used for all the CV measurements. The DC bias was varied from -3 V to +3 V for most cases with a small AC signal of 30 mV. To measure $C_{GS} = -\partial Q_G/\partial V_S \mid V_G, V_D, V_{SUBS}$, the low terminal of the setup was connected to the gate which monitors the current and the high terminal to the source where the bias is applied. CV measurements on TFETs show perfectly symmetric nature due to ambi-polar characteristics [3] of TFETs as shown in Fig.1. TFET capacitances were also found to be reciprocal in nature which means $C_{ij} = C_{ji}$ for $i,j=\{G,S,D\}$ at $V_{DS} = 0$ V (Fig. 1a).

Drain-gate capacitance was measured at different values of V_{DS} in Fig. 1b. Similarly source-gate capacitance was measured at different values of V_{DS} in Fig. 1c. We observe that gate-source capacitance (C_{GS}) under strong inversion remains negligible compared to gate-gate capacitance (C_{GG}) even at high $V_{DS} = 1.5$ V. So, it can be said that source has negligible contribution to the inversion charges, and that only depletion charges contribute to the source capacitance. Hence C_{GG} is dominated by the gate-drain capacitance (at least for low injection levels smaller than the depletion charges, as in the studied devices). This is also the reason for the strong miller effect in TFETs [4]. Further to the measurements, TCAD simulations were done to create a fictitious device TFET A with 2 orders of magnitude more current than a template 100 nm device in Fig. 2(top). As shown in the capacitance simulation in Fig. 2(bottom), the TFETA device with almost mA current level still has negligible C_{GS} compared to C_{GD}. TFET A shows a slightly higher C_{SG} and slightly lower C_{DG} compared to TFET B. This plot clearly shows that the additionally injected carriers from the source are still negligible compared to the inversion charge. So we can say that tunneling generated carriers has little or negligible role to play in the charge distribution of a tunnel FET. The following assumptions can now be taken reliably: (a) 100-0 charge partitioning scheme with 100% to drain. (b) Source depletion charge entirely dictates the source-gate capacitance.

Model Description: A 100nm double gate device with 3nm HfO$_2$ gate dielectric and 20nm silicon body thickness was used in the simulations. Source/drain doping was 1×10^{20} cm^{-3} and abrupt junctions were assumed. As described in [5], eqn. (1) is used to compute the gate charge Q_G. The gate charge Q_G is then used to compute the surface potential ψ_s in eqn. (2). The surface potential ψ_s

is used with Gauss law to approximate inversion charges in eqn. (3). The drain-body depletion charge under the gate is also estimated by Gauss law in eqn. (4). The total drain charge is computed in eqn. (5) and finally the source charge is evaluated in eqn. (6) by following charge conservation rule.

$$V_G^* - E_{fn}(x) \approx \frac{Q_G}{C_{OX}} + U_T \cdot \ln\left(\frac{Q_G^2}{2\varepsilon_{Si}qU_Tn_i} + \frac{2Q_G}{qn_iT_{SI}}\right) \quad (1)$$

$$V_G^* - \psi_s(l_g/2) = Q_G/C_{OX} \quad (2)$$

$$Q_{D\text{-}INV} = C_{OX}(V_{GS} - V_{FB} - \psi_s)(L_G W_G) \quad (3)$$

$$Q_{D\text{-}DEPL} = \epsilon_{Si}(V_{DS} + V_{BI} - \psi_s)(T_{SI} W_G)/\lambda \quad (4)$$

$$Q_{DRAIN} = -(Q_{D\text{-}INV}) + Q_{D\text{-}DEPL} \quad (5)$$

$$Q_{SOURCE} = -(Q_{DRAIN} + Q_G) \quad (6)$$

Where λ (lambda) is the characteristic length of a double gate structure [7], V_{BI} corresponds to the built in potential in the source-channel *pn* junction, U_T is the thermal voltage, C_{OX} is the gate-oxide capacitance, V_{GS} is the gate-source bias, V_{FB} is the flat band voltage, L_G and W_G are the length and width of the gate, V_{DS} is the gate-source bias, T_{SI} is the Silicon layer thickness. Capacitances are then evaluated as derivatives of the computed charges. Earlier work on capacitance behaviour of a TFET such as [8] is based on BSIM, whereas [9] also compute charges but do not discuss capacitances voltage behaviour.

Comparison with TCAD simulations: The drain, source and gate charges with respect to V_{GS} are shown in Fig. 3. As discussed earlier source charges remain negligible compared to drain charges even at $V_{DS} = 1.5$ V. The capacitance curves with respect to gate-source voltage (top) and drain-source voltage (bottom) are shown in Fig. 4. We see that in spite of the approximations introduced the model estimates the trans-capacitances with relative accuracy both above and below the inversion threshold. Finally as shown in Fig. 5, the model also works for varying drain voltages for C_{GG}, C_{SG} and C_{DG} @ $V_{DS} = 0.5$V, 1.0V, 1.5V. C_{SG} particularly shows large discrepancy due to the assumption of a constant depletion width for all T_{SI}. However the shape and the variation with V_{DS} and V_{GS} is captured in this model.

Conclusions: Based on capacitance measurements a first order capacitance model for DG-TFETs has been proposed. The model matches numerical simulations well and shows the same behaviour as measurements.

References: [1] Synopsys Senataurus TCAD, ver. 2012.06. [2] F. Mayer, et al, IEDM Technical Digest, 2008 [3] A. M. Ionescu and H. Riel, Nature, Nov. 2011 [4] S. Mookerjea, et al., Electron Device Lett. IEEE 30, 1102 (2009) [5] A. Biswas et al. ULIS 2014 [6] Mansun Chan et al, NSTI-Nanotech 2004 [7] K. Suzuki et al., IEEE TED (1993). [8] Y. Yang, et al., IEEE Electron Device Lett. 31, 752 (2010). [9] L. Zhang et al. IEEE TED, vol. 61, no. 2, 2014

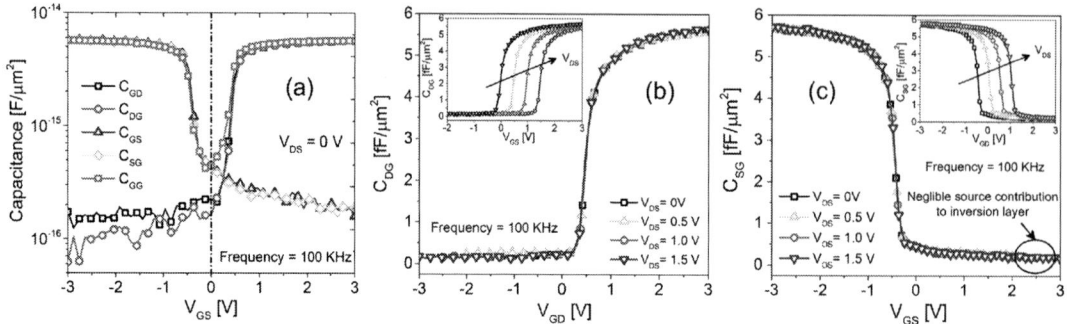

Figure 1: (a) Measured C_{GS}, C_{GD} and C_{GG} with respect to V_{GS}. Perfect symmetry is observed highlighting the ambi-polar nature of TFETs. (b) Measured Gate-Drain capacitance for different V_{DS}. Inset shows measured C_{GD} curves with respect to V_{GS}. (c) Measured Gate-Source capacitance for different V_{DS}. Source contribution to inversion electrons is negligible even at $V_{DS} = 1.5$ V with device turned ON. Inset shows measured C_{GS} curves with respect to V_{GD}. Fully depleted SOI TFET with $L_G = 200$ nm, $T_{SI} = 21$ nm and 6 nm SiO_2 dielectric.

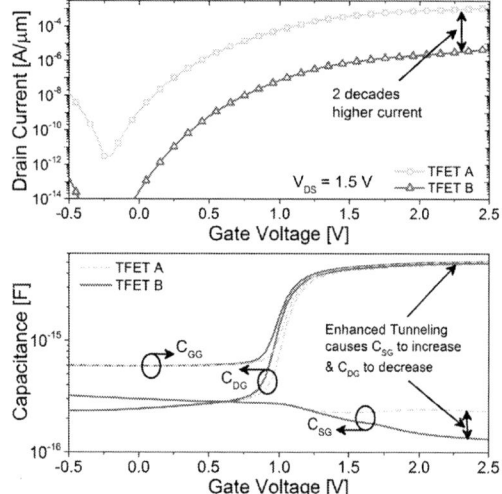

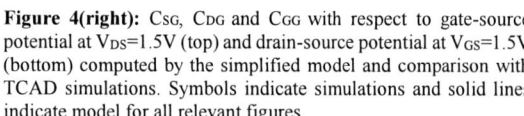

Figure 2(above): (top) TFET A simulated with artificially enhanced tunneling injection to have 2 order of magnitude more current than TFET B. (Bottom) TFET A shows a slightly higher C_{SG} and slightly lower C_{DG} compared to TFET B. C_{SG} remains negligible compared to C_{DG} in TFET A, verifying that BTBT has little or negligible influence in the charge distribution of a TFET.

Figure 4(right): C_{SG}, C_{DG} and C_{GG} with respect to gate-source potential at V_{DS}=1.5V (top) and drain-source potential at V_{GS}=1.5V (bottom) computed by the simplified model and comparison with TCAD simulations. Symbols indicate simulations and solid lines indicate model for all relevant figures.

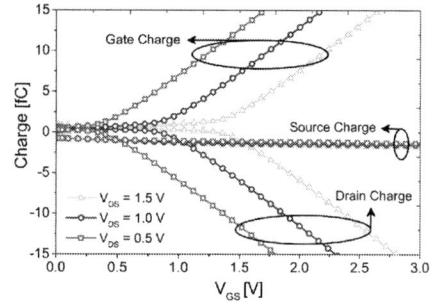

Figure 3(above): Modelled gate, drain and source charges of a double gate tunnel FET as a function of gate and drain voltage.

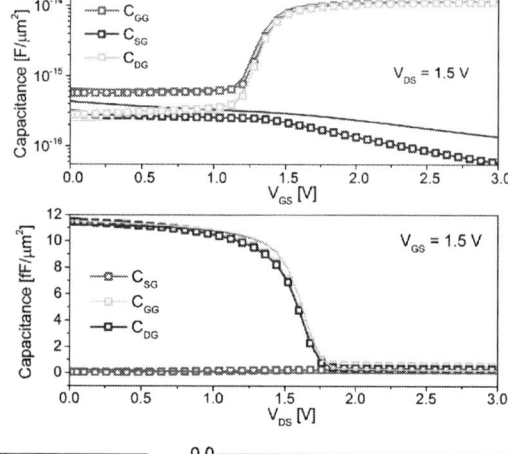

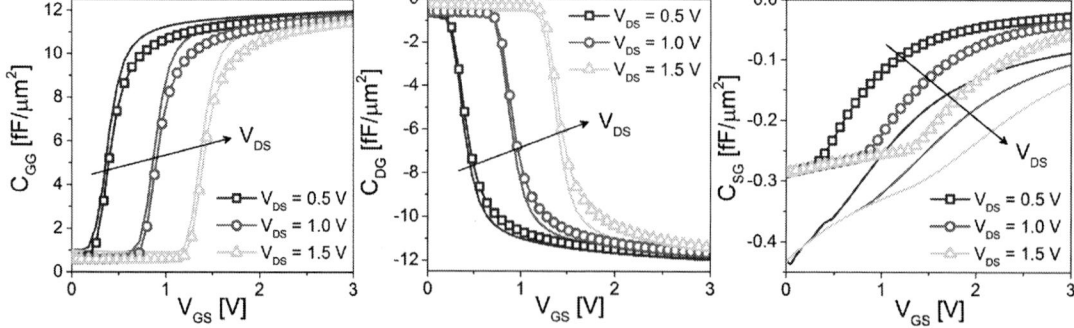

Figure 5: Modelled C_{GG}, C_{DG} and C_{SG} curves with respect to V_{GS} for three different drain voltages and comparison with TCAD simulations. Modelled C_{SG} curves with respect to V_{GS} although do not show a good match, predicts the trend correctly.

Physics-based Model for the Conductive Filament at the Low Resistance State of Thin SiO₂ Films

Rintaro Yamaguchi[1], Shingo Sato[1,2], and Yasuhisa Omura[1,2]

[1]Dpt. Electronics, Kansai University, Japan
[2]ORDIST, Kansai University, Japan
Email: omuray@kansai-u.ac.jp

Abstract - **This paper proposes the possible physics-based model for the conductive filament (CF) at the low-resistance state (LRS) of thin SiO₂ films that were formed by sputtering technique. The closed and analytical current models proposed here are examined by experimental results.**

The about 60-nm-thick SiO₂ films are deposited by RF sputtering technique on the β-FeSi₂ film (200nm)/n-Si substrate [1]. The Si substrate is the bottom electrode (BE) and Al top electrode (TE) was deposited by evaporation technique on the SiO₂ films. The schematic of fabricated capacitor is shown in Fig. 1(a). Energy band diagram of materials of the capacitor is shown in Fig. 1(b), where the energy band diagram assumed at the LRS is also illustrated.

The low resistance state was formed by applying a positive constant voltage (8V) to the top electrode; the current compliance is 10 μA. Current vs. voltage characteristics of the SiO₂ film is shown in Fig. 2 for a voltage range from +0.7 V to -0.7 V at room temperature. The capacitor shows an asymmetric current vs. voltage characteristic at room temperature. Therefore we assumed the energy band diagram (dotted lines) for the LRS shown in Fig. 1(b).

We propose two-different transport models for the LRS as shown in Fig. 3.

(1) Positive bias voltage to the top electrode (see Fig. 3(a))

When a positive bias voltage is applied to the top electrode, we assume that electrons are injected from the β-FeSi₂ film. Some electrons are captured at the traps of the conductive filament and/or the surrounding SiO₂ film region, and they work as negative charges; that is, the electron transport is dominated by the space-charge-limited (SCL) conduction. The SCL current density is expressed theoretically as [2]

$$J_{SCLC} = \frac{9}{8} \frac{\varepsilon_{CF} \mu_n V^2}{t_{ox}^3} \exp(0.89 \eta_c \sqrt{E}), \qquad (1)$$

where ε_{CF} is the permittivity of conductive filament, t_{ox} is the SiO₂ film thickness, and η_c is SCL current correction factor.

(2) Negative bias voltage to the top electrode (see Fig. 3(b))

When a negative bias voltage is applied to the top electrode, we assume that electrons are injected from the Al electrode and holes are injected from the β-FeSi₂ film; some electrons recombine with holes in the conductive filament. The electrons are supplied by the Fowler-Nordheim tunneling mechanism [3] and they are transported by the Poole-Frenkel emission mechanism [3]. Theoretical expressions for the current density are given as

$$J_{FN} = (\frac{q^3 E^2}{8\pi\hbar\phi}) \exp\left[\frac{-4(2m)^{1/2}\phi^{3/2}}{3\hbar q E}\right], \qquad (2)$$

$$J_{PF} = E \exp\left[\frac{-q(\phi - \sqrt{qE/4\pi\varepsilon_{CF}})}{k_B T}\right], \qquad (3)$$

where ϕ is barrier height in the conductive filament.

We examined the model proposed here by comparing with experimental result shown in Fig. 2, where the broken line shows the calculation result; it is assumed that the cross sectional area of the conductive filament is 50 nm², $\phi = 0.2$ eV, $\varepsilon_{CF} = 11\varepsilon_o$, and $\eta_c = -0.003$. The model basically reproduces the experimental result.

REFERENCES

[1] R. Yamaguchi, S. Sato, Y. Omura and K. Nakamura, "Characterization and Modeling of Resistive-Transition Phenomena and Electronic Structure of Sputter-Deposition SiO2 Films", *Tech. Dig., WOLTE-11* (Grenoble, July, 2014), pp.69-72.

[2] D. F. Barbe, "Space-charge-limited current enhanced by Frenkel effect,"
J. Phys. D, Appl. Phys., vol. 4, no. 11, pp. 1812–1815, 1971.

[3] S. M. Sze and K. K. Ng, "Physics of Semiconductor Devices", 3ʳᵈ ed. (Wiley, 2007), p. 227.

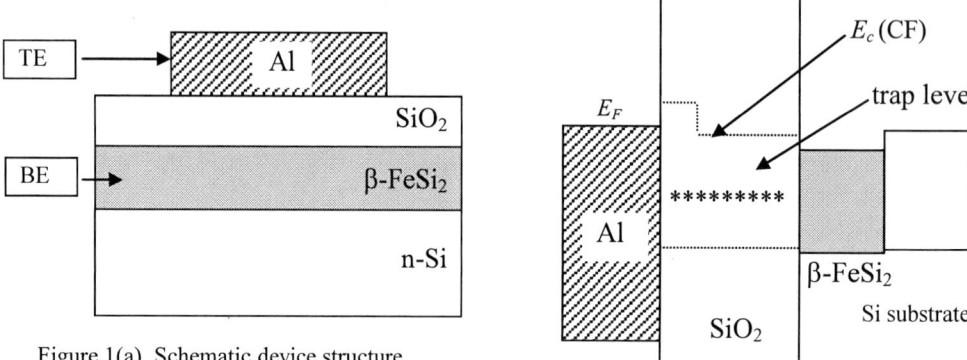

Figure 1(a). Schematic device structure.

Figure 1(b). Energy band diagram assumed.

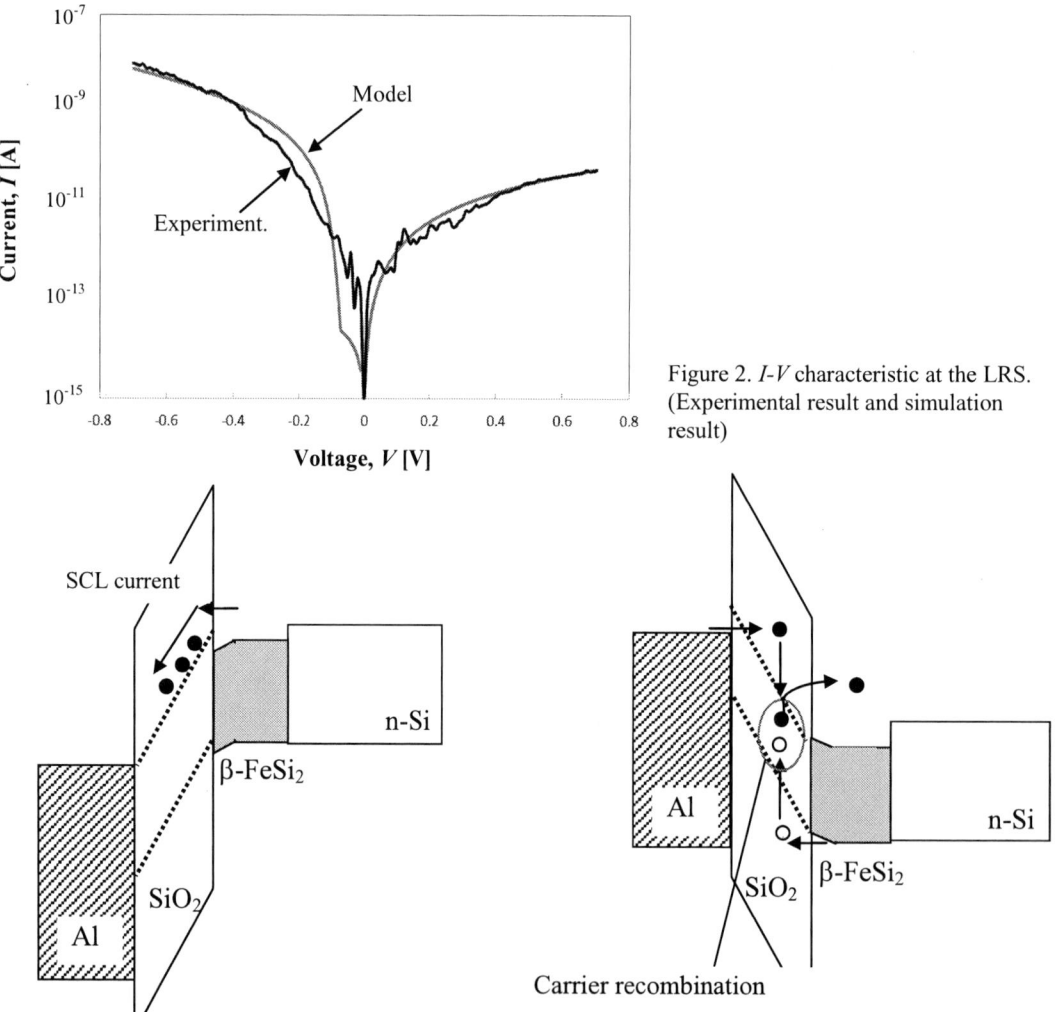

Figure 2. *I-V* characteristic at the LRS. (Experimental result and simulation result)

Figure 3(a). Band diagram at the positive bias to TE.

Figure 3(b). Band diagram at the negative bias to TE.

Performance Evaluation of Si Ultra-Thin Body (1 nm) Junctionless FET with L_G = 1 nm and L_G = 3 nm

Yi-Ruei Jhan, Yan-Bo Liu, and Yung-Chun Wu*

Department of Engineering and System Science, National Tsing Hua University, Hsinchu, Taiwan

*Phone: +886-3-5715131 ext: 35802, E-mail: jhanyiruei@gmail.com

Abstract–A Si ultra-thin body (UTB) junctionless field-effect transistor (UTB-JLFET) with L_G = 1 nm and L_G = 3 nm have been demonstrated by solving the coupled drift-diffusion (DD) and density-gradient (DG) model. The simulation results show that the Si can be used in ultra-short channel device as long as UTB is employed. As UTB is employed, ultra-short channel device does not need to follow an empirical rule of T_{ch} = L_G / 3. Furthermore, UTB-JLFET 6T-SRAM cell has reasonable static noise margin (SNM) value of 138 mV. The circuit performances reveal UTB-JLFET can be used for sub-5 nm CMOS technology nodes.

I. Introduction

Recently, the conventional MOSFETs face a lot of challenges such as random dopant fluctuation, thermal budgets and short channel effect (SCE). The JLFET can avoid aforementioned issues because the channel region of JLFET has high doping concentrations and the same dopant type as source/drain regions [1]-[3]. But the JLFET has turn-off problem due to high doping concentration in channel region. To solve this problem, JLFET need UTB structure to reach fully depleted channel region in off-state. The UTB structure can provide low off-state current owing to large bandgap which induce by quantum confinement effect. In this work, Si has been employed to investigate electrical performance of UTB-JLFET with L_G = 1 nm and L_G = 3 nm by 3-D simulations.

II. Device Structure

Fig. 1 shows the architecture of the UTB-JLFET and the parameters of this study. The Synopsys TCAD simulator was used to perform 3-D simulations, which included the coupled drift-diffusion (DD) and density-gradient (DG) model [4]. A Si were used in the simulated channel material. The channel width is 10 nm. **Fig. 2(a) and 2(b)** show conduction band energy (E_C) diagrams of Si UTB-JLFET with L_G = 1 nm in off-state and on-state, respectively. In off-state, the UTB structure builds a high E_C level at gated region because of quantum mechanism bandgap shift. This energy barrier can block electrons which pass through channel by thermal injection and direct tunneling. In on-state, owing to absence of N-P junction barrier of JLFET, the electrons pass through channel by ballistic transport.

III. Results and Discussion

Fig. 3(a) and 3(b) show the transfer characteristics of UTB-JLFET with L_G = 1 nm and L_G = 3 nm, respectively. Owing to the ultra-thin channel, this device has high I_{ON}/I_{OFF} current ratio of 10^5 at V_G = 1 V. The SS are 100 mV/decade of PFET and 99 mV/dec of NFET with L_G = 1 nm, respectively. The DIBL are 225 mV/V of PFET and 222 mV/V of NFET with L_G = 1 nm, respectively. Even though this ultra-short channel device does not follow an empirical rule of T_{ch} = L_G / 3, the electrical properties can meet the industry requirements because of quantum confinement effect. The saturation current are 1.23×10^{-3} A/µm and 8.21×10^{-4} A/µm for NFET and PFET with L_G = 1 nm, respectively. **Fig. 4(a) and 4(b)** plot the I_D-V_D characteristics of UTB-JLFET with L_G = 1 nm and L_G = 3 nm, respectively. UTB-JLFET shows low parasitic resistance and large drain saturation current ($\sim$2.5 mA/µm).

IV. Circuit Performances

Fig. 5(a) and 5(b) plots the timing characteristics of an inverter circuit and static transfer characteristic curves of Si UTB-JLFET with L_G = 1 nm 6T-SRAM cells, respectively. **Fig. 6(a) and 6(b)** plots the timing characteristics of an inverter circuit and static transfer characteristic curves of Si UTB-JLFET with L_G = 3 nm 6T-SRAM cells, respectively. UTB-JLFET shows lower delay time and larger static noise margin (SNM). UTB-JLFET with L_G = 3 nm has large SNM value of 138 mV. This result can prove that ultra-short channel device does not need to follow an empirical rule of T_{ch} = L_G / 3 if UTB is employed.

V. Conclusion

In summary, Si UTB-JLFET with L_G = 1 nm and L_G = 3 nm is successfully demonstrated. The off-state leakage current can be reduced by quantum confinement effect. As UTB is employed, Si UTB-JLFET with L_G = 1 nm and L_G = 3 nm have reasonable SNM that can meet the industry requirements. Finally, circuit performances reveal UTB-JLFET can be used in advanced logic ICs applications.

VI. Reference

[1] J. P. Colinge *et al.*, Nature Nanotech., vol. 5, 2010, pp. 225. [2] S. Migita, *et al.*, IEDM, 2012, pp. 191. [3] K. H. Goh *et al.*, IEDM, 2013, pp. 433. [4] Synopsys TCAD, Version G-2012.06.

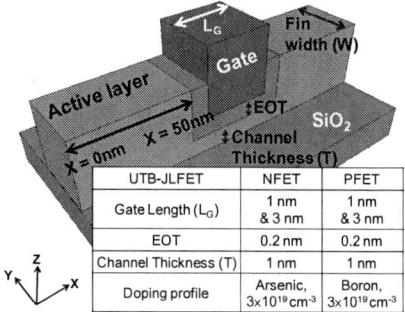

Fig. 1. Device structure and important parameters of simulated UTB-JLFET with coupled DD and DG model.

UTB-JLFET	NFET	PFET
Gate Length (L_G)	1 nm & 3 nm	1 nm & 3 nm
EOT	0.2 nm	0.2 nm
Channel Thickness (T)	1 nm	1 nm
Doping profile	Arsenic, $3\times10^{19}\,cm^{-3}$	Boron, $3\times10^{19}\,cm^{-3}$

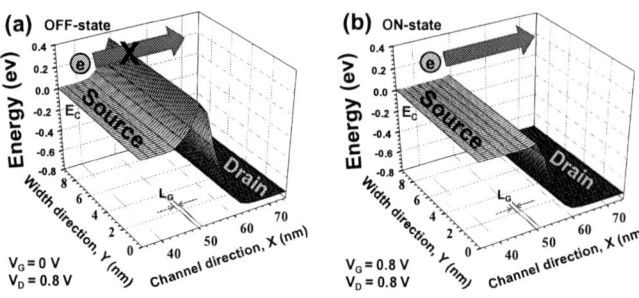

Fig. 2. Conduction band energy (E_C) diagrams of UTB-JLFET with $L_G = 1$ nm and $T = 1$ nm in (a) off-state and (b) on-state. In off-state, a high E_C level at channel region and depletion width of 2.5 nm at both source and drain side.

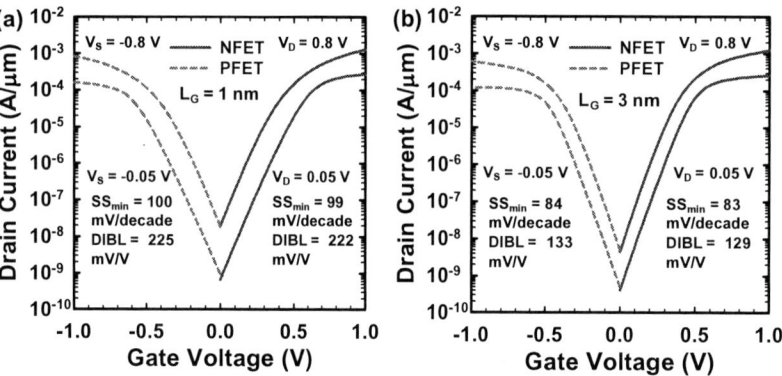

Fig. 3. I_D-V_G of UTB-JLFET with channel thickness (T) is 1 nm for (a) $L_G = 1$ nm and (b) $L_G = 3$ nm. Both UTB-JLFETs with $L_G = 1$ nm and $L_G = 3$ nm show normally-off performances because of UTB structure. The SS are 100 mV/decade of PFET and 99 mV/dec of NFET with $L_G = 1$ nm, respectively. The DIBL are 225 mV/V of PFET and 222 mV/V of NFET with $L_G = 1$ nm, respectively. The saturation current are 1.23×10^{-3} A/μm and 8.21×10^{-4} A/μm for NFET and PFET with $L_G = 1$ nm, respectively. The saturation current are 1.24×10^{-3} A/μm and 6.1×10^{-4} A/μm for NFET and PFET with $L_G = 3$ nm, respectively.

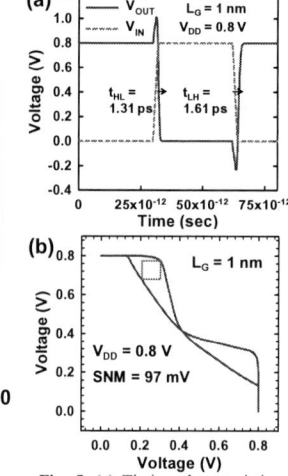

Fig. 5. (a) Timing characteristics of CMOS inverter for $L_G = 1$ nm. (b) Static transfer characteristic curves of 6T-SRAM cells.

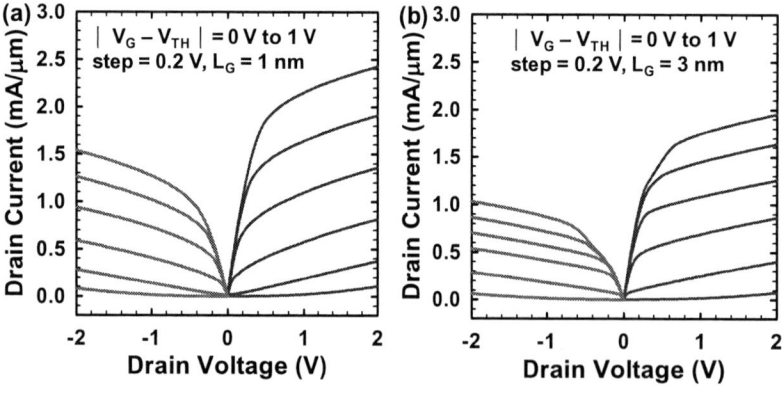

Fig. 4. I_D-V_D characteristics of UTB-JLFET with (a) $L_G = 1$ nm and (b) $L_G = 3$ nm. UTB-JLFET with $L_G = 1$ nm shows low parasitic resistance and large drain saturation current (~2.5 mA/μm). The maximum drive currents (@ | V_D | = 2 V, | V_G-V_{TH} | = 1 V) are 2.42 mA/μm and 1.54 mA/μm for NFET and PFET with $L_G = 1$ nm, respectively. The maximum drive currents (@ | V_D | = 2 V, | V_G-V_{TH} | = 1 V) are 1.95 mA/μm and 1.03 mA/μm for NFET and PFET with $L_G = 3$ nm, respectively. The drive current is proportional to the gate voltage.

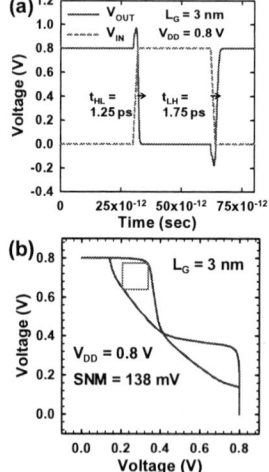

Fig. 6. (a) Timing characteristics of CMOS inverter for $L_G = 3$ nm. (b) Static transfer characteristic curves of 6T-SRAM cells.

Design and analysis of electric-field-assisted nonlocal silicon-channel spin devices

D.Kitagata[1], T.Akushichi[1], Y.Takamura[2], Y.Shuto[1], S.Sugahara[1]
[1]Imaging Science and Engineering Laboratory, Tokyo Institute of Technology, Yokohama, Japan.
[2]Department of Physical Electronics, Tokyo Institute of Technology, Tokyo, Japan
Tel: +81(45)924-5456, Fax: +81(45)924-5456, Email: kitagata.d@isl.titech.ac.jp

INTRODUCTION: Over recent years, spin-MOSFETs (Fig.1) [1] have attracted considerable attention as a key transistor for low-standby-power integrated circuits. To realize spin MOSFETs, understanding and controlling of spin dynamics in the Si channel is indispensable. The Hanle effect of spin-polarized electrons transported in the channel of spin devices is a powerful tool for evaluating spin dynamics. Using the period (B_π) of Hanle-effect oscillating signals (see Fig. 2), spin lifetime can be calculated accurately [2]. However, there exists a challenge in the generation/measurement of Hanle-effect oscillating signals. The generally used four-terminal nonlocal (4TNL) technique cannot generate sufficient oscillating signals, since this technique employs diffusive spin transport that has a short diffusion length. This technique also causes the widely spread distribution of spin transport time, resulting in the inaccuracy of B_π [2]. The width of the ferromagnetic contact needs to be carefully designed, since Hanle-effect signals are distorted owing to phase randomizing caused by spin polarized electrons passing through the channel beneath the contact. Inadequate ferromagnetic contact width weakens the satellite peak intensity of Hanle-effect oscillating signals and affects the accuracy of B_π. Recently, we proposed a new electric-field-assisted (EFA)-4TNL technique based on drift spin transport [2], which makes it possible to measure correct B_π and evaluate spin lifetime accurately (Fig.3). In the EFA-4TNL technique, the contact width effect can be negligible when an appropriate electric field is applied [2]. In this paper, we computationally analyze Hanle-effect signals of the EFA-4TNL devices and establish an optimization scheme of applied electric field and channel length design of the EFA-4TNL devices. Bulk channel and bottom-gated MOS channel EFA-4TNL devices are investigated (Fig. 3).

ANALYSIS METHOD: Figures 4 (a)-(c) show schematic illustrations of simplified models for spin injection and detection mechanism, which represent the line injection / line detection (LILD), line injection / domain detection (LIDD), and domain injection / domain detection (DIDD) models, respectively. In the LILD model, the contact width effect is ruled out. In the LIDD and DIDD models, the contact width effect is included only for the detection contact and for both the injection and detection contacts, respectively. These models were analyzed by appropriate convolution procedures using the solution of the drift-diffusion equation with the boundary conditions. The details were described in Ref. 2.

SIMULATION: Table 1 shows parameters used in the following simulations. For the bulk device, the highly doped Si substrate is used to obtain a sufficient bias current, and thus the mobility of the bulk device is not so high. On the other hand, the bias current can be easily obtained by forming MOS inversion layer for the MOS device. The universal effective electron mobility in the inversion channel [3] is used for the MOS device. Spin lifetime (τ_{sf}) was treated as a parameter, since experimentally obtained τ_{sf} values vary widely depending on measurement methods and they are still under investigation. In the following simulation, firstly, τ_{sf} was fixed to a relatively low value (100ps), and then it is varied from 100ps to 10ns.

Figure 5 shows Hanle-effect signals for the LIDD and DIDD models, in which Hanle-effect signals for the LILD model are also shown. The top and bottom panels in both the figures show the cases for E_{acc} = 0 and 7 kV/cm, respectively, in which L_{eff} is set to 0.4 μm and d is set to 10, 50, 100% of L_{eff}. By the application of E_{acc}, the signal intensities are enhanced

and the multiple oscillations become visible for both the cases. Figures 6 (a) and (b) show B_π as a function of E_{acc} for the LIDD and DIDD models, respectively, in which L_{eff} is set to 0.4 μm and d is varied. For both the cases, B_π is deviated with increasing d from that for the LILD model. However, under a particular E_{acc}, there exists a crosspoint between the B_π - E_{acc} curves for the LILD and LIDD (DIDD) models. This means that the contact width effect can be eliminated by the application of an appropriate E_{acc}. Figure 7 (a) shows E_{acc} at the crosspoint when L_{eff} is varied. E_{acc} at the crosspoint is varied with L_{eff}. Thus, for a given L_{eff}, an appropriate E_{acc} that satisfies the crosspoint condition can be chosen. Figure 7 (b) shows the first and second peak intensities of the Hanle-effect signals as a function of L_{eff}, in which E_{acc} is varied. The first peak intensity is rapidly reduced with increasing L_{eff}, and the second peak intensity is maximized at a specific L_{eff} value. The L_{eff} value that maximizes the second peak intensity is almost unchanged, even when E_{acc} increases. This restricts the design of L_{eff} and thus E_{acc}. Figure 8 shows the first and second peak intensities for the LIDD and DIDD models, in which d is varied. As noted previously, the first peak intensity monotonously decreases with increasing L_{eff}, and the second peak intensity has a peak near L_{eff} = 0.4 μm. The second peak intensity decreases with increasing d, while the first peak intensity is not varied by d. Figures 9 (a) and (b) show the second peak intensity as a function of d for the LIDD and DIDD models, respectively. The second peak intensity monotonously decreases with decreasing d. Therefore, the narrower d is more preferable. An acceptable value of d would be determined by allowable device process. Then, E_{acc} and L_{eff} can be self-consistently optimized, as shown in Figs. 6-8. Figure 10 shows the optimized E_{acc} and L_{eff} (that are denoted by E_{acc}^{opt} and L_{eff}^{opt}, respectively) and acceptable d (that is denoted by d^{min}) for the DIDD model, in which τ_{sf} is varied from 100ps to 10ns and d^{min} is determined as the value that the second peak intensity decreases by 10 % of that for the LILD model. The results for the MOS device are also shown in these figures. Although E_{acc}^{opt} decreases with increasing τ_{sf} and L_{eff}^{opt} and d^{min} increase with increasing τ_{sf}, the ranges of these values are feasible for both the bulk and MOS devices. For the MOS device, smaller E_{acc}^{opt} and larger L_{eff}^{opt} and d^{min} are allowed in comparison with the bulk device. This means that the device fabrication and the measurements of B_π are easier for the MOS devices.

CONCLUSION: The EFA-4TNL devices with L_{eff}^{opt} and d^{min} disigned in this study are feasible using present microfabrication lithography technique and E_{acc}^{opt} is easily applied by an ordinary bias technique. A wide range of spin lifetime can be applicable for the EFA-4TNL device design. Using the EFA-4TNL devices, the spin lifetime can be evaluated accurately from measuring B_π of Hanle-effect oscillating signals. It is worthy to note that B_π has the same universality as the effective electron mobility in MOS inversion channels [4]. Therefore, spin lifetime can be evaluated with respect to each individual scattering process using the EFA-4TNL devices with the MOS channel. The resulting finding would lead to important benefits for the design of spin-MOSFETs.

REFERENCES:[1]S. Sugahara, IEE Proc. Circuits, Devices & Systems, **152**, 355 (2005) [2]Y. Takamura, *et al*, to be published J. Appl. Phys. [3]S. Takagi *et al*., IEEE Trans. Electron Devices **41**, 2357 (1994) [4]Y. Takamura, S. Sugahara, J. Appl. Phys. , **111**, 07C323 (2012)

Table 1 Simulation parameters.

Accelerating electric field : E_{acc}	0–10 kV/cm
Effective channel length : L_{eff}	0–10 μm
Width of ferromagnetic contact : d	10–100% of L_{eff}
Mobility : μ_{eff} for bulk device	100 cm^2/Vs
for MOS device	600 cm^2/Vs
Spin lifetime : τ_{sf}	100 ps–10 ns
Temperature : T	300 K

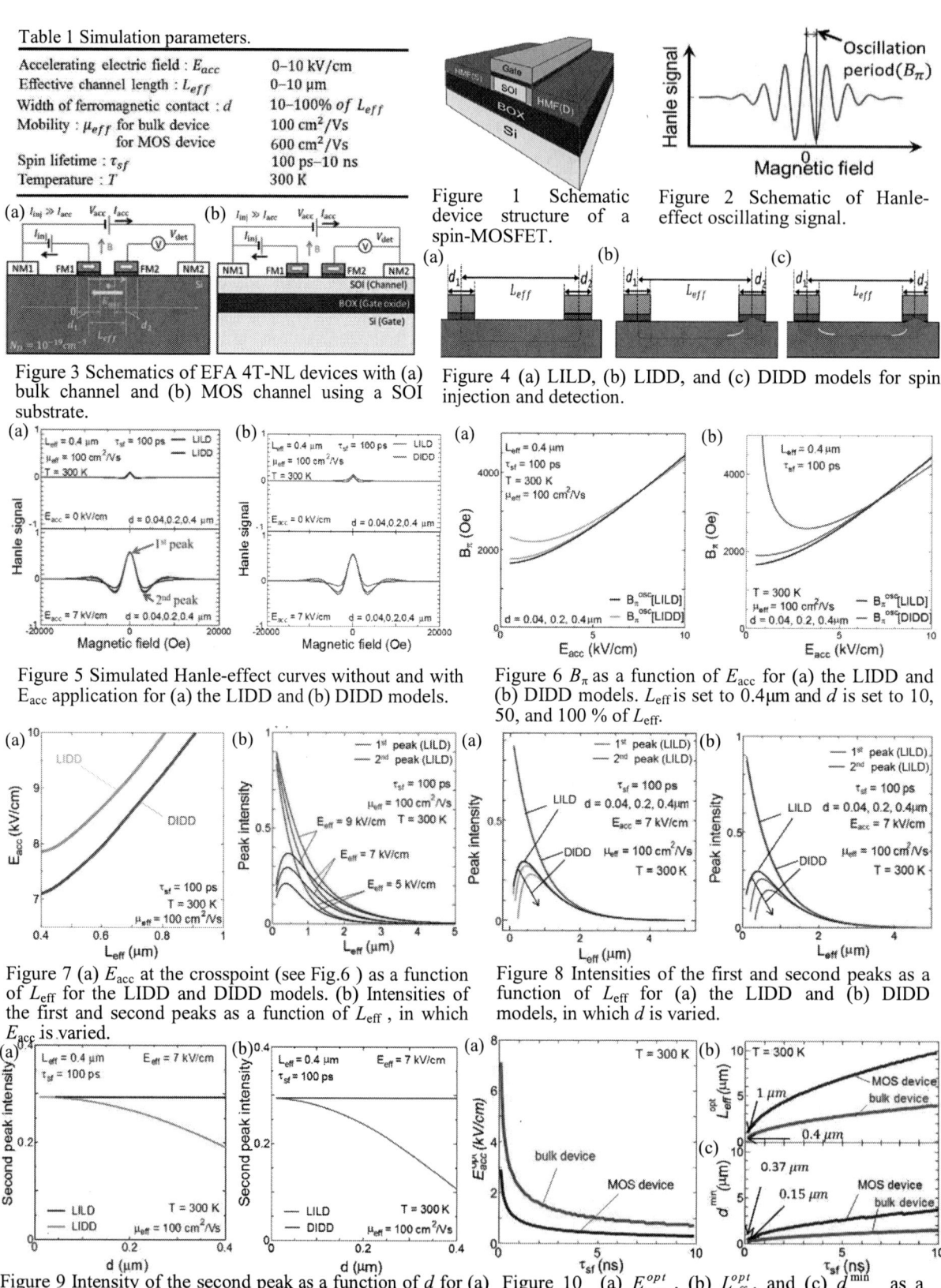

Figure 1 Schematic device structure of a spin-MOSFET.

Figure 2 Schematic of Hanle-effect oscillating signal.

Figure 3 Schematics of EFA 4T-NL devices with (a) bulk channel and (b) MOS channel using a SOI substrate.

Figure 4 (a) LILD, (b) LIDD, and (c) DIDD models for spin injection and detection.

Figure 5 Simulated Hanle-effect curves without and with E_{acc} application for (a) the LILD and (b) DIDD models.

Figure 6 B_π as a function of E_{acc} for (a) the LILD and (b) DIDD models. L_{eff} is set to 0.4μm and d is set to 10, 50, and 100 % of L_{eff}.

Figure 7 (a) E_{acc} at the crosspoint (see Fig.6) as a function of L_{eff} for the LILD and DIDD models. (b) Intensities of the first and second peaks as a function of L_{eff}, in which E_{acc} is varied.

Figure 8 Intensities of the first and second peaks as a function of L_{eff} for (a) the LILD and (b) DIDD models, in which d is varied.

Figure 9 Intensity of the second peak as a function of d for (a) the LILD and (b) DIDD models. Blue lines in both the figures indicate the second peak intensity for the LILD model.

Figure 10 (a) E_{acc}^{opt}, (b) L_{eff}^{opt}, and (c) d^{min} as a function of τ_{sf} for both the bulk and MOS devices.

Silicon-Compatible Resonant Plasma-Wave Transistor with 2D Silicene Channel for High-Performance Terahertz Electromagnetic Wave Emitters

Jong Yul Park, Sung-Ho Kim, and Kyung Rok Kim*

School of Electrical and Computer Engineering, Ulsan National Institute of Science and Technology, Korea

*e-mail: krkim@unist.ac.kr

Abstract — **In this work, we propose a novel Si-compatible resonant plasma-wave transistor (R-PWT) with 2D silicene channel for a high-performance terahertz (THz) electromagnetic (EM) wave emitter. High resonance quality i.e. narrow emission spectra can be obtained by high mobility of 2D silicene channel (μ= 2×10^5 cm^2·V^{-1}·s^{-1}) since nanoscale channel length L can be much smaller than the maximum channel length L_{max}= 1395 nm even under relatively low gate voltage.**

I. INTRODUCTION

Since terahertz (THz) electromagnetic (EM) wave, which lies in the "THz gap" of f= 0.1~10 THz, has properties to penetrate non-conducting materials and be absorbed in conductors with characteristic spectral responses, THz technology can be applied to security and food inspection. As a THz EM wave emitter, resonant plasma-wave transistor (R-PWT) is promising because R-PWT can modulate frequency easily and has fairly smaller size than optical devices which are based on large foot-print facilities. With aid of silicone (Si) technology, it is possible to fabricate R-PWT with low cost and large scale integration. However, it is still challenging to make efficient Si R-PWT due to its low channel electron mobility (μ). To overcome this problem, silicene, which has very high μ comparable to graphene (μ~ 2×10^5 cm^2·V^{-1}·s^{-1}) [1], can be chosen as the channel material of R-PWT.

II. DEVICE STRUCTURE OF SILICENE R-PWT

Because silicene has buckled honeycomb structure as shown in Fig. 1(a) and 1(b), the modulation of band gap is easier than graphene [2] so that silicene is proper to be used as a channel material of field effect transistor (FET). However, since silicene forms covalent bond with SiO$_2$, which causes mobility degradation, it is recommended to cover silicene with hexagonal BN having same honeycomb structure and smooth surface [3]. R-PWT is based on conventional FET structure with boundary conditions of Z_{GS}= 0 and Z_{GD}= ∞ (see Fig. 1(c)). With these boundary conditions, the oscillation of collective electron density (i.e. plasma-wave n= $n_0(x)$ + $n_1(x,t)$) occurs in the channel

with amplification due to the imaginary part of angular frequency ω= $\omega'+i\omega''$ as follows [4]:

$$\omega'' = \frac{s^2 - v_0^2}{2Ls} \ln\left|\frac{s+v_0}{s-v_0}\right| - \frac{1}{2\tau_p} \tag{1}$$

where s is plasma-wave velocity, v_0 is electron drift velocity, L is channel length, and τ_p is momentum relxation time.

III. PLASMA-WAVE RESONANCE IN SILICENE R-PWT

With physical conditions in Table I, a resonance window is confined on s-v_0 plot (design window) as shown in Fig. 2. If we assume μ= 1×10^5 cm^2·V^{-1}·s^{-1} for silicene, THz emitter operates at L= 450 nm which is fairly high L compared to Si R-PWT (maximum channel length L_{max}= 12 nm) [4]. If the values of s and v_0 are on the resonance window, oscillatory component of plasma-wave (n_1), which are divided down (n_{1+}) and upstream (n_{1-}), starts to resonate with drain reflection coefficient $R_{n,d}$= $(s+v_0)/(s-v_0)$ (see Fig. 3(a)). Figure 3(b) illustrates that resonance acts like coupled harmonic oscillator, which emits undesired frequency. That is because of amplitude discrepancy between n_+ and n_- due to $R_{n,d}$. The high quality of the resonance is determined by the reflection condition of high $(s-v_0)\tau_p/L$ for fixed L as shown in Fig. 3(c) and 3(d).

Figure 4 illustrates 3D design window with L as the z-axis, which shows resonance windows for silicene mobilities μ= 1×10^5, 1.5×10^5, and 2×10^5 cm^2·V^{-1}·s^{-1}. Due to its high μ, THz emitter operates when $L\leq L_{max}$= 1395 nm. However, it is needed to fabricate $L<< L_{max}$ because R-PWT with short L easily emits single-resonance peak than of longer L. As shown in Fig. 4, the resonance windows for relatively short L have the wide ranges of low v_0 so that it is possible to suppress the amplitude discrepancy by using low v_0, which yields low $R_{n,d}$ and results in simple harmonic oscillator when $L << L_{max}$ (see Fig. 5). The other reason is that it is necessary to bias high voltage U_0 to induce high s= $(eU_0/m)^{1/2}$ for R-PWT with longer L. For example, U_0~ 10 V should be biased to emit f= 2 THz at L= 1200 nm for t_{ox}= 15 nm, which may cause gate oxide breakdown. Also, high U_0 induces strongly degenerated 2DEG in which plasma-wave cannot exist due to Pauli exclusion principle.

ACKNOWLEDGEMENT

This work was supported by the Pioneer Research Center Program through the National Research Foundation of Korea funded by the Ministry of Science, ICT & Future Planning (Grant No. 2012-0009594).

REFERENCES

[1] B. Bishnoi, and B. Ghosh, "Spin transport in buckled bilayer silicene," *Comput. Mater. Sci.*, vol. 85, pp. 16-19, April 2014.

[2] G. L. Lay, "2D materials: silicene transistors," *Nature Nanotechnology*, vol. 10, pp. 202-203, February 2015.

[3] Z. Ni, Q. Liu, K. Tang, J. Zheng, J. Zhou, R, Qin, Z. Gao, D. yu, and J. Lu, "Tunable bandgap in silicene and germanene," *Nano Letters*, vol. 12, no. 1, pp. 113-118, November 2011.

[4] J. Y. Park, S.–H. Kim, S.-M, Hong, and K. R. Kim, "Physical analysis and design of resonant plasma-wave transistors for terahertz emitters," *IEEE Trans. THz Sci. Technol.*, vol. 5, pp. 244-250, March 2015.

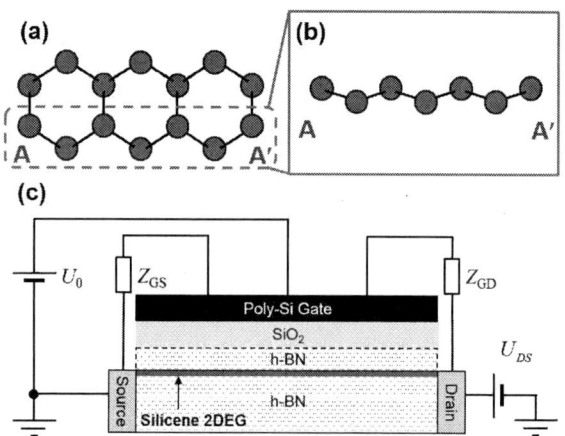

Figure 1. (a) Top and (b) side view of silicene which is used as channel material of (c) R-PWT.

TABLE I.   PYSICAL CONDITIONS FOR R-PWT

Physical Conditions	Criteria
reflection	$(s-v_0)\tau_p/L > 1$
instability	$v_0 < v_{inj} < s$
increment	$\omega'' > 0$
frequency	$f < 10$ THz
underdamped	$\omega\tau_p > 1$

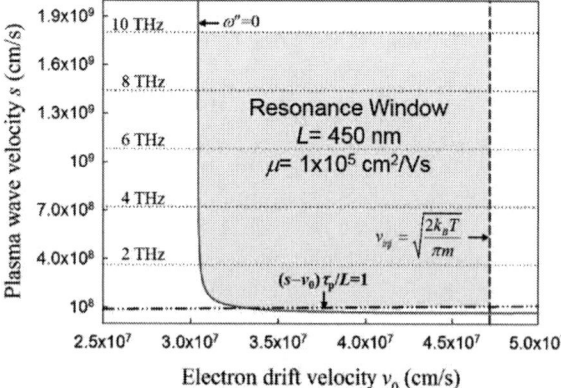

Figure 2. Resonance window for silicene mobility $\mu = 1\times10^5$ cm²·V⁻¹·s⁻¹ and electron effective mass $m = 0.013m_0$ at $L = 450$ nm. The underdamped condition ($\omega\tau_p > 1$) is neglected because the reflection condition (($s-v_0)\tau_p/L$) is always dominant than the underdamped condition.

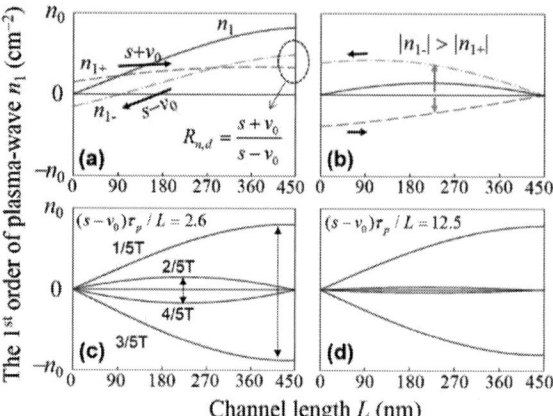

Figure 3. Plasma-waves in the channel ($s = 2\times10^8$ and $v_0 = 4\times10^7$ cm/s) at (a) $1/5T$ and (b) $2/5T$ for silicene $\mu = 1\times10^5$ cm²·V⁻¹·s⁻¹ and $m = 0.013m_0$ at $L = 450$ nm. (c) In this case, resonance occurs in the form of coupled harmonic oscillator with $(s-v_0)\tau_p/L = 2.6$. (d) By increasing $s = 8\times10^7$ cm/s, waveforms are close to simple harmonic oscillator with $(s-v_0)\tau_p/L = 12.5$.

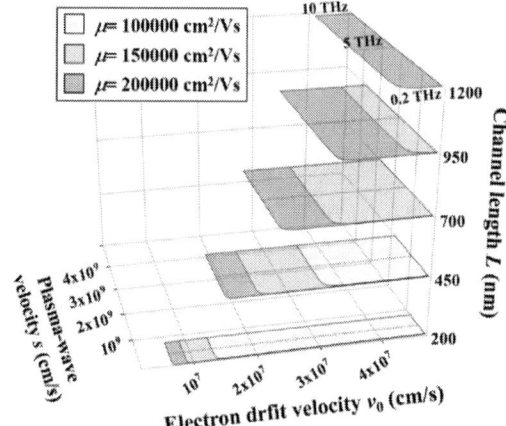

Figure 4. 3D design window for assumed silicene mobilities.

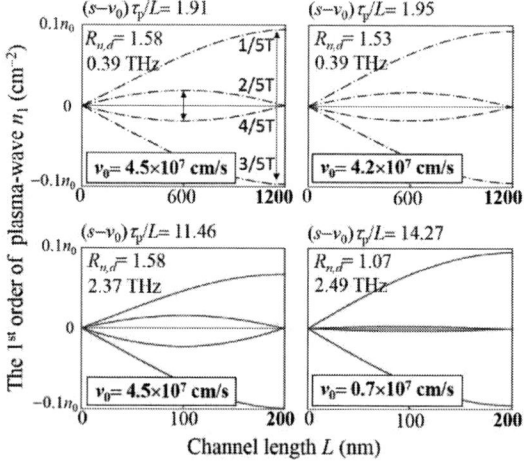

Figure 5. Plasma-wave resonance ($s = 2\times10^8$ cm/s) in the silicene channel ($\mu = 2\times10^5$ cm²·V⁻¹·s⁻¹). In the case of $L = 1200$ nm, the range of v_0 on resonance window is narrow so that it is hard to enhance the quality of resonance except changing frequency and increasing s with taking risk. On the contrary, nearly single emission peak can be observed at $L = 200$ nm with small change of frequency for low v_0.

Novel Trigate Field-Plated Poly-Si TFT with Improved Leakage Current and High On/Off Current Ratio

Yong-Hong Syu, Hsin-Hui Hu*, Jhen-Yu Tsai, Kai-Ming Wang, Jia-Jin Tsai

Department of Electronic Engineering, National Taipei University of Technology, Taipei 106, Taiwan, R.O.C.
Tel: +886-2-27712171 ext. 2287; Fax: +886-2-27317120; *E-mail: hhhu@ntut.edu.tw

Abstract

Abstract - A metal field plate thin-film transistor combined with extended drift region is fabricated. In this study, the influence of different channel wire width (W_0) and extended length (L_{EX}) on off-state leakage current and on-state current are investigated. When extending the drift region, the electric field near drain is suppressed and the gate induced drain leakage (GIDL) is reduced. In addition, high on-state current can be observed when L_{EX}=1.6 μm due to high electron density in the drift region.

I. Introduction

Polycrystalline silicon thin-film transistors (poly-Si TFTs) have been widely studied and suitable for system-on-panel (SOP) applications [1-2]. A variety of high-voltage TFT (HVTFT) structures such as lightly doped drain and offset drain have been proposed to improve the breakdown voltage and leakage current but degraded the on-state current [3-4]. Metal field plate (MFP) TFT is one of the various HVTFTs which can also improve on-state characteristic [5]. However, MFP need to supply an additional external bias. In our previous study [6], extended drift region of gate-all-around (GAA) HVTFT combined the MFP presents high breakdown voltage and low specific on-resistance. This study investigates the off-state leakage current and on-state current of MFP TFT with various extended length (L_{EX}). The simulated electric field distribution and electron density are discussed to assist in the analysis of GIDL and on-state current, respectively.

II. Experiments

Fig. 1 shows the schematic cross section of MFP TFT. The MFP is connected to the source electrode. Fig. 2 presents top-view SEM image of MFP TFT with multiple channel wires structure and extended drift region. The L_{EX} is defined in Fig. 2(c) and changed from 0 to 2.6 μm (denoted as EX0, EX1.6 and EX2.6) in this study. The measured dimensions of MFP TFT are listed in Table I. The effective channel width W_{eff} is fixed at 1.6 μm for comparison. This device's channel is a solid-phase-crystallized poly-Si film. A 4-nm thick HfO$_2$ and 50-nm thick TiN are deposited to form the gate electrode. Dopants activation is performed by using low-temperature MWA at 3 kW for 300 s.

III. Results and Discussion

The following measurements are performed by source field plate (V_{FP}=V_S=0 V). When W_0 is narrower, the device has a better gate controllability as shown in Fig 3. Table II lists the parameters in comparison with each device. While the L_{EX} extends from 0 to 1.6 μm, the on-state current increases.

However, as the L_{EX} extends to 2.6 μm, the device with EX2.6 has lower on-state current than EX0 and EX1.6. Fig. 4 shows various V_{DS} with different L_{EX} for MC012. The GIDL can be suppressed through increasing L_{EX} in saturation region.

According to previous research [7-8], the GIDL effect of HfO$_2$ TFTs are more apparent in the off-state due to higher electric field near the drain side. Fig. 5(a) shows the magnitude of electric field distribution of MC012 with EX0 at V_{DS}=2 V and V_{GS}=0 V. The high electric field near the drain side results from the thinner EOT of the high-k dielectric. As shown in Fig. 5(b), electric field at drain junction of MC012 is suppressing after extension of the L_{EX} from 0 μm to 2.6 μm. The extended drift region under MFP restrains GIDL significantly. To clarify the influence of L_{EX} on on-state current, electron density is simulated at V_{GS}-V_{TH}=1 V as shown in Fig. 6. When the L_{EX} increases from 0 to 1.6 μm, more electrons concentrating at drain junction leads to higher on-state current. In Fig. 6(c), zero bias of MFP made electron density be limited under MFP and therefore lessen the on-state current. In Fig. 7, electron density at drain junction of MC012 with EX0 and EX1.6 are similar but MC012 with EX2.6 is much lower. It explains the results of on-state current in Fig. 3.

IV. Conclusions

Conventional multi-channel devices present superior output electrical characteristics. However, at high drain voltage, GIDL is increased due to high electric field near the drain. It is successful to suppress the high electric field and reduce GIDL through extending the drift region under MFP. MC012 with EX1.6 shows not only lower GIDL but also better electrical characteristics such as I_{ON}/I_{OFF}, and SS than single channel (SC) device.

Acknowledgement

The authors would like to thank the staff of National Nano Device Laboratories. This work was supported by the Ministry of Science and Technology of Taiwan under Contract MOST 103-2221-E-027-120.

References

[1] S. D. S. Malhi, *et al.*, *IEEE J Solid-State Circuits*, **20**, p .178, 1985.
[2] H. Wang, *et al.*, *IEEE EDL*, **21**, p .439, 2000.
[3] M. Hack, *et al.*, *IEDM Tech. Dig.*, p .252, 1988.
[4] K. Tanaka, *et al.*, *IEEE EDL*, **9**, p .23, 1988
[5] T. Y. Huang, *et al.*, *IEEE EDL*, **11**, p. 244, 1990.
[6] J. Y. Tsai, *et al.*, *IEEE Trans. ED*, **62**, p. 822, 2015.
[7] C. P. Lin, *et al.*, *IEEE EDL.*, **27**, p. 360, 2006
[8] C. M. Lee, *et al.*, *IEEE EDL.*, **32**, p. 327, 2011

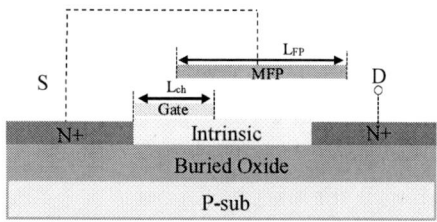

Fig.1 Schematic cross-section of MFP TFT.

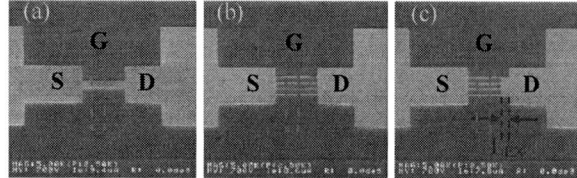

Fig.2 SEM graph of (a) SC, (b) MC02 with EX0, and (c) MC02 with EX1.6.

TABLE I

DEVICE DIMENSIONS OF SC, MC02, AND MC012. HEIGHT H OF EACH CHANNEL IS 0.1 μm. CHANNEL LENGTH OF EACH DEVICE IS 0.7 μm. EFFECTIVE WIDTH $W_{eff} = (W_0+2H) \times N_f$ IS FIXED AT 1.6 μm

Device Name	Each channel width, W_0	Channel numbers, N_f	Effective width, W_{eff} (μm)
SC	1.4 μm	1	1.6
MC02	200 nm	4	1.6
MC012	120 nm	5	1.6

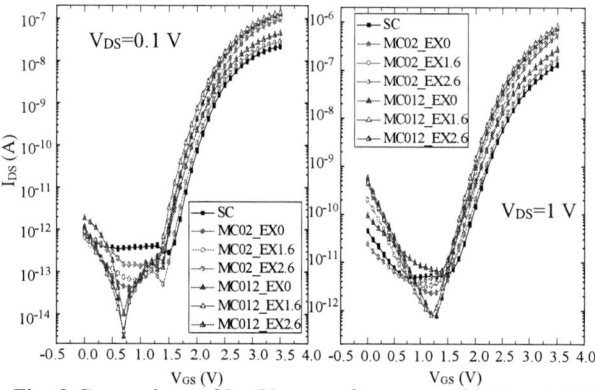

Fig. 3 Comparison of I_{DS}-V_{GS} transfer curves with V_{DS}=0.1 V and V_{DS}= 1 V for each device.

TABLE II

DEVICE PARAMETERS OF EACH DEVICE, SS AND I_{ON}/I_{OFF} (OFF-STATE: I_{OFF} IS SMALLEST AND ON-STATE: V_{GS}=3.5 V) WERE EXTRACTED AT V_{DS} = 0.1V.

Parameters		I_{ON}/I_{OFF}	SS (mV/Dec)
SC		5.5×10^4	204
MC02	EX0	1.93×10^6	186
	EX1.6	2.96×10^6	181
	EX2.6	5.63×10^5	182
MC012	EX0	3.15×10^7	121
	EX1.6	3.41×10^7	127
	EX2.6	4.34×10^6	130

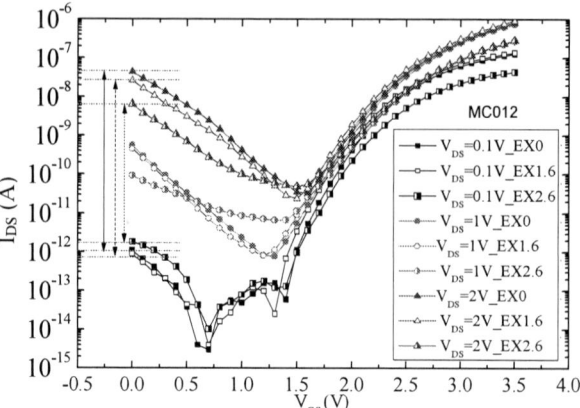

Fig.4 Comparison of I_{DS}-V_{GS} transfer curves with different V_{DS}.

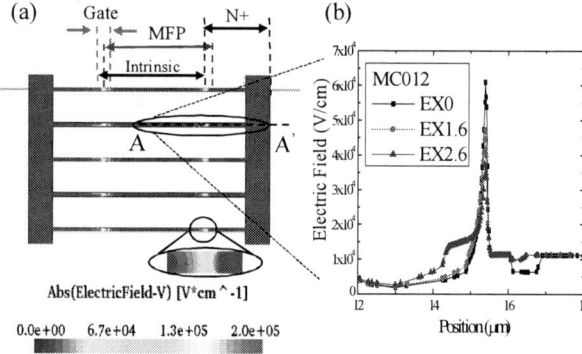

Fig. 5 (a) The magnitude of electric field distribution at V_{DS}= 2 V and V_{GS} = 0 V, and (b) along cutline AA' at drain junction for MC012.

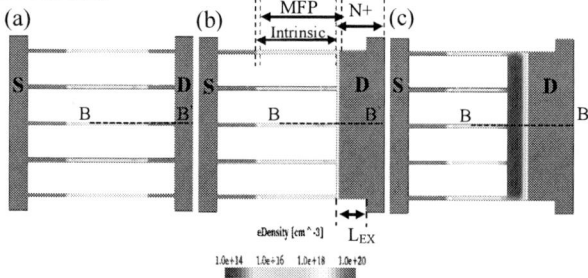

Fig. 6 The electron density of (a) MC012 with EX0, (b) MC012 with EX1.6, and (c) MC012 with EX2.6 when V_{GS}-V_{TH}=1 V and V_{DS}=0.1 V.

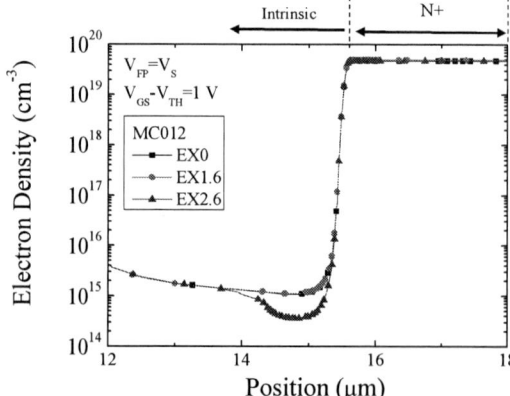

Fig. 7 Electron density near drain junction along cutline BB' for MC012.

Frequency-Dependent Response of Nanoscale Thermocouples Using Temperature Oscillations Produced by Nanoscale Heaters

Gergo P. Szakmany, Alexei O. Orlov, Gary H. Bernstein, and Wolfgang Porod

Center for Nanoscience and Technology, Department of Electrical Engineering,
University of Notre Dame, Notre Dame, IN 46556, USA
Email: gszakman@nd.edu

Abstract — **Frequency-dependent response of nanoscale thermocouples (NTCs) is studied by using a temperature oscillation produced by a nanoscale heater. The thermal response of the NTC and the heater is measured as a function of heater-current frequency by the 2ω and the 3ω method, respectively.**

I. INTRODUCTION

The response time of nanoscale thermocouples is widely studied using laser pulses or with chopper-modulated CW lasers; however, the frequency-dependent response, i.e., the NTC response as a function of the frequency of an oscillating heat source, was not investigated. In this work, we measure the frequency-dependent response of the NTC by using a temperature oscillation produced by a nanoscale heater that is energized by a sinusoidal current at ω frequency. Since the temperature oscillation of the heater decays with increasing frequency, the measured response of the NTC must also decay with increasing frequency. Therefore, in order to decouple the effect of the heater behavior from that of the NTC, we simultaneously measured the voltage oscillation of the NTC (2ω method) and the temperature oscillation of the heater (3ω method) as a function of heater-current frequency.

II. MEASUREMENTS & RESULTS

Figure 1 shows the structure used to characterize the frequency-dependent response of the NTCs. The nanoscale Cr/Ni thermocouples are constructed from lithographically defined nanowires made by lift-off on a Si/SiO2 substrate. The heater and the NTC are thermally connected, but electrically insulated by a 20-nm-thick Al2O3 layer deposited by atomic layer deposition (ALD). Note, two separate, but nominally identical, NTCs were fabricated on the heater.

In order to measure the frequency-dependent response of the NTC, an AC current at various frequencies, ω, is passed through the heater. This produces a temperature oscillation at 2ω due to Joule heating [1], and as a result, the voltage response of the NTC and the resistance of the heater vary with 2ω. We used the 3ω method to monitor the temperature oscillation of the heater as a function of heater-current frequency. Since the voltage across the heater is the product of the input current at ω and the resistance change at 2ω, the voltage across the heater contains a 3ω frequency component that is proportional to the temperature oscillation [2].

Figure 2 shows the schematic of the measurement setup. We used the signal generator from the lock-in amplifier in series with a 560 Ohm resistor as the current source. The 1st and 3rd harmonic components of the signal across the heater are measured by one channel of the digital lock-in amplifier. The voltage response of the NTC is amplified by a differential amplifier, and measured by the second channel of the lock-in at 2ω. The measured 2ω signal of the NTC was normalized to the pre-amplifier frequency response to accommodate the frequency-dependent gain.

Figure 3 shows the 2ω NTC response and the 3ω heater response as a function of heater-current frequency from 10 kHz to 8 MHz. It is seen that the frequency dependence of both signals are nearly identical, indicating that the reduction of the NTC response is determined by the temperature oscillation of the heater at least up to 4 MHz, and is not the inherent response of the thermocouple. Above this frequency the NTC response cannot be accurately determined because the 3ω response of the heater is distorted, and the temperature oscillation of the heater can no longer be detected. We conclude that the NTC is able to detect thermal oscillations with frequencies of at least 8 MHz, since the temperature oscillates at twice the heater-current frequency.

REFERENCES

[1] H. H. Roh, J. S. Lee, D. L. Kim, J. Park, K. Kim, O. Kwon, *et al.*, "Novel nanoscale thermal property imaging technique: The 2ω method. I. Principle and the 2ω signal measurement," *J. Vac. Sci. Technol., B,* vol. 24, no. 5, pp. 2398-2404, 2006.

[2] D. G. Cahill, H. E. Fischer, T. Klitsner, E. T. Swartz, and R. O. Pohl, "Thermal conductivity of thin films: Measurements and understanding," *Journal of Vacuum Science & Technology A,* vol. 7, no. 3, pp. 1259-1266, 1989.

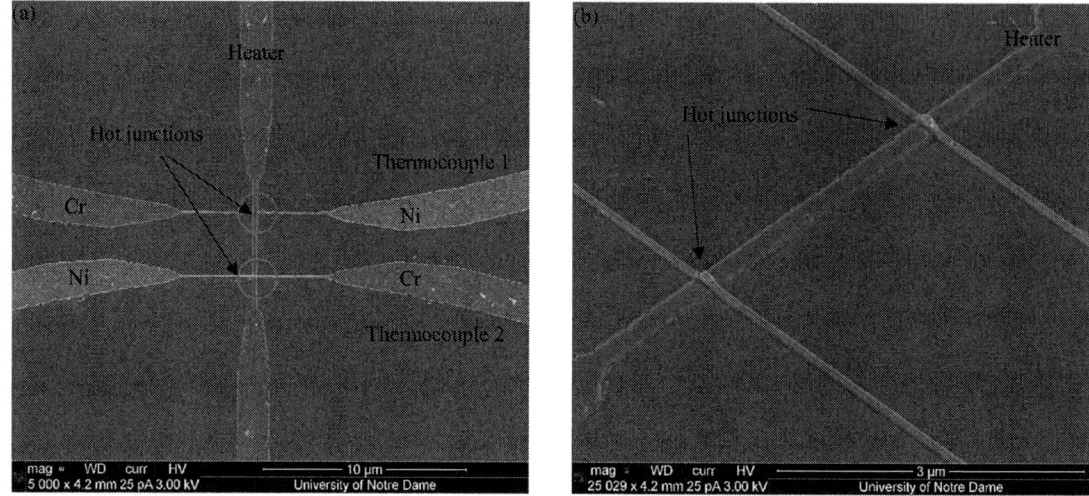

Figure 1. Scanning electron micrograph of our structure used for the frequency-dependent measurements of the NTCs. The nanoscale heater increases the temperature of the hot junctions of the thermocouples. Two nominally identical thermocouples were fabricated on each heater.

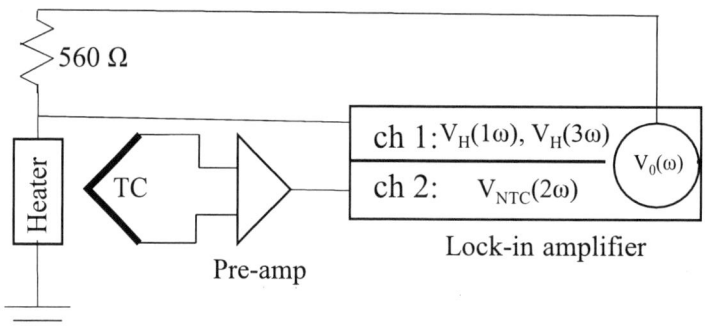

Figure 2. Schematic of the experimental setup. The lock-in voltage generator excites the heater at ω frequencies. The NTC response at 2ω and the signal across the heater at both ω and 3ω are detected.

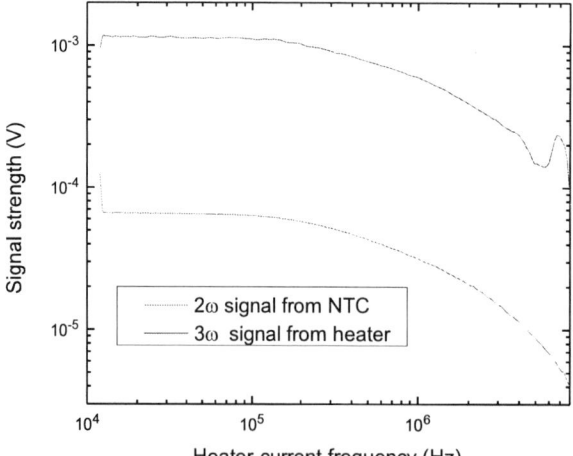

Figure 3. 2ω and 3ω thermal responses of the NTC and the heater, respectively, as a function of the heater-current frequency.

New features in Planar SiGe Channel Tunnel FETs Performance and Operation

C. Le Royer[1], L. Hutin[1], S. Martinie[1], P. Nguyen[1], S. Barraud[1], F. Glowacki[1], S. Cristoloveanu[2], and M. Vinet[1]

[1]CEA, LETI, MINATEC Campus, F-38054 Grenoble, France
[2]MEP-LAHC, INP-Grenoble, MINATEC, France
Email: cyrille.leroyer@cea.fr

Abstract — **We report the characterization of SiGe Tunnel FETs (TFETs) fabricated on SGOI with a standard CMOS process. The large gain in saturation gain (x20) is due to the threshold voltage shift and to enhanced intrinsic band-to-band tunneling injection (both related to the narrow band gap of SiGe channels). We also investigate the ambipolar signature from the $I_D(V_{DS})$ of TFETs which we compare to MOSFETs. A simple protocol is proposed and validated to get a rapid insight in injection mechanism at the two junctions of any FET device.**

I. INTRODUCTION

A Tunnel FET is an unconventional field-effect transistor relying on band-to-band tunneling (BTBT) injection rather than thermal emission above a diffusion barrier, thus enabling it in principle to beat the kT/q.ln(10) sub-threshold swing (SS) limit [1-3]. The TFET ambipolarity allows to get n and p mode operations from a single device (**Fig. 1**). However the reported performance of fabricated TFETs (I_{ON}) are poor compared to MOSFET ones. SiGe materials offer new paths for CMOS and TFET optimization due to their narrow band gap. **In this work, we investigate i) the integration of SiGe channels for both TFET and MOSFET devices and ii) the electrical signature of ambipolar tunneling signature from the output characteristics of TFETs.**

II. SiGe CHANNEL BENEFITS FOR TFETs

A. TFET and CMOS cointegration

We have fabricated planar P and N-MOSFETs and TFETs (**Fig. 2**) on SiGe-on-insulator (SGOI) structures (10 to 12nm $Si_{1-x}Ge_x$ with x=0/20/25%) using Ge enrichment and a standard High K Metal Gate CMOS process (detailed flow in [3]).

B. Impact of the Ge content on $I_D(V_{GS})$, V_{th}, I_{ON}

Increasing the Ge fraction in the SiGe channel from 0 to 25% leads to threshold voltage shift (**Fig. 3, 7**) and increased p-mode TFET drain current (**Fig. 4, 5**) because of enhanced BTBT injection. We have checked the absence of DIBL in TFETs operation (**Fig. 8**) and that $I_D(V_G)$ curves are independent on gate length down to 50nm (in agreement with previous data [1], as the tunneling mechanism (BTBT) is only related to the injection junction (source N+ / channel, here). The extracted $I_{ON}(x_{Ge})$ exponential dependence (**Fig. 5**) is related to the SiGe bandgap evolution as function of x_{Ge} (**Fig. 6**): BTBT injection [4] depends on the bandgap [5] and thus on the Ge fraction, as illustrated by (1).

$$I_{BTBT} = A \cdot V_G{}^2 \cdot e^{-B/V_G} \text{ with } B \propto \left[E_G(x)\right]^{3/2} \quad (1)$$

III. AMBIPOLARITY IN SiGe TFETs

A. Ambipolar tunneling signature

Biasing the exact same TFET device under a "p" or "n mode" scheme leads indeed to a visible transistor effect (**Fig.9-a&b**). In addition to the current plateaus increasing with V_{GS} (linked to BTBT occurring at the primary junction), parasitic current tails tend to appear at high V_P (resp. V_N), shifting by the amount of the V_G increment, and corresponding to unwanted tunnelling occurring at the secondary junction (**Fig. 1**). However, in spite of a different type of injection a similar aspect can also be obtained from "dual $I_D(V_D)$" curves in Schottky-Barrier FETs. SiGe channel SB FETs with interfacial n-type doping exhibit clearly visible V_G-dependent plateaus under both "pMOSFET" and "nMOSFET" biases (**Fig.9-c&d**).

B. Antisymmetry

Running regular $I_D(V_D)$ test or "dual $I_D(V_D)$" test after swapping electrodes yields basically the same result for MOSFETs. This no longer holds true with TFETs: **Fig.10-a&b** show overlays of the natural and "swapped" p mode and n mode TFET biasing schemes. Interestingly, both **"swapped" sets of curves indicate that the "Drain" current is mostly independent on V_G** as the characteristics appear to merge. From TCAD simulations [6] which reproduce the measured curves, we deduce that in the "swapped modes", the electrostatic potential in the channel appears to be pinned, independent of V_G. The TFET behaves in the "swapped modes" like a two-terminal resistor.

REFERENCES

[1] F. Mayer *et al.*, IEDM 2008, p. 163.
[2] K. Tomioka *et al.*, VLSI tech symp. 2012, p. 47.

[3] A. Villalon *et al.*, VLSI tech symp. 2012, p. 49.
[4] E. Kane, Appl. Phys., vol. 32, N° 1, 1961.
[5] M. Rieger and P. Vogl, Phys. Rev. B 48 (1993).
[6] R. P. Oeflein *et al.*, EuroSOI-ULIS 2015, p. 145.

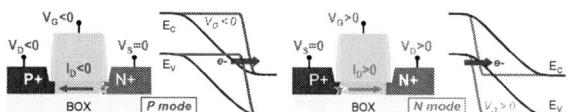

Fig.1: TFET bias scheme for p and n mode operations with corresponding band diagram illustrating the OFF and ON states.

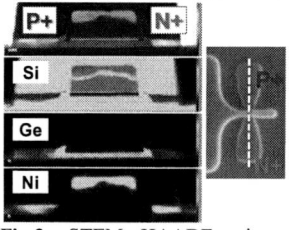

Fig.2: STEM HAADF micrograph and EDX elemental mappings (Si, Ge, Ni) on a SGOI TFET (with top view SEM).

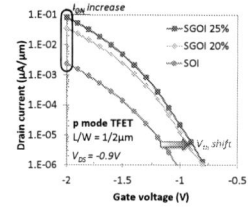

Fig.3: Measured $I_D(V_{GS})$ of wide TFETs with SOI and SGOI channels.

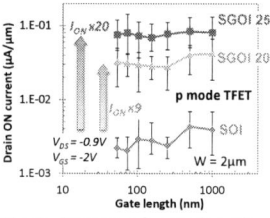

Fig.4: Impact of gate length on extracted ON currents ($I_{ON} = I_D$ @ $V_{GS} = -2$ V) for wide SOI and SGOI p mode TFETs.

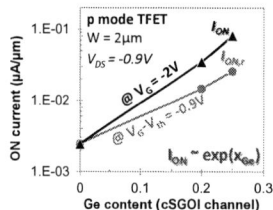

Fig.5: Impact of the Ge content in the SiGe channel on the TFET ON currents with two extraction conditions (p mode).

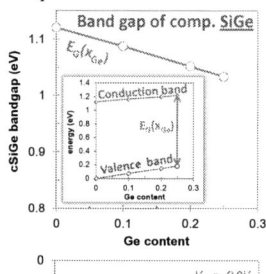

Fig.6: Impact of the Ge content on the compressive SiGe bandgap with inset showing the evolution of the conduction and valence bands (data from [5]).

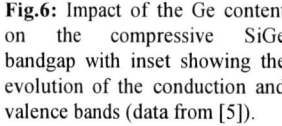

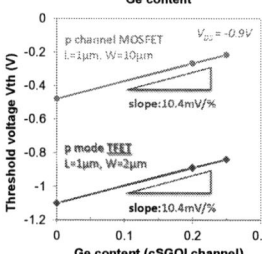

Fig.7: Impact of the Ge content in the strained SiGe channel TFET on the cSiGe bandgap.

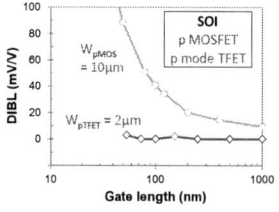

Fig.8: Impact of gate length on the DIBL values of wide pMOSFETs and TFETs (V_{DS}=-0.05V/-0.9V), with ~11nm Si thick channels.

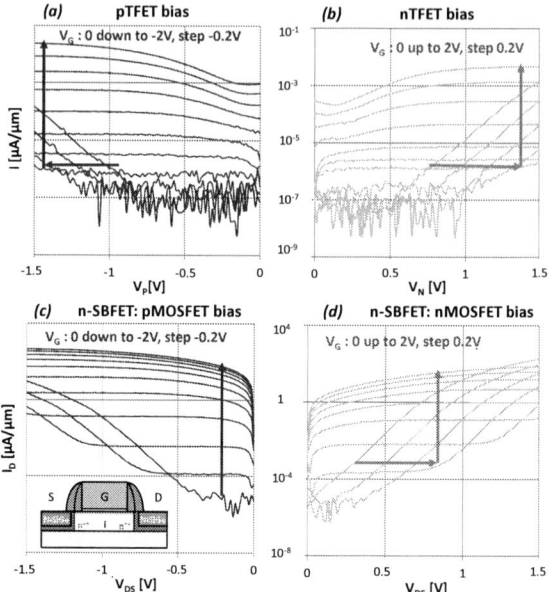

Fig.9: a/b): Measured log-scale $I_P(V_P)$ in p mode and $I_N(V_N)$ in n mode of a $Si_{0.8}Ge_{0.2}$ channel TFET (L_G=100nm); **c/d):** Measured log-scale $I_D(V_{DS})$ under pMOSFET and nMOSFET bias of a $Si_{0.8}Ge_{0.2}$ channel Schottky Barrier FET with n-type interfacial doping (L_G=50nm).

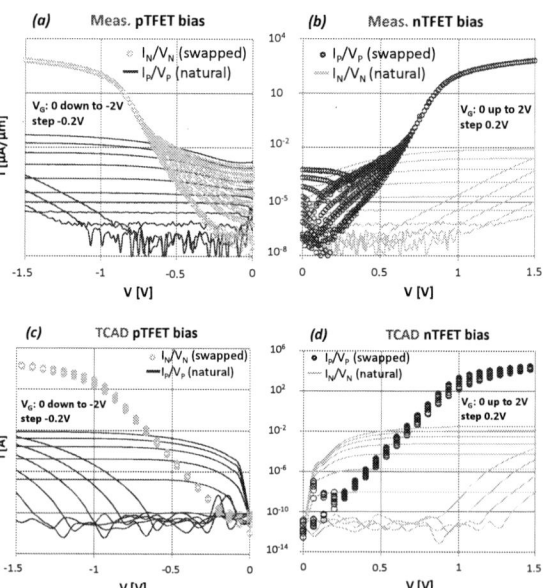

Fig.10: a/b) Measured "dual I_D-V_D" curves on a $Si_{0.8}Ge_{0.2}$ channel TFET (L_G=100nm) in "natural" biasing scheme (solid lines), and swapping the N and P terminals (open circles). Gate leakage is apparent at low V in the latter case; **c/d):** Same sets of curves using TCAD simulation (not calibrated for transport), showing a similar trend with the lack of V_G dependence in swapped mode.

Design of Complementary Tilt-gate TFETs with SiGe/Si and III-V Integrations Feasible for Ultra-low-power Applications

E. R. Hsieh[1], Y. S. Lin[2], Y. B. Zhao[1], C. H. Liu[2], C. H. Chien[1], and Steve S. Chung[1]

[1]Department of Electronics Engineering & Institute of Electronics, National Chiao Tung University, Taiwan, [2] Department of Mechatronic Technology, National Taiwan Normal University Taiwan

Abstract- A new concept of the structure design with an alignment between the maximum band-to-band tunneling rate and electric field has been proposed to enhance the performance of TFETs. It was found that the specific gate of TFET to form an obtuse shape can dramatically improve the on-current of TFET, with over 4 order improvement in comparison to planar ones. This complementary TFET (CTFET) was also demonstrated by SRAM as a benchmark, with SiGe/Si integrated with III-V on Si substrate. In order to increase WNM and RSNM of CTFET SRAM, a new scheme has been adopted, in which SRAM has been successfully demonstrated with operating bias down to 0.3V.

1. Introduction

As Moore's law continues, the trend of applications has been turned from high-performance before 2010s to the low-power after 2010s. To meet the demand, CMOS circuit designers push harder to lower down the operation voltage. However, due to the limit of device physics, the sub-threshold swing (S.S) can not be down to below 60mV/decade, which constraints V_{dd}. Tunneling FET(TFET) has received appreciably attention because Zener band-to-band tunneling (B2BT) allows S.S. smaller than 60mV/decades, which enables V_{dd} to scale down further. However, small on-current still be a critical issue, while a mismatch of the bandgap created by hetero- junction is a standard technique to boost I_{on}, which can be realized by SiGe based materials, e.g., pTFET [1,2], and III-V based materials, e.g., nTFET [3,4]. Another strategy is to enlarge the tunneling area since the tunneling current is proportional to the area [5]. In this paper, we are interested in how to optimize the B2BT rate by exactly aligning the max. electric field and the max. B2BT rate as well as providing larger tunneling area. Finally, the goal is to design complementary TFETs suitable for low voltage/power operation and the demonstration of SRAM cell as a benchmark.

2. Description of Devices

Three structures have been designed to investigate the characteristics of CTFETs, including heterojunction-planar TFET, homojunction planar TFET, and tilt-gate TFET, whose gate is formed as an angle between the source and drain, Fig. 1. The hetero junction is constructed by SiGe/Si for pTFET; InAs/GaSb for nTFET. Fig. 2 is a suggested process flow to implement tilt-TFET by III-V-OI transferring technique, following by selective epitaxial hetero-materials on holding wafer.

3. Results and Discussion

A. Engineering Between Field and B2BT Alignment

First, Fig. 3 compares $I_d V_{gs}$ of the devices, showing 90^0-gate TFET, which has reached 4 orders of I_{on}, compared to the control, with around 10mV/decade of S. S. much steeper than that of control, Fig. 4. In order to understand what mechanisms induce such small S.S., Fig. 5 shows the comparisons of the distributions for the electric field and B2BT rate. It was found that, the max. field distribution of 90^0 gate TFET is broader and extended into channel (Fig. 5a), but the planar one is much steeper and localized at the source edge (Fig. 5b). Thus, where the max. of B2BT rate takes place is much easier to match the region of electric field for the 90^0-gate TFET; otherwise,it is difficult to align the maximum

field at source edge (Fig. 5b). Thus, where the max. of B2BT rate takes place is much easier to be matched the region of electric field for the 90^0-gate TFETs; otherwise, it is difficult to align the max. field and rates at the same position in the planar TFET. As a result, B2BT rate of 90^0-gate TFET is 5-order larger than that of the planar and contributes to larger current accordingly

B. Design of Tilt-Complementary TFET(CTFET)

Since the maxium field of a tilt TFET becomes more broad, let's examine the case with different angles. Fig. 6 compares $I_d V_{gs}$ characteristics of 140^0, 90^0, and 40^0-gate TFETs. It was found that the wider the angle is, the higher the on-current we can get, since the wider the angle is, the more broader the region of max. electric field becomes, i.e., in Fig. 7, it expands the area of tunneling region and yields to a large on-current finally. Fig. 8 summarizes I_{on} and S. S. for different angles of tilt TFETs. As expected, the current increases and S.S. decreases as the angles increased. However, a turning point is observed when the angle is larger than a certain degree, since the max. field is diluted as it becomes broader. Finally, we compare $I_d V_{gs}$ of 140^0-gate CTFETs and CMOS devices at the same 20nm gate length. The performance of 140^0-gate CTFETs is not only comparable to that of LOP CMOS devices, but also its off-current is smaller than that of LSTP CMOS devices.

C. Demonstration of Tilt-TFET SRAM

Fig. 10 compares the capacitances of tilt CTFETs and CMOS devices, in which C_{gd} of CTFETs is much larger than that of CMOS devices because TFET is asymmetry structure. C_{gc} will share minority charge with C_{gd} in strong inversion and will pull C_{gd} up, which hurts the delay of CTFET inverter. Fortunately, C_{gd} is decreased as V_{dd} reduces so one can operate CTFET at low V_{dd} to avoid this issue. Fig. 11 demonstrates this concept. When V_{dd} is high, the delay of CTFET inverter is worse than that of CMOS one. However, when V_{dd} is lower, the delay of CTFET inverter becomes better than that of CMOS one. Moreover, since TFET is an asymmetry device, it is difficult to realize the access gate by TFET, especially for 6T SRAM. Figs. 12 and 13 provide a new scheme, to replace single gate of nTFET by a bi-directional pass gate by connecting two identical nTFETs in parallel. It was found that the write noise margin (WNM) of new CTFET SRAM is better than that of CMOS ones, Fig. 14. Furthermore, a new strategy to improve read static noise margin(RSNM) is proposed by grounding the top world line(WL) but pulling up the bottom WL. It enables non-destructive read and increases RSNM, Fig. 15.

In summary, a new concept of alignment between the max. electric field and B2BT rate to enhance the performance of TFET has been demonstrated. A new structure with tilt-gate complementary TFET has been proposed to achieve this goal. A bi-directional pass gate has been applied to CTFET SRAM to improve the WNM and RSNM, with operation voltage down to 0.3V, Fig. 16 This shows great potential of the new structure TFET for ultra-low power applications.

Acknowledgments This work was support in part by the Ministry of Science & Technology, Taiwan, under contract no. *NSC102- 2221-E009-094* and *NCTU-UC Berkeley* I-RiCE program, under *MOST-104-2911-I-009-301*.
References: [1] A. Villalon, VLSI, p. 49, 2012. [2] A. M. Walke, IEEE TED, p. 707, 2014. [3] K. E. Moselund, EDL, p. 1454, 2012.[4] R. Li, EDL, p. 364, 2012. [5] Q. Huang, IEDM, p. 178 , 2012.

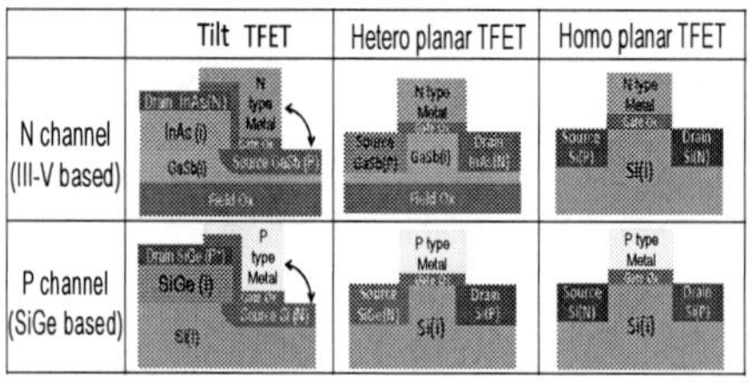

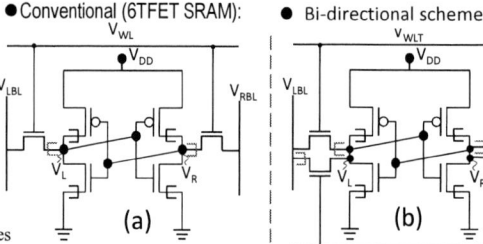

Fig. 1 Three kinds of TFETs are designed and simulated in this work, including heterojunction planar TFET, homo-junction planar TFET(the control), and the tilt-gate TFET with an angle between the drain and source.

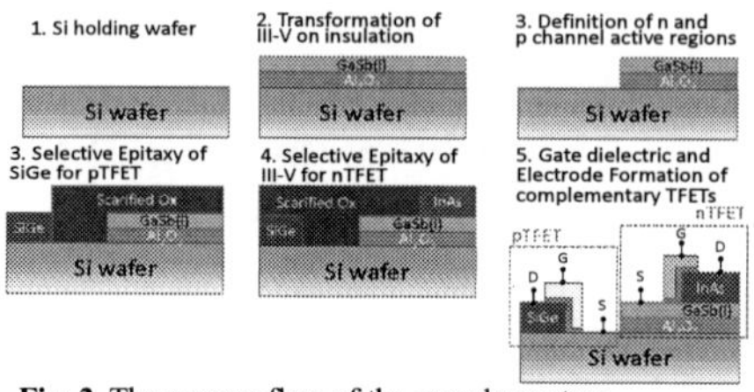

Fig. 2 The process flow of the complementary TFET by SiGe/Si integrated with III-V on Si substrate, which is designed by selective epitaxial growth of hetero-material on Si wafer through III-V on insulator transferring technique.

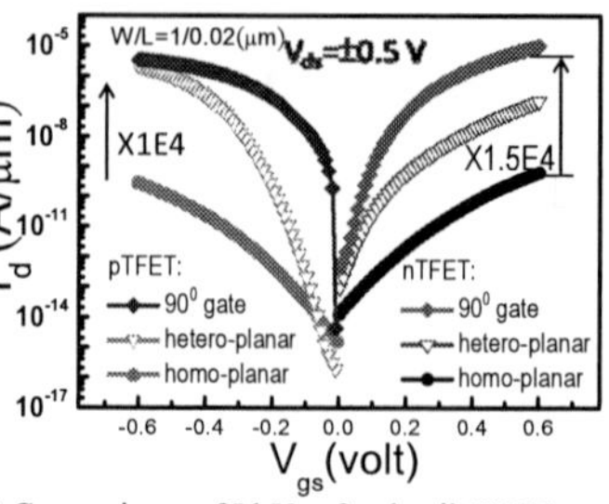

Fig. 3 Comparisons of Id-Vgs for the tilt TFETs and the control devices, showing that the tilt one exhibits better performance, especially the S.S.

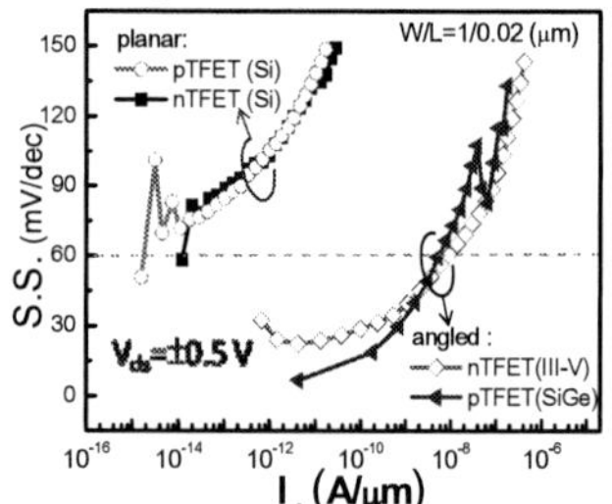

Fig. 4 Comparisons of S.S. for the tilt and the controls. It shows that the S. S. of tilt devices have been improved a lot, much below 60mV/decade.

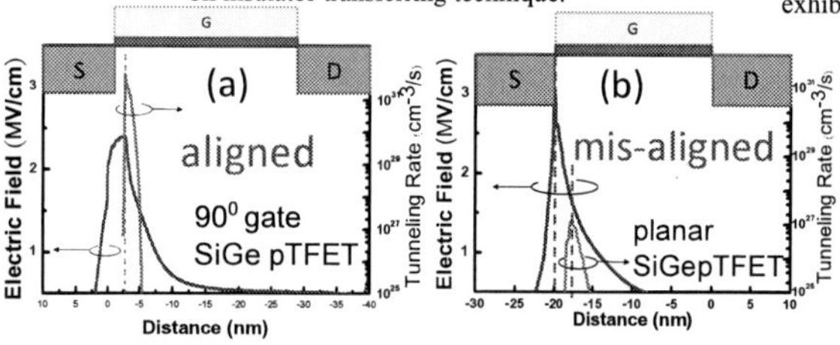

Fig. 5 (a) The improvement of S.S. is owing to the location of max. electric field and the max. band-to-band tunneling rate, i.e., both are aligned at the same position. whereas, (b) in the conventional planar TFET, the max. B2BT does not align with the max. field.

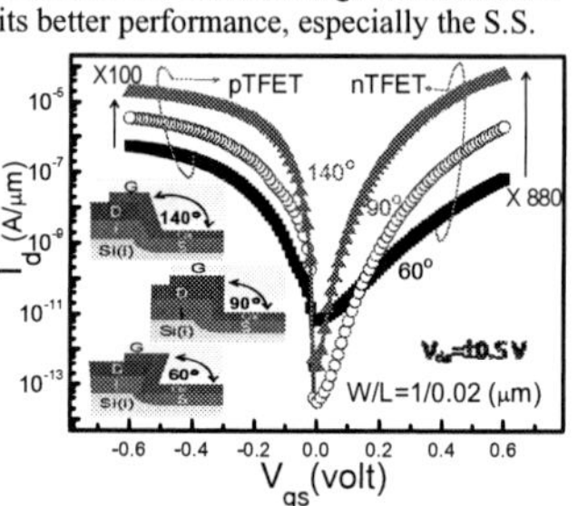

Fig. 6 Comparisons of the I_dV_{gs} for different TFET devices, for the 40^0, 90^0, and 140^0 tilt-gate, showing larger tilt angle shows better performance.

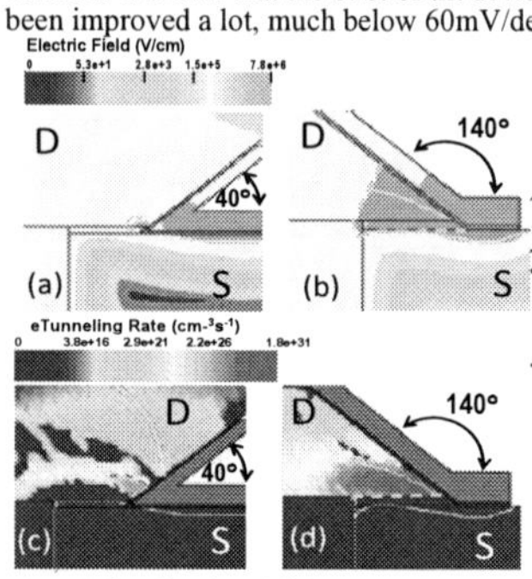

Fig. 7 The simulated fields (a,b) and B2BT (c,d) distributions to explain why the 140^0TFET can enlarge the tunneling current. It was found that the 140^0 one provides much wider and stronger field region than the 40^0 one.

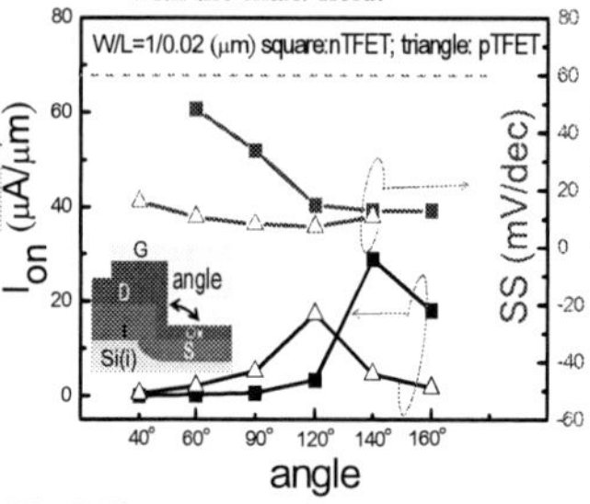

Fig. 8 The summary of on-currents and S.S. for different tilt TFETs. It is noticed that when the angle is wider than a limit, the current decreases because the field is lessened as its distribution is wider than a limit value.

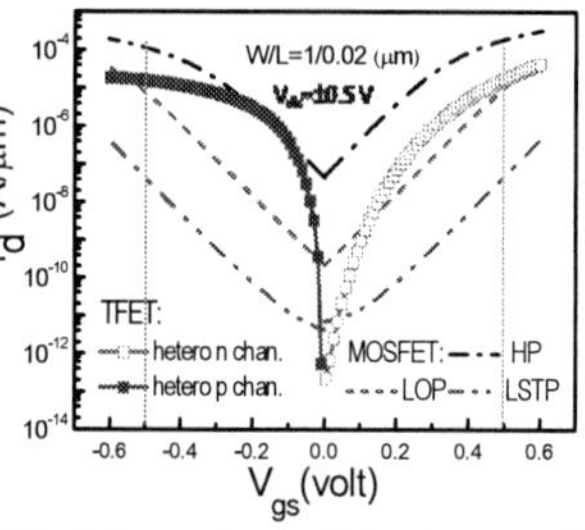

Fig. 9 Comparisons of I_dV_{gs} for 140^0 TFETs and different-operation-mode MOSFETs. This TFET is superior to the LOP and LSTP MOSFETs when the operation voltage is below 0.6V.

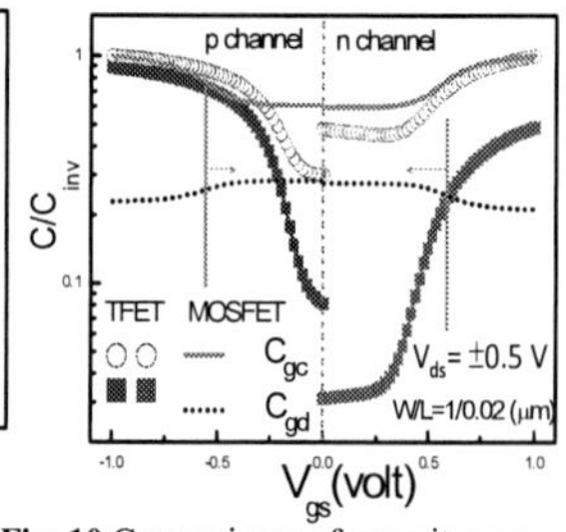

Fig. 10 Comparisons of capacitances for 140^0TFET and LOP MOSFETs. It was found that the C_{gd} of TFET is much larger than that of MISFETs, but it is rapidly decreased as the operation voltage reduced.

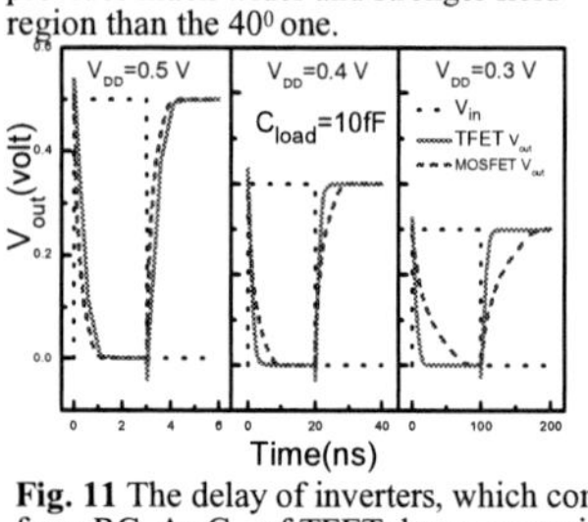

Fig. 11 The delay of inverters, which comes from RC. As C_{gd} of TFET decreases rapidly, the voltage transient behaviors of TFET at small V_{dd} is much steeper than that of MOSFETs.

● Conventional (6TFET SRAM):   ● Bi-directional scheme (8TFET)

Fig. 12 A new structure of SRAM is proposed to replace the traditional 6T one (a) by inserting a bi-directional pass gate (b) as the access component.

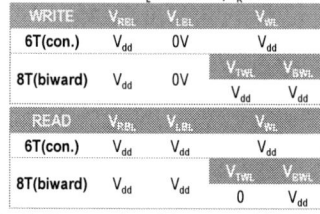

V_L is stored in 1; V_R is stored in 0			
WRITE	V_{RBL}	V_{LBL}	V_{WL}
6T(con.)	V_{dd}	0V	V_{dd}
8T(biward)	V_{dd}	0V	V_{TWL} V_{dd} / V_{BWL} V_{dd}
READ	V_{RBL}	V_{LBL}	V_{WL}
6T(con.)	V_{dd}	V_{dd}	V_{dd}
8T(biward)	V_{dd}	V_{dd}	V_{TWL} 0V / V_{BWL} V_{dd}

Fig. 13 The operation of the proposed SRAM. Taking the advantage of bi-directional pass gate, writing can be carried by pulling up VTWL&VBWL; for reading, only VBWL is pulled up to provide a un-destructive operation.

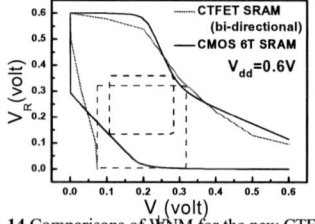

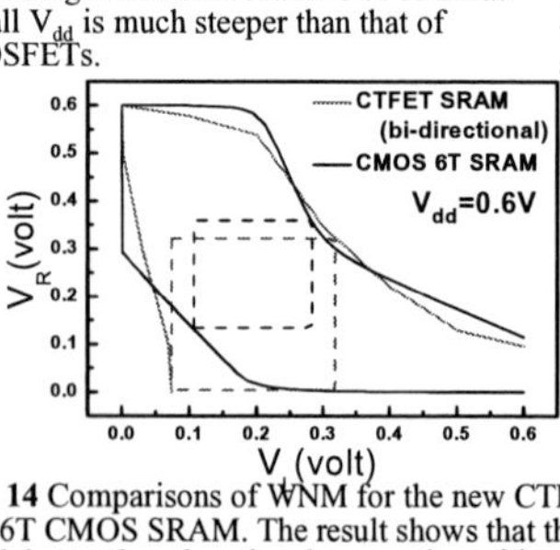

Fig. 14 Comparisons of WNM for the new CTFET SRAM and 6T CMOS SRAM. The result shows that the new one is much better than the other, because the writing current can easily pass through the bi-directional access gate.

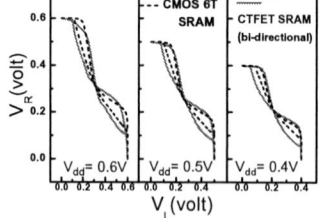

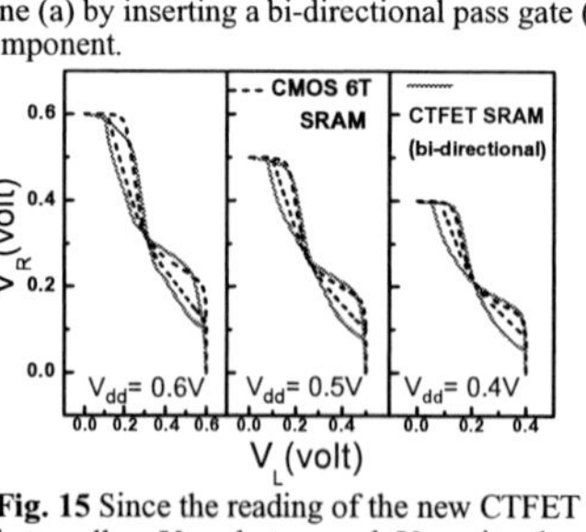

Fig. 15 Since the reading of the new CTFET SRAM is to pull up V_{BWL} but ground V_{TWL}, it enlarges the RSNM by non-destructive read.

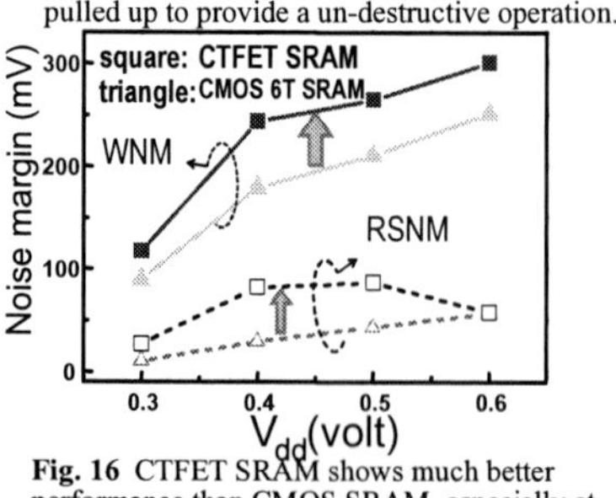

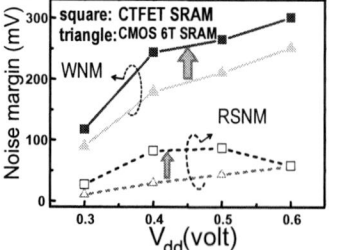

Fig. 16 CTFET SRAM shows much better performance than CMOS SRAM, especially at very low operation voltage.

Dopant-Assisted Tunnel-Current Enhancement in Two-Dimensional Esaki Diodes

H.N. Tan[1], D. Moraru[2], K. Tyszka[1,3], A. Sapteka[4], S. Purwiyanti[4], L. T. Anh[5], M. Manoharan[5],
T. Mizuno[1], R. Jablonski[3], D. Hartanto[4], H. Mizuta[5,6], and M. Tabe[1]

[1]Research Institute of Electronics, Shizuoka University, Japan (Email: f0330299@ipc.shizuoka.ac.jp)
[2]Faculty of Engineering, Shizuoka University, Japan
[3]Institute of Metrology and Biomedical Engineering, Warsaw University of Technology, Poland
[4]Department of Electrical Engineering, University of Indonesia, Indonesia
[5]School of Materials Science, Japan Advanced Institute of Science and Technology, Japan
[6]Nano Group, Faculty of Physics and Applied Sciences, University of Southampton, United Kingdom

Abstract — We study ultrathin (2D) lateral Si Esaki tunneling diodes, and find that anomalous current peaks and humps are observed to be superimposed on the ordinary negative differential conductance (NDC). The remarkable enhancement of interband tunneling current is primarily ascribed to resonant tunneling via gap-states created by large potential fluctuation due to prominent inhomogeneity of dopant distribution (dopant-clusters) in the 2D depletion region.

1. Introduction

Recently, interband tunneling originated in Esaki tunnel diodes [1,2] has attracted increasing interest for low power consumption devices. In fact, tunnel field-effect transistors (TFETs) [3] use interband tunneling to achieve steep sub-threshold slopes. However, since tunneling current is relatively low in Si because of indirect bandgap nature, a number of ways to enhance the current are proposed by involving hetero-structures [4], deep-level impurities or isoelectronic traps [5].

In this paper, we find that interband tunneling current is strongly enhanced in ultrathin (2D) tunnel diodes without introducing extrinsic impurities. This remarkable current enhancement is ascribed to the resonance via dopant-clusters formed due to the 2D-characteristic inhomogeneity of dopant distribution.

2. 2D Esaki tunneling diodes

We fabricated Esaki tunneling diodes in silicon-on-insulator (SOI) layers. The *pn* junction is formed in a constriction region, with a final width on the order of ~100 nm and thickness ~5 nm, as shown in **Fig. 1(a)**. The conventional selective-doping processes were used to define the *n*-type region (doped with phosphorus, P, $N_D \approx 2.7 \times 10^{20}$ cm^{-3}) and the *p*-type region (doped with boron, B, $N_A \approx 0.9 \times 10^{20}$ cm^{-3}) with an overlap area. Doping profile is shown in **Fig. 1(b)**.

3. Low-temperature *I-V* characteristics

I-V characteristics measured in forward bias regime at T=5.5 K are shown in **Fig. 2(a)**. It is seen that current rapidly increases with increasing V. In the low voltage region (<0.25 V), however, several specific features can be noticed, as marked in **Fig. 2(b)**. First, fine current humps at ~20 and ~60 mV are observed, which are due to phonon-assisted interband tunneling [6], in addition to well-known negative differential conductance (NDC) due to interband tunneling current [1,2]. Secondly, we find two extra features, i.e., the enhanced-current peak (A) and hump (B). Such interesting features are often observed in our 2D diodes, although the current peak and hump structures are different from each other.

4. Analysis of enhanced-current features

In early work on Esaki diodes [7], excess current, which is observed in higher voltage region than the interband tunneling peak voltage (~0.1V), is ascribed primarily to a two-step process, i.e., tunneling and recombination (SRH mechanism) involving a gap-state. The hump (B), which also lies in the excess current region, is likely to be caused by tunneling-recombination process [**Fig. 2(d)**]. In contrast, since the narrow peak (A) appears at $V \approx 0.1$ V superimposed on the interband tunneling peak, the peak (A) is ascribed to resonant tunneling through an aligned energy gap-state without energy dissipation [**Fig. 2(c)**].

In order to confirm this model, we studied the substrate bias (V_{sub}) dependence of the current peak, as shown in **Fig. 3**. V_{sub} is expected to work to shift the gap-state energies present in the depletion region. The features A and B significantly depend on V_{sub}, while other regions are insensitive to V_{sub}. It is found that the behaviors of A and B are different and shift oppositely as a function of V_{sub} (see **Fig. 4(c)**). The difference is consistently explained by the spatial locations of the gap states relative to the density of states (DOS) in the leads, as schematically illustrated in **Figs. 4(a)** and **4(b)**.

Finally, we studied the origin of the gap-states by simulation and conclude that the energy states are ascribed to potential fluctuation produced by localized dopant clusters, as schematically shown in **Fig. 5(a)**. Such large potential fluctuation is generated especially for 2D systems, but is significantly screened for 3D systems. Simulated potential profiles for the ultrathin co-doped *pn* junction, as shown in **Fig. 5(b)**, confirm the presence of huge potential fluctuation. The simulation results are quite consistent with our model.

5. Conclusions

We studied ultrathin Esaki diodes fabricated in silicon-on-insulator layers. We find interband tunneling current with strong enhancement due to energy states in the depletion region. These states are ascribed to localized dopant clusters that prominently appear in two-dimensional structures. These findings give an insight into the impact of low-dimensionality on the operation of interband tunneling devices.

Acknowledgements

This work was supported by MEXT Kakenhi (23226009, 25630144) & Coop. Res. Project (RIE, Shizuoka Univ.).

References

[1] L. Esaki, Phys. Rev. **109**, 603 (1958).
[2] J. Wallentin et al., Nano Lett. **10**, 974 (2010).
[3] A.M. Ionescu and H. Riel, Nature **479**, 329 (2011).
[4] C.D. Bessire et al., Nano Lett. **11**, 4195 (2011).
[5] T. Mori et al., VLSI Tech. Dig., 978 (2014).
[6] A.G. Chynoweth et al., Phys. Rev. **125** (1962).
[7] C.T. Sah el al., Phys. Rev. **123**, 1594 (1961).

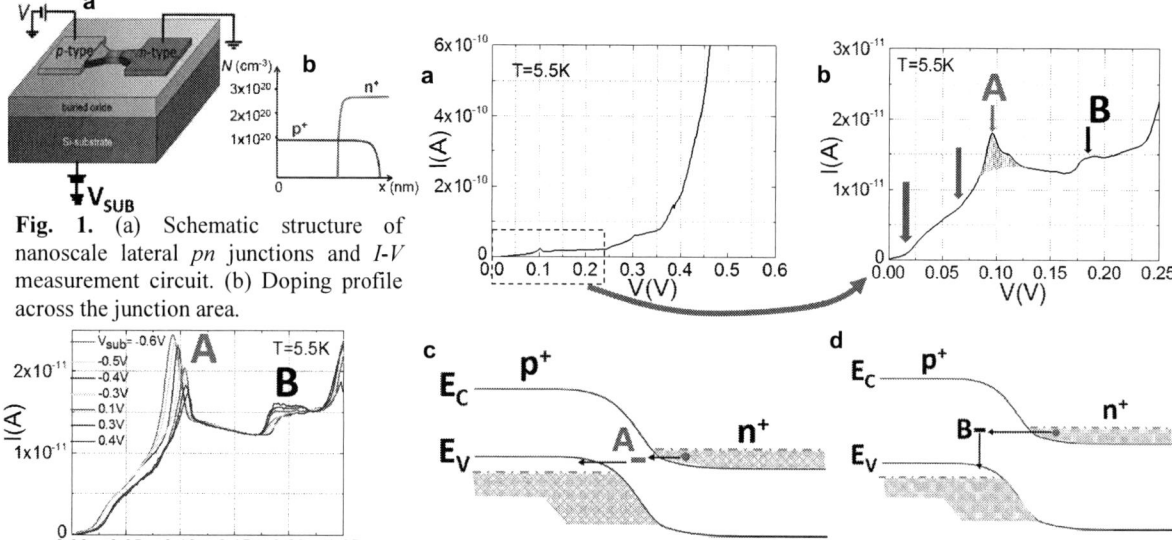

Fig. 1. (a) Schematic structure of nanoscale lateral *pn* junctions and *I-V* measurement circuit. (b) Doping profile across the junction area.

Fig. 2. (a) Low temperature (5.5 K) *I-V* characteristics in forward bias region in a wider voltage range. (b) Zoom-in on the low bias region, with marks indicating main features (including peak A and hump B). (c)-(d) Band diagrams schematically showing energy states responsible for features A and B.

Fig. 3. *I-V* characteristics as a function of V_{sub}, showing the evolution of features A and B.

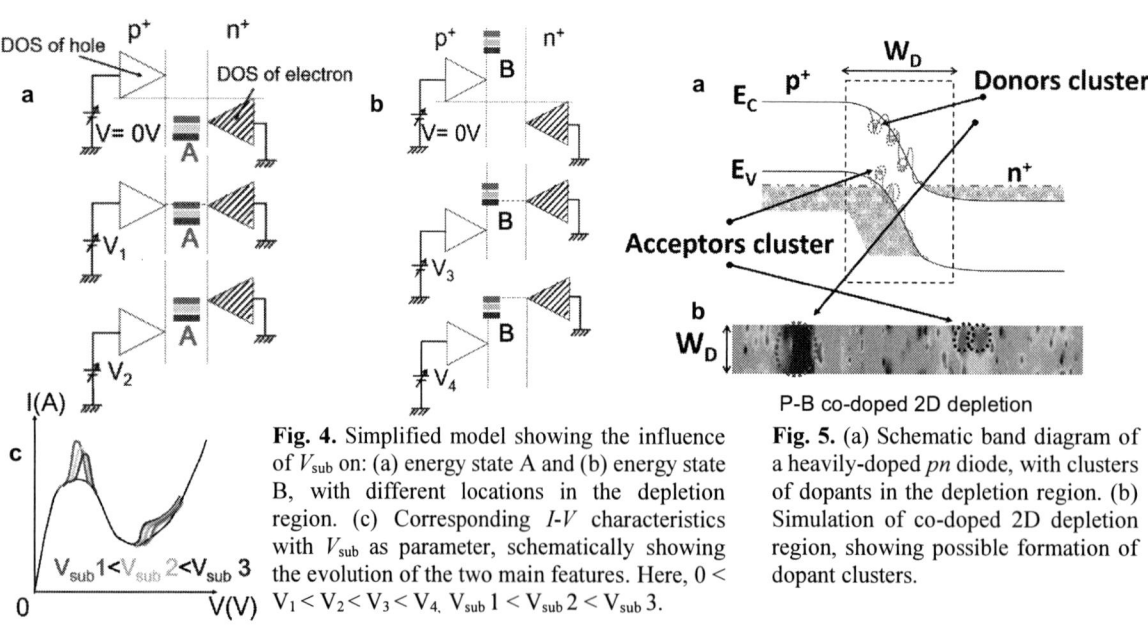

Fig. 4. Simplified model showing the influence of V_{sub} on: (a) energy state A and (b) energy state B, with different locations in the depletion region. (c) Corresponding *I-V* characteristics with V_{sub} as parameter, schematically showing the evolution of the two main features. Here, $0 < V_1 < V_2 < V_3 < V_4$, $V_{sub}1 < V_{sub}2 < V_{sub}3$.

Fig. 5. (a) Schematic band diagram of a heavily-doped *pn* diode, with clusters of dopants in the depletion region. (b) Simulation of co-doped 2D depletion region, showing possible formation of dopant clusters.

Fabrication of high-quality $Co_2FeSi_{0.5}Al_{0.5}$/CoFe/MgO/Si spin injectors for Si-channel spin devices

T. Kondo[1], Y. Kawame[1], Y. Takamura[2], Y.Shuto[1], S. Sugahara[1]

[1]Imaging Science and Engineering Laboratory, Tokyo Institute of Technology, Yokohama, Japan.
[2]Department of Physical Electronics, Tokyo Institute of Technology, Tokyo, Japan.
Tel: +81 (45) 924-5456, Fax: +81 (45) 924-5456, Email: tsuyoshi.k@isl.titech.ac.jp

INTRODUCTION: In recent years, spin-MOSFETs [1] have received much attention for low-power and high-performance integrated circuits. Figure 1 shows the schematic device structure of a spin-MOSFET that employs half-metallic ferromagnet tunnel contacts for the source and drain. To realize spin-MOSFETs, highly efficient spin-injector/-detector for the Si channel is required. Half-metallic full-Heusler alloy $Co_2FeSi_{0.5}Al_{0.5}$ (CFSA) is promising for a ferromagnetic electrode material owing to its high Curie temperature (>>RT) and extremely high spin polarization. Although CFSA has the completely ordered $L2_1$, partially ordered $B2$, and disordered $A2$ structures, its half-metallicity remains even for the partially disordered $B2$ structure [2]. This structural robustness is beneficial for device process of spin-MOSFETs. Recently, we demonstrated spin accumulation in a Si channel using a CFSA/MgO/Si tunnel contact (Fig. 2) [3]. The Hanle-effect signal for spin accumulation in the channel was successfully observed. However, its intensity was weak, i.e., the spin injection efficiency was low. The roughening of the CFSA/MgO interface was also observed for the tunnel contact, which could be the one of the origins of the low spin injection efficiency.

In this paper, we proposed CFSA/CoFe/MgO/Si tunnel contacts and successfully demonstrated the quality improvement of the tunnel contacts by introducing the CoFe interlayer into the contact. The CoFe interlayer is known to enhance the TMR rario of full-Heusler-alloy/Mg-O-compound MTJs [4].

EXPERIMENTAL: CFSA(30nm)/CoFe(1nm)/MgO(1-3nm)/n^+-Si(4×10^{19}cm^{-3}) tunnel contacts were prepared using a multi-chamber system without breaking ultra-high vacuum (Fig. 3). The control sample of CFSA/MgO/ n^+-Si tunnel contacts was also fabricated. The Mg thin film was deposited at RT after thermal cleaning of the silicon substrate. Then, the MgO tunnel barrier was formed by radical oxidation of the Mg layer and the resulting film was annealed in radical-oxygen atmosphere at 400°C for 30 min. The CoFe layer was deposited at RT and the CFSA layer was deposited at 400°C. Structural characteristics of the CFSA/CoFe/MgO/Si tunnel contacts were evaluated using X-ray diffraction (XRD) and transmission electron microscope (TEM).

RESULT AND DISCUSSION: For the XRD evaluation of the orientation of the CFSA films, the (220) and (400) fundamental lines were used, and the ordered structure was analyzed by the (220) fundamental lines and (111) and (200) superlattice lines. Figure 4 shows the definition of axes with respect to the sample stage for the XRD measurements. Figure 5(a) shows the XRD spectra of the CFSA/CoFe/MgO (3nm)/Si and CFSA/MgO(3nm)/Si samples for the 2θ wide scan. The CFSA(220) diffraction peak for the CoFe-introduced sample was weakened in comparison with the sample without the CoFe interlayer. Figure 5(b) shows the Ψ-rocking curve for the (220) diffraction. Both the CFSA films with/without the CoFe interlayer showed the coexistence of (110)- and (100)-orientations. Figures 5(c) and (d) show the CFSA(220) and CFSA(400) fundamental lines. The CFSA(220) fundamental line of the CFSA/CoFe/MgO/Si sample was much weaker than that of the CFSA/MgO/Si sample, and the CFSA(400) fundamental line of the sample with the CoFe interlayer was stronger than that of the sample

without the CoFe interlayer. These results indicated that the CoFe interlayer improved the (100) orientation of the CFSA film. Figure 6 shows the superlattice lines for the CFSA films with/without the CoFe interlayer. For both the CFSA films, the (200) superlattice and (220) fundamental lines were clearly observed and the (111) superlattice line was not detected. Therefore, both the CFSA films have the $B2$-ordered structure. The degree of $B2$ order was estimated at 64% for the CFSA/MgO/Si sample and 51% for CFSA/CoFe/MgO/Si sample, in which the extended Webster model was used for the estimation [5].

Figures 7(a)-(d) show the effect of the MgO thickness on the orientation of the CFSA/CoFe/MgO/Si and CFSA/MgO/Si samples. For the CFSA/MgO/Si sample, the intensity of the CFSA(400) fundamental line decreased with decreasing the MgO thickness. In particular, the (100)-orientation component ((400) diffraction peak) almost disappeared for the CFSA/MgO/Si sample, when the thickness of the MgO layer is reduced to less than 2nm. However, for the CFSA/CoFe/MgO/Si sample, the (100)-orientation component ((400) diffraction peak) remained, even when the thickness of the MgO layer was reduced to 1nm. Therefore, the CoFe interlayer was effective for the (100)-orientation of the CFSA film. Figure 8 shows the intensity ratio of (400) and (220) diffraction lines for both the CFSA films. Although these ratios were reduced with decreasing the MgO thickness, the intensity ratio ((100)-orientation) of the CFSA/CoFe/MgO/Si samples was larger (stronger) than that of CFSA/MgO/Si samples. Figures 9(a) and (b) show the CFSA(200) superlattice lines (that represent the $B2$-ordering) for the CFSA/MgO/Si and CFSA/CoFe/MgO/Si samples, respectively. Although the intensity of the (200) superlattice lines decreased with decreaseing the thickness of the MgO barrier for both the CFSA films, this tendency was relatively weak for the CFSA/CoFe/MgO/Si sample.

Figures 10(a)-(d) show TEM images of the CFSA/MgO(3nm)/Si and CFSA/CoFe(1nm)/MgO(3nm)/Si samples. For the CFSA/MgO/Si sample, the roughening of the CFSA/MgO interface was clearly observed. However, it was effectively suppressed for the CoFe-introduced sample as shown in Fig. 10(b). The crystallization of the MgO layer near the CFSA/MgO interface was observed for both the samples. This crystallized MgO layers would also play a important role for the crtstallographic properties of the CFSA films. By improving the crystallinity of the thin MgO layers (≤2nm), high quality CFSA films would be also formed even on the thin MgO layers.

CONCLUSION: High quality CFSA/CoFe/MgO/Si tunnel contacts were successfully fabricated. The CoFe interlayer was effective at improving crystallinity and interface roughening of the contacts. The presented technique would be promising for enhancing the spin injection efficiency of full-Heusler-alloy-based tunnel contacts.

REFERENCES: [1] S. Sugahara, IEE Proc. Circuits Devices Syst. **152**, 355 (2005). [2] T.M.Nakatani *et al.*, J. Appl. Phys. **102**, 033916 (2007). [3] Y. Kawame, *et al.*, to be published J. Appl. Phys. [4] T.Scheike, *et al.*, Appl. Phys. Lett. **105**, 242407 (2014). [5] Y. Takamura, *et al.*, J. Appl. Phys. **105**, 07B109 (2009).

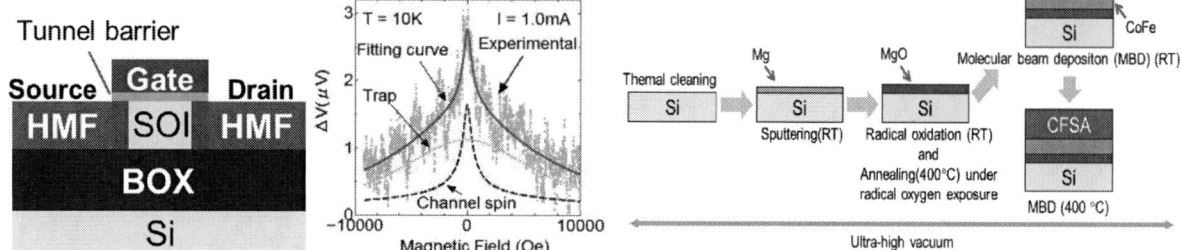

Fig. 1. Schematic illustration of a spin-MOSFET.

Fig. 2. Observed Hanle-effect signal using a CFSA/MgO/Si contact.

Fig. 3. Fabrication process for CFSA/CoFe/MgO/Si samples.

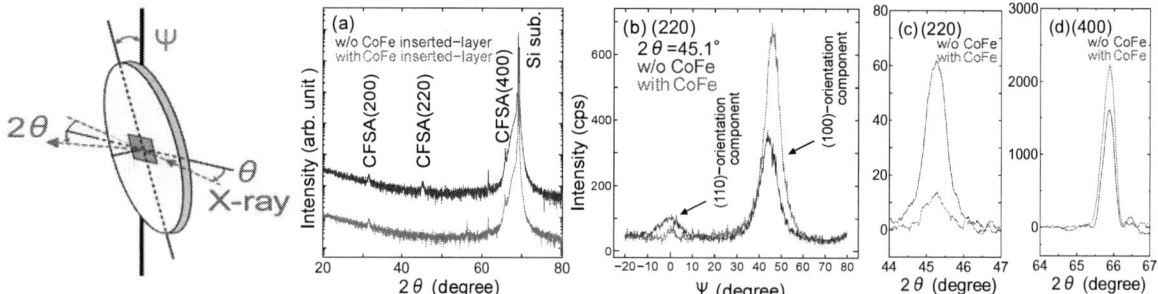

Fig. 4. Definition of sample stage axes for XRD measurements.

Fig. 5. (a) XRD θ/2θ wide scan, (b) Ψ-rocking curve for (220) diffraction, and (c) (220) and (d) (400) fundamental lines for CFSA/MgO(3nm)/Si and CFSA/CoFe(1nm)/MgO(3nm)/Si samples.

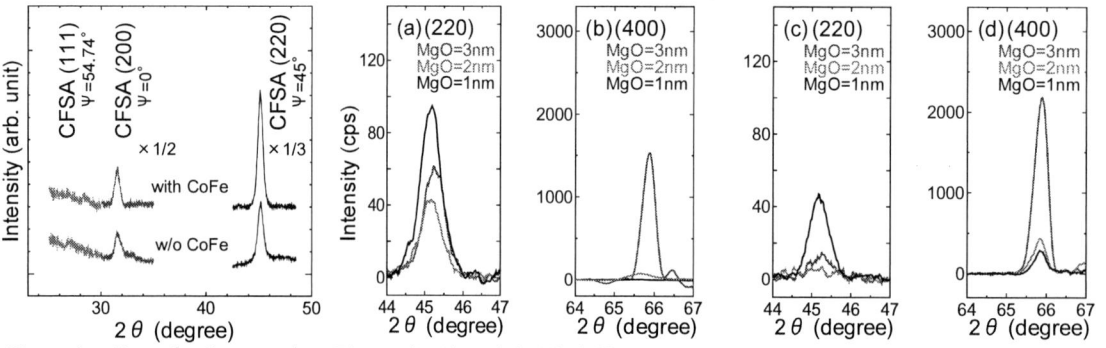

Fig. 6. Superlattice and fundamental lines for the CFSA/CoFe/MgO/Si and CFSA/MgO/Si samples.

Fig. 7. (220) and (400) diffraction intensities as a function of the (a, b) CFSA/MgO/Si and (c, d) CFSA/CoFe(1nm)/MgO/Si samples. The MgO thickness was varied from 1nm to 3nm.

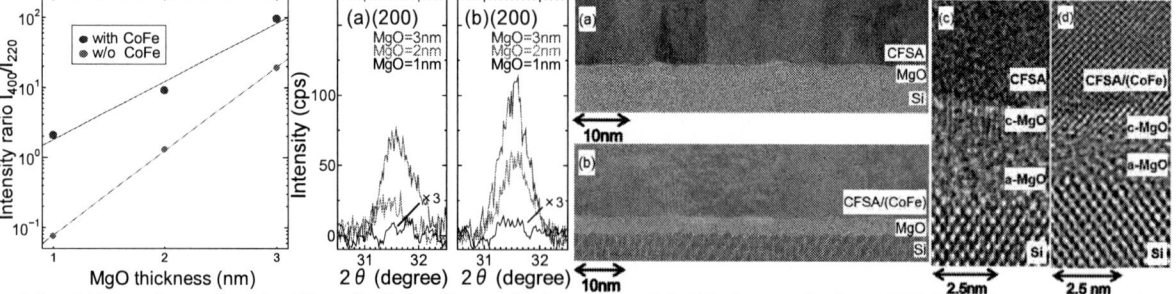

Fig. 8 Intensity ratio of (400) and (220) fundamental lines for the CFSA/MgO/Si and CFSA/CoFe/MgO/Si samples.

Fig. 9 Intensity of the (200) superlattice line for the CFSA/MgO/Si and CFSA/CoFe/MgO/Si samples.

Fig. 10 High-resolution TEM images of (a) the CFSA/MgO(3nm)/Si samples and (b) the CFSA/CoFe(1nm)/MgO(3nm)/Si samples and these magnified images ((c) and (d)).

98

Spin-based Quantum Computing in Silicon

Andrew Dzurak

UNSW, School of Electrical Engineering, Sydney, Australia

Email: a.dzurak@unsw.edu.au

Abstract — **This talk will review recent progress in the development of spin qubits in silicon. In particular it will consider qubit systems based on single phosphorus atoms, in which either the bound donor electron or the ^{31}P nuclear spin can serve as a qubit, and also silicon metal-oxide-semiconductor (SiMOS) quantum dots within which single confined electrons can act as a qubit. In both cases the qubit control fidelity can be above above 99%, as required for some quantum error correction codes.**

Spin qubits in silicon are excellent candidates for scalable quantum information processing [1] due to their long coherence times and the enormous investment in silicon MOS technology. I will discuss qubits based upon the electron and nuclear spins associated with single phosphorus (P) dopant atoms in silicon [2-5] and also more recent work based upon electron spins confined in Si-MOS quantum dots [6-9]. In each case, single-shot electron spin readout is performed using an adjacent single electron transistor and the process of spin-to-charge conversion, showing spin lifetimes of order seconds [2, 7] for the electrons and many minutes for the nuclear spins [4]. Control of individual electron and nuclear spins is achieved by spin resonance using an on-chip microwave transmission line [3].

In isotopically enriched Si-28 all of these spin qubits show control fidelities F_C above 99%, consistent with some fault-tolerant QC error correction codes. Specifically the P donor electron spin qubit has $F_C^e > 99.6\%$ [5], the ^{31}P nuclear spin qubit has $F_C^n > 99.99\%$ [5], and the Si-MOS quantum dot electron spin qubit has $F_C^e > 99.6\%$ [8]. Using dynamical decoupling sequences the coherence times for the P atom qubits can reach $T_{2e}^{CPMG} = 0.5$ s for the electron and $T_{2n}^{CPMG} = 30$ s for the nuclear spin.

In the SiMOS quantum dot qubits the electron g*-factor can be tuned using a gate voltage, leading to a Stark shift in the qubit operation frequency of > 10 MHz [8], allowing individual addressability of many qubits. Most recently we have demonstrated the exchange coupling of two SiMOS qubits to realize CNOT gates [9] for which over 100 two-qubit gates can be performed within a coherence time of 8 μs. Figure 1 shows the device structure for this 2-qubit SiMOS quantum logic gate, together with data showing Rabi spin control of each of the two qubits operated independently.

REFERENCES

[1] D.D. Awschalom et al., "Quantum Spintronics", *Science* 339, 1174 (2013).

[2] A. Morello et al., "Single-shot readout of an electron spin in silicon", *Nature* 467, 687 (2010).

[3] J.J. Pla et al., "A single-atom electron spin qubit in silicon", *Nature* 489, 541 (2012).

[4] J.J. Pla et al., "High-fidelity readout and control of a nuclear spin qubit in Si", *Nature* 496, 334 (2013).

[5] J.T. Muhonen et al., "Storing quantum information for 30 seconds in a nanoelectronic device", *Nature Nanotechnology* 9, 986 (2014).

[6] S.J. Angus et al., "Gate-defined quantum dots in intrinsic silicon", *Nano Lett.* 7, 2051 (2007).

[7] C.H. Yang et al., "Spin-valley lifetimes in a silicon quantum dot with tunable valley splitting", *Nature Communications* 4, 2069 (2013).

[8] M. Veldhorst et al., "An addressable quantum dot qubit with fault-tolerant control fidelity", *Nature Nanotechnology* 9, 981 (2014).

[9] M. Veldhorst et al., "A two-qubit logic gate in silicon", arXiv:1411.5760.

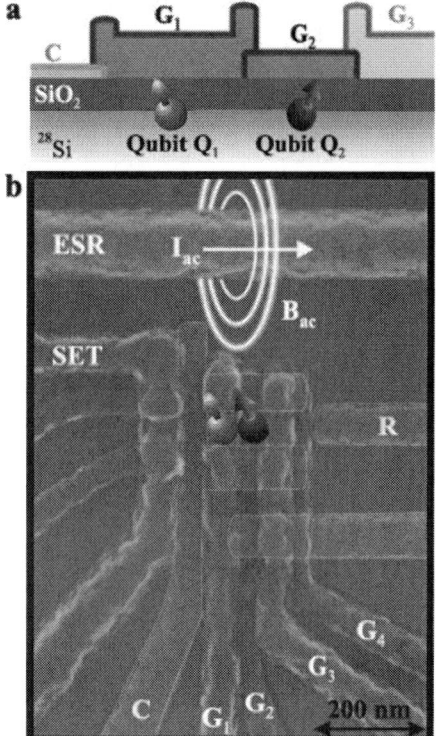

Figure 1: Silicon two-qubit logic device, incorporating SET readout and selective qubit control. **a** Schematic and **b** SEM coloured image of the device. The quantum dot structure (labels C and G) can be operated as a single or double quantum dot by appropriate biasing of gate electrodes G1-G4, where we choose here to confine the dots underneath G1 and G2. The confinement gate C runs underneath the gates G1-G3 and confines the quantum dot on all sides except on the reservoir side (R). **c** Stability diagram of the double quantum dot obtained by monitoring the capacitively coupled SET. Qubit Q1 and qubit Q2 are realized by depleting dot 1 and dot 2 to the last electron. **d** The quantum dot qubits can be individually controlled by electrically tuning the ESR resonance frequency using the Stark shift. Clear Rabi oscillations for both qubits are observed. [Taken from: M. Veldhorst *et al.*, "*A two-qubit logic gate in silicon*", arXiv:1411.5760]

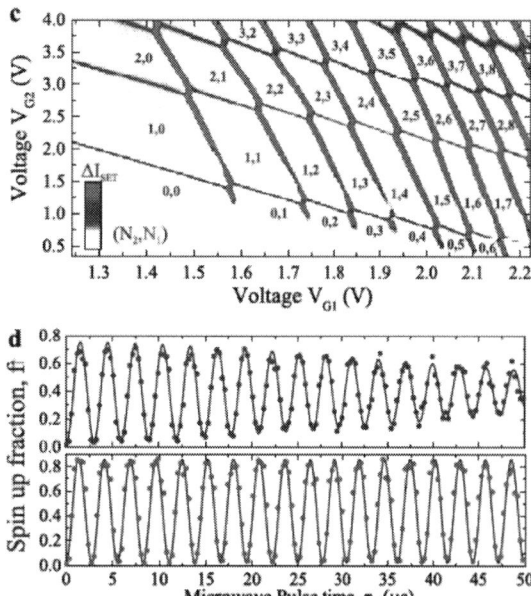

Variation of Coulomb diamonds and excited states caused by electric field in Si single-electron transistor

Hikaru Satoh[1], Takafumi Uchida[1], Atsushi Tsurumaki-Fukuchi[1],
Masashi Arita[1], Akira Fujiwara[2], and Yasuo Takahashi[1]

[1]Graduate school of Information Science and Technology, Hokkaido University, Sapporo, 060-0814, Japan
[2]NTT Basic Research Laboratories, NTT Corporation, 3-1 Morinosato Wakamiya, Atsugi, 243-0198, Japan
E-mail: hika-s@frontier.hokudai.ac.jp

Abstract — **Small Si single-electron transistors (SETs) show interesting nature due to their complicated energy-level structure. We evaluate Coulomb diamonds and excited states of dual-gate SETs by changing an electric field in the SET island. Even though the number of electron in the SET island is constant, the electric field applied by the dual gates remarkably changes the electron addition energies. In addition, interesting variation of excited states due to the electric field are also observed.**

I. INTRODUCTION

Si single-electron transistors (SETs) that have a small Si island are advantageous because of the possibility of combinational usage with CMOS circuits. One of the specific feature of SETs is that they can inherently have many gate electrodes, which provides higher functionalities [1]. In addition, small Si islands show complicated characteristics completely different from metal SETs described by the orthodox theory. It was shown that gate electric field dependence of the peak positions of the Coulomb blockade oscillation significantly depends on the number of electrons [2, 3]. On the other hand, it is also guessed that the electric field varies the wave function of electrons in the small Si island and thereby can affect its excited-state spectrum. Here, we evaluated Coulomb diamonds and excited states of dual-gate SETs by changing the electric field in the SET island.

II. EXPERIMENTAL DETAILS

The schematic device structure fabricated on a silicon-on-insulator (SOI) wafer is shown in Fig. 1. We fabricated with widths of 36−50 nm and lengths of 50−100 nm Si nanowires in the Si layer of the 25 nm thick SOI wafer by electron-beam lithography and dry etching as shown Fig. 1. Oxidization at 1000 °C in dry oxygen ambient automatically formed Si nanodots sandwiched by tunnel barriers in the nanowires by means of pattern-dependent oxidation (PADOX) [4]. After the 50-nm-thick SiO_2 deposition, a P-dope poly-Si gate electrode was formed on the wire.

Electron transport property was measured at a temperature of about 8K. We used the poly-Si gate (top gate) and substrate (back gate) as dual-gates so as to vary the vertical electric field applied to the SET Si island.

III. RESULTS AND DISCUSSION

Contour plots of differential conductance as a function of top-gate voltage (V_G) and drain voltage (V_D), which elucidate the nature of SET, are shown in Figs. 2, 3 and 4 for three SETs named A, B and C, respectively. The parameter is the back gate voltage (V_B). Each pair of the contour plots shows the region of the same number of electron in the island considering the effect that the application of positive V_B shifts the characteristics to negative V_G direction. As shown in Figs. 2(a) and (b), electron addition energies (Coulomb diamond sizes) and excited states of SET A were remarkably changed by V_B application. Figs. 2(c) and (d) shows the addition energies of the three Coulomb diamonds and energy of excited states evaluated from the Coulomb diamond edge as a function of V_B, respectively. The fact that the variations of excited states are relatively smaller than those of the addition energies may imply small variation in wave functions.

The size of the Coulomb diamonds were not almost unchanged by V_B in SETs B and C. The diamond sizes were both almost identical between the devices without being affected by the number of electron in the island. This indicates that the island size was almost constant. However, a significant difference was observed in behaviors of excited states as shown in Figs. 3 and 4. In SET B, two of the four excited states were degenerated by increasing V_B. Conversely, an excited state in SET C separated into two at high V_B. These behaviors are thought to be caused by the change in the wave functions of electrons in the island by electric fields. The existence of many excited states due to the complicated energy-level structure can be understood as the origin of this fact. In comparison with these results, larger variations of addition energies with small change in excited states observed in Fig. 2 seems strange because the addition energy corresponds to grand states mainly determined by the island size, which is not expected to be largely changed by the field.

IV. SUMMARY

The properties of dual-gate Si SETs were evaluated when the electric field in the SET island was varied. Large variations of the addition energy and excited states caused by the electric field were observed, which are attributed to the large change of wave function of electrons in the Si island [5].

ACKNOWLEDGEMENT
This work was partly supported by the KAKENHI by MEXT and JSPS (24360128, 25420279, and 26630141).

REFERENCES
[1] T. Kaizawa *et al.*, IEEE Trans. Nanotech, **8**, 535 (2009).
[2] S. Horiguchi *et al.*, Jpn. J. Appl. Phys., **43**, 2036 (2004).
[3] S. Horiguchi *et al.*, Thin Solid Films, **520**, 3349 (2012).
[4] Y. Takahashi *et al.*, Electron. Lett., **31**, 136 (1995)
[5] T. Uchida *et al.*, J. Appl. Phys., **117**, 084316 (2015).

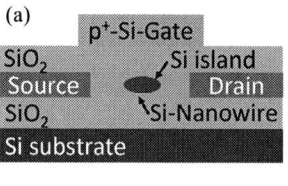

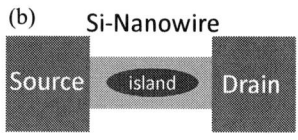

Fig. 1. Schematic of the device structure; (a) cross-section and (b) top view of Si-SET.

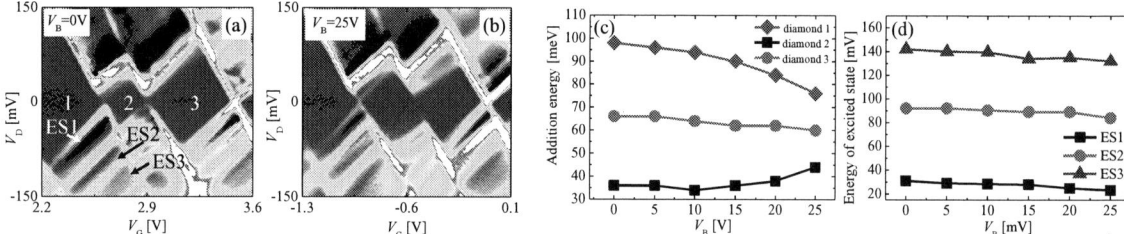

Fig. 2. The measured stability diagram of SET A shown as contour plot of dI_D/dV_D as a function of V_G and V_D at (a) $V_B=0V$ and (b) $V_B=25V$. It is noted that the number of electrons in the SET island is the same in the two figures. (c) Addition energies evaluated from the Coulomb diamond heights. (d) Energy of excited states evaluated from the distance from the edge of the diamond.

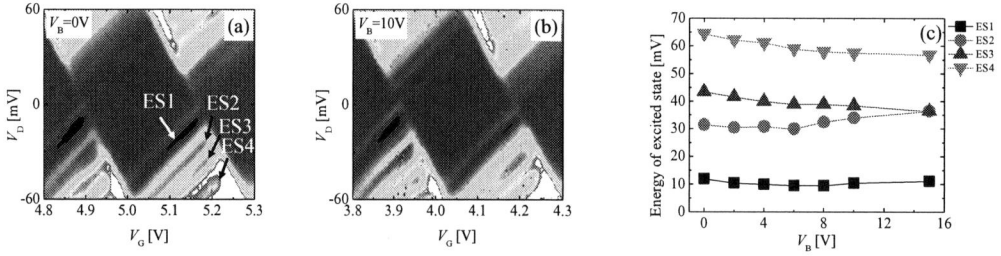

Fig. 3. Stability diagrams of SET B measured at (a) $V_B=0V$. (b) $V_B=10V$. (c) Energy of excited states.

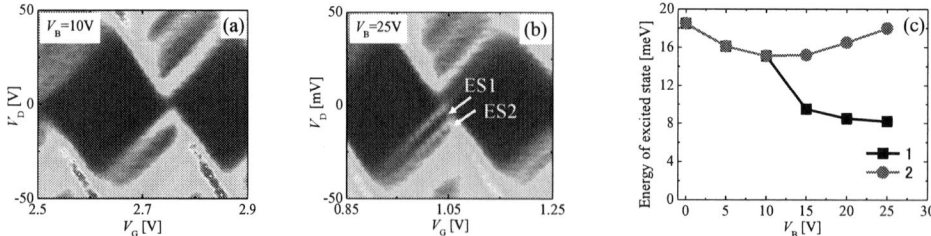

Fig. 4. Stability diagrams of SET C measured at (a) $V_B=10V$. (b) $V_B=25V$. (c) Energy of excited states.

Study of charged island formation in nanoscale Si single-electron transistors using dual port reflectometric spectroscopy

Alexei O. Orlov[1], Patrick Fay[1], Gregory L. Snider[1], Xavier Jehl[2], Romain Lavieville[3],
Sylvain Barraud[3], and Marc Sanquer[2]

[1]Department of Electrical Engineering, University of Notre Dame, USA.
[2]DSM-INAC, CEA-Grenoble, France
[3]DRT-Leti, CEA-Grenoble, France

Email: aorlov@nd.edu

Abstract — **Recently developed dual port reflectometric spectroscopy (DPRS) not only enables detection of single electron charging events in nanoscaled MOS single-electron transistors (SETs) that are not detectable using conventional transport spectroscopy, but DPRS also provides the ability to spatially localize charging events and discriminate between charging events involving defects and those associated with the SET island. Here we present an experimental study of charging processes in Si single-hole transistors (SHTs) from deep depletion (prior to electrostatic formation of the island) through the initial stages of SHT island formation and population with the first few holes.**

I. INTRODUCTION

Single-electron transistors are one of the most promising electron devices that benefit from ultimate downscaling. One of the intriguing questions defining the operation of semiconductor SETs is the determination of the border at which the population of the island begins and whether this transition corresponds to the onset of charge transport through the island, or if the initial stages of island population are dominated by population of the island as "charge pockets" decoupled from the source and drain. These charges have great influence on the threshold voltage variability of nanoscale transistors and possibly on phase coherence for CMOS-based Q-bits. In addition, it is also important to understand the degree to which charging of the defects surrounding SET island may interfere with SET operation.

II. EXPERIMENTAL RESULTS

A. Samples and experimental setup

The devices studied here were fabricated using fully depleted SOI [1] (Fig. 1a,b); a Si nanowire is surrounded by the gate on three of the four facets, and spacers above two undoped regions near the source and drain form the tunnel barriers. Transport in these devices at temperatures <20K is dominated by Coulomb blockade and the population of the central island is controlled by the gate. To investigate the charging processes in the SET we use dual-port reflectometric spectroscopy (DPRS) [2] (Fig. 1c) combined with conventional ("DC") single-electron spectroscopy (SES) measurements.

B. Experimental results and discussion

Figure 2 shows the results obtained for a single hole transistor (SHT) using a combination of DPRS and SES techniques. We were able to detect the first current-carrying state in the device, i.e. switching between 0 and 1 hole in the island (note that SES technique produces measurable results only at $|V_{ds}|>50$ mV). The DPRS technique is able to detect a number of additional charging events at gate voltages $V_g>-600$ mV that are apparently distinct from this first current-carrying state. For example, charging of single dopants at the edges of the source and drain, below the spacers results in two distinctly different lines: tilted lines with a slope $\alpha=dV_{ds}/dV_g\sim1$ result from single-electron (SE) charging near the drain, while vertical lines only visible in the gate coupled maps result from SE charging near the source. There is also a number of lines with $\alpha>1$ corresponding to charging extrinsic to the SHT, most likely due to Coulomb charges in the gate stack. The full vector analysis of the reflected signal contains information on spatial location of charging processes and their principal impact (dissipative vs dispersive processes).

ACKNOWLEDGEMENTS

This work was supported by NSF Foundation grant DMR-1207394 (A. Orlov and P. Fay); S. Barraud, X. Jehl, and M. Sanquer acknowledge financial support from the EU through FP7 initiative under Project TOLOP (318397) and SiAM (610637).

REFERENCES

[1] S. Barraud *et al.*, *Elec. Dev. Lett.*, 33, 1526 (2012).
[2] B. J. Villis *et al.*, *Appl. Phys. Lett.*, 104, 233503 (2014).

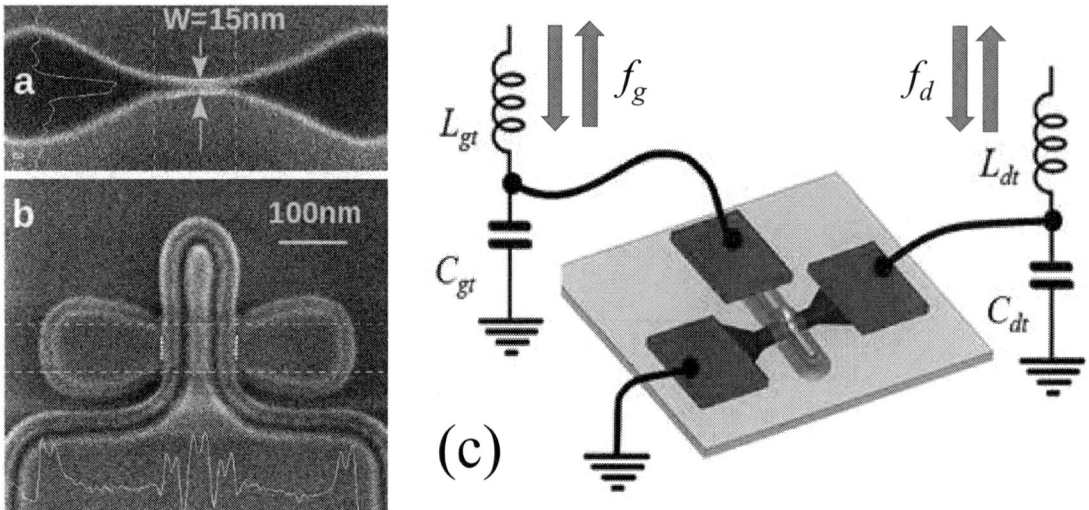

Figure 1. (a,b) SEM micrographs of a device nominally identical to the one studied. Width of the nanowire is indicated in (a). The length of the SHT island, visible in (b) is defined by the space covered by the gate, ~20 nm. The thickness of the nanowire is 11 nm (c) Sketch of the dual channel reflectometry experiment. LCs represent the simplified resonators coupled to the gate and to the drain of the device. Blue arrows indicate RF incoming and reflected signals at two different frequencies, f_g = 319 MHz and f_d =453 MHz.

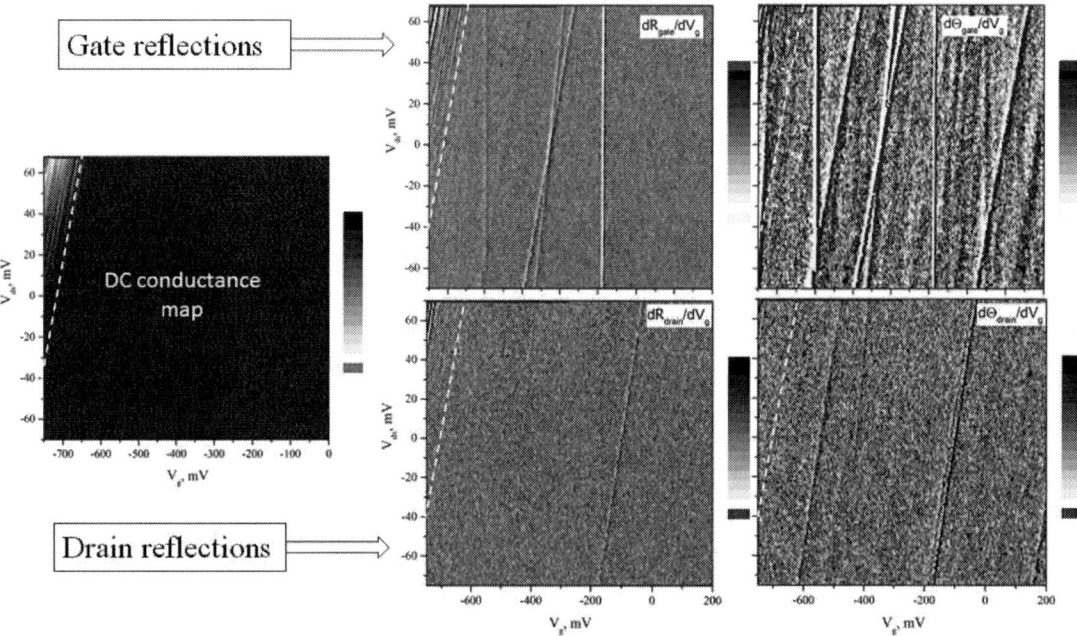

Figure 2. Experimental results of the dual port reflectometry measurements. Dashed white line delineates the onset of SHT current-carrying population. In this sample the transition 0->1st hole in the island corresponds to the carrier transport through the dot. Gray scale map of conductance is shown on the left, black color corresponds to unmeasurably low conductance (<0.1 nS).

Two-component gray scale maps are shown in the right. Here, the upper two maps show the two components of the gate reflected signals (magnitude on the left, phase on the right), f_g=319 MHz. The lower two maps show the two components of the drain reflected signals (magnitude on the left, phase on the right), f_d=453 MHz. The sample is cooled to about 3.8K.

Low Temperature Charge Pumping in SOI Gated PIN Diode

T. Watanabe[1], M. Hori[1], T. Saruwatari[1], A. Fujiwara[2], and Y. Ono[1*]

[1] University of Toyama, 3140 Gofuku, Toyama 930-8555, Japan
[2] NTT Basic Research Laboratories, 3-1 Morinosato Wakamiya, Atsugi, Kanagawa, 243-0198, Japan
Email: *yukiono@eng.u-toyama.ac.jp

I. INTRODUCTION

The recent progress of Si-based quantum electronics indicates the importance of silicon-on-insulator (SOI) materials, owing their unique characteristics, e.g., the ability to control the valley degree of freedom [1, 2]. Since such characteristics appear to be related to the quality of the Si/oxide interfaces, it is important to analyze the interface states in SOIs. However, the charge pumping (CP), the most reliable method for the interface-state analysis [3-5], is not well established for the SOI devices, in particular at low temperatures.

In this study, we perform systematic measurements of the CP current in an SOI gated PIN diode at 7 K. Mapping of the CP current in the front-gate/back-gate voltages plane reveals that the CP current is reduced when the dual (front and back) channels are formed. This anomaly is a unique feature of the low-temperature CP in SOI materials and indicates the importance of the interplay between the interface states and the remote channel.

II. DEVICE STRUCTURE

Top and cross-sectional schematic views of the device are shown in Fig. 1. The device consists of six terminals, the n-type and p-type sources and drains, and the front and back (substrate) gates. Owing to this multi-terminal structure, the device allows us to measure the drain current (I_D) vs. gate voltage (V_G) characteristics for both n- and p-channel MOSFETs and analyze the relationship between the CP current and the electron- and hole-channel formation.

III. RESUTS AND DISCUSSIONS

A. DC characteristics of n- and p-MOSFETs

Figs. 2(a) and 2(b) show grayscale plots of the transconductances, dI_D/dV_f and dI_D/dV_b, respectively, in the V_f-V_b plane, where V_f and V_b denote the voltages to the front and back gates, respectively. The measurement temperature is 7 K. These transconductance plots accentuate the gate-voltage boundary between the front and back channels, as shown by the schematic diagram in Fig. 2(c).

B. CP currents

The CP applies the voltage pulse to the gate so that the electron and hole channels are alternately formed. The repetition of the gate pulse results in a recombination current I_{cp} flowing between the n- and p-type electrodes upon the presence of the interface states. The I_{cp} is thus proportional to the pulse frequency f and interface state density N_{it}; i.e., $I_{cp} \propto N_{it} f$.

For the present CP measurements, we apply the pulse voltage to the front gate, and the I_{cp} is measured as a function of the base voltage $V_{f\text{-base}}$ of the front-gate pulse with a fixed back-gate voltage V_b.

Fig. 3(a) shows the I_{cp} measured at 7 K as a function of $V_{f\text{-base}}$ for three different V_b values. The voltage trajectory and the pulse shape are shown in Fig. 3(b). As shown in Fig. 3(a), the maximum I_{cp} value decreases when a positive or negative V_b is applied.

Fig. 4(a) shows the mapping of I_{cp} in the $V_{f\text{-base}}$-V_b plane. The I_{cp} at $V_{f\text{-base}} = -2.84$ V as a function of V_b is also shown (right figure). The voltage conditions for obtaining a non-zero value of I_{cp} can be divided into three regions: (1), (2), and (3) (See Fig. 4(b)). We found that the boundary line between (1) and (2) and that between (2) and (3) coincide with lines (III) and (IV) in Fig. 2(b), respectively. This indicates that the reduction of the CP currents in (1) and (3) is related to the formation of the dual channels during the CP sequence.

On the basis of these results, one possible mechanism for CP-current reduction is the following. In regions (1) and (3), the formation of the dual channels pins the back-interface potential to 0 V (i.e., to the Fermi level of the source). This Fermi-level pinning results in a strong vertical electric field in the Si layer. This strong field then induces the electron (or hole) emission, reducing the recombination rate and thus the CP current.

The present results suggest that biasing to the back gate elucidates the correlation between the electron/hole emission and recombination in CP sequences of the SOI devices.

IV. SUMMARY

The CP current in SOIs is systematically investigated at 7 K. It decreases when biasing to the back gate, and mapping of the CP current in the gate-voltages plane indicates that the reduction is related to the dual gate formation. The present results reveal the importance of the interplay between the interface states and the remote channel for the low-temperature CP in SOIs.

This work was partially supported by JSPS KAKENHI Grant Nos. 23226009, 24360128, 25289098, 25600015, and 25706003. We thank Prof. Toshiaki Tsuchiya for his valuable discussion.

REFERENCES

[1] K. Takashina Y. Ono, A. Fujiwara, Y. Takahashi, Y. Hirayama, Phys. Rev. Lett. **96**, 23681 (2006).

[2] J. Noborisaka, K. Nishiguchi, A. Fujiwara, Sci. Rep. **4**, 6950 (2014).

[3] G. Groeseneken, H. E. Maes, N. Beltran, R. F. DeKeersmaecker, IEEE Trans. Electron Devices **31**, 42 (1984).

[4] T. Ouisse, S. Cristoloveanu, T. Elewa, H. Haddara, G. Borel, D. E. Ioannou, IEEE Trans. Electron Devices **38**, 1432 (1991).

[5] Y. Li, T.-P. Ma, IEEE Trans. Electron Devices **45**, 1329 (1998).

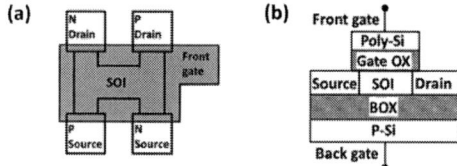

Fig. 1. Top (a) and cross-sectional (b) schematic views of device. Area of channel is 380 μm^2. Si layer is 30 nm thick, and oxide for front and back gates is 100 and 400 nm thick, respectively.

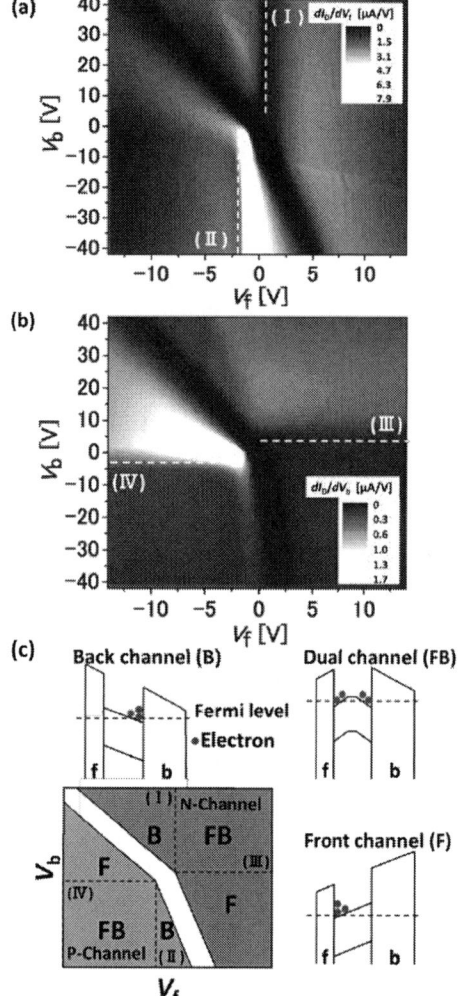

Fig. 2. Grayscale plot of dI_D/dV_f (a) and dI_D/dV_b (b) in V_f-V_b plane measured at 7 K. I_D was measured with n-drain voltage $V_{nd} = 0.1$ V, p-drain voltage $V_{pd} = -0.1$ V, and both n- and p-sources grounded. Dashed lines (I)–(IV) in (a) and (b) show boundary between single (electron or hole) channel and dual-channel regions. (c) Schematic of channel formation. "F," "B," and "FB" denote front, back, and dual (front and back) channels, respectively. Energy band diagrams for electron channels are also shown. In the band diagrams, "f" and "b" denote front- and back-gate oxides, respectively, and horizontal dashed lines indicate source Fermi level.

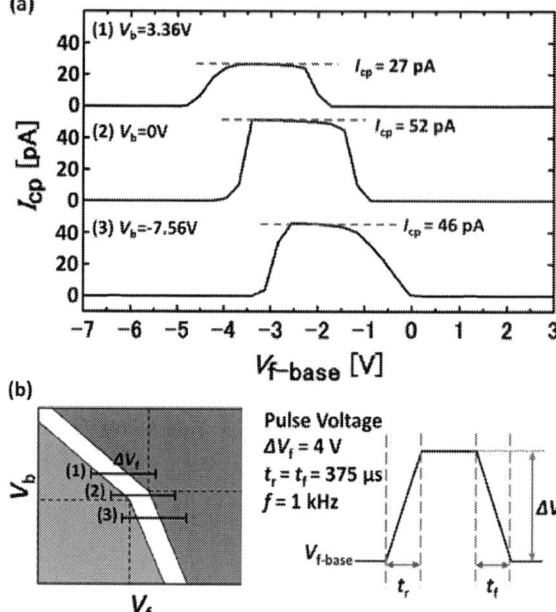

Fig. 3. (a) I_{cp} as a function of $V_{f\text{-base}}$ for three V_b values, measured at $f = 1$ kHz and $T = 7$ K. Pulse voltage amplitude ΔV_f and rise/fall times t_r/t_f are respectively 4 V and 375 μs. N_{it} estimated from data (2) was 8.6×10^{10} cm^{-2}. (b) Gate voltage trajectory and the gate pulse voltage.

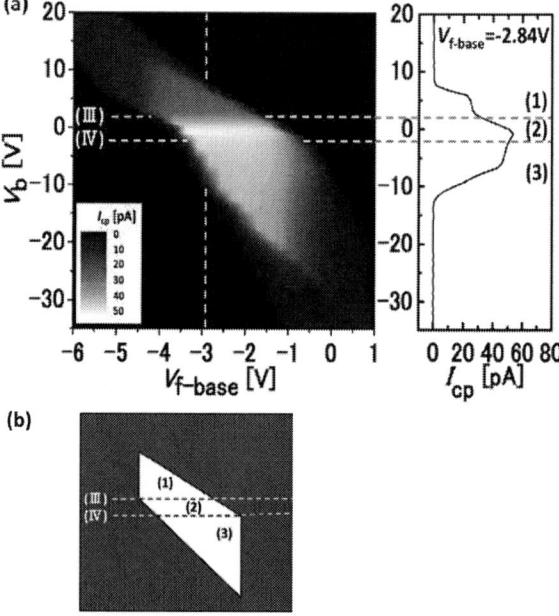

Fig. 4. (a)-*left* Grayscale plot of I_{cp} in $V_{f\text{-base}}$ and V_b plane, measured at $f = 1$ kHz and $T = 7$ K. Lines (III) and (IV) correspond to those in Fig. 2(b). (a)-*right* I_{cp} as a function of V_b at $V_{f\text{-base}} = -2.84$ V (vertical dashed line in the grayscale plot). (b) Diagram of I_{cp} in $V_{f\text{-base}}$ and V_b plane.

Charge sensing of p-channel double quantum dots fabricated on (110) silicon substrate

K. Iwasaki[*], T. Kodera and S. Oda

[1] Department of Physical Electronics, Quantum Nanoelectronics Research Center, Tokyo Institute of Technology, 2-12-1, Ookayama, Megruro-ku, Tokyo, Japan

[*]Tel/Fax: +81-3-5734-2542, E-mail: iwasaki.k.aj@m.titech.ac.jp

Abstract — We fabricate and characterize p-channel double quantum dots (DQDs). The DQDs are formed on a silicon-on-insulator (SOI) wafer and integrated with a single hole transistor (SHT) as a charge sensor. We observe charge stability diagram of the DQDs in the characteristic of the charge sensor. Furthermore, few-hole regime in the DQDs is clearly obtained.

INTRODUCTION

Silicon double quantum dots (Si DQDs) have been well studied toward electron spin quantum bits (qubits) which are expected to have long coherence time and suitability for integration. However, there are few reports on p-channel Si DQDs for hole spin qubits [1,2]. Holes in Si quantum dots (QDs) have a wave function with a p-like orbital and then hole spins have a weaker hyperfine interaction with nuclear spins than electron spins. Therefore, longer coherence time of hole spins is expected than that of electron spins. Furthermore, on (110) Si substrate, the effective mass of holes is smaller than that on (100) Si substrate [3]. As a result, the quantum confinement effect of holes is larger, which leads to the larger energy separation between ground state and excited state in QDs. This is advantageous for the stability of the quantum states in the QDs.

Single hole transistors (SHTs) have been well studied and operated at room temperature for beyond CMOS application [4]. In this work, we utilize the SHTs as charge sensors (CSs) for detecting the charge states in the DQDs. The sensitivity of the SHTs is so high that hole transitions in the DQDs can be detected even though no current is observed through the DQDs due to small tunneling rate. We demonstrate charge sensing experiments using physically-defined p-channel Si DQDs with SHT on (110) substrate and clearly observe few-hole regime in the DQDs with SHT charge sensor (CS).

DEVICE FABRICATION

DQDs are fabricated on (110) Si-on-insulator (SOI) substrate, by using electron beam lithography, SF_6 dry etching, and thermal oxidation. Figure 1 shows a scanning electron microscope (SEM) image of the device, composed of Si DQDs with an SHT CS and three side gates (SG_l, SG_r, SG_{CS}). The SGs are used for controlling the energy levels in the QDs. Accepter of Boron (B^+) is implanted at source (S) and drain (D) regions in SOI layer. The doping concentration is 1.1×10^{20} cm^{-3}. Holes are induced in the QDs by applying negative voltage to a top gate (TG: not shown in Fig. 1) like a metal-oxide-semiconductor field effect transistor (MOSFET).

MEASUREMENT RESULTS AND DISCUSSION

Figure 2 (a) shows current thorough CS I_{CS} as a function of V_{TG} when $V_{SGl} = V_{SGr} = V_{SGCS} = 0$ V at 4.2 K. Coulomb oscillation is observed. This result indicates that single hole tunneling through the CS occurs at 4.2 K and the CS works as an SHT. Figure 2 (b) shows the derivatives of CS current with SG_l voltage, dI_{CS}/dV_{SGl}, as a function of V_{SGl} and V_{SGr}. In Fig.2 (b), we observed two slopes of charge transition lines (green and orange lines), reflecting from different capacitances between each QD and each side gate. The each line corresponds to the change in number of holes in each QD in the DQDs. This means that we demonstrate the charge detection of the DQDs with the SHT CS. Furthermore, the transition lines are not observed at the upper right region in Fig.2 (c), which indicates that the few-hole regime of DQDs is clearly observed.

CONCLUSIONS

We fabricate p-channel DQDs integrated with an SHT as a CS and investigate the transport properties. At 4.2 K, honeycomb-like charge stability diagrams of the

DQDs are clearly obtained by the SHT CS. We also obtain the few-hole regime, which is a state that is prefered to hole spin qubits.

ACKNOWLEDGEMENT

This work was financially supported by Kakenhi Grants-in-Aid (Nos. 26709023, 26630151, and 26249048), the Murata Science Foundation, and Project for Developing Innovation Systems of the Ministry of Education, Culture, Sports, Science and Technology (MEXT).

REFERENCES

[1] K. Yamada, T. Kodera, T. Kambara and S. Oda, "Fabrication and characterization of p-channel Si double quantum dots", Appl. Phys. Lett. **105**, 113110 (2014).

[2] P. C. Spruijtenburg, J. Ridderbos, F. Mueller, A. W. Leenstra, M. Brauns, A. A. I. Aarnink, W. G. van der Wiel and F. A. Zwanenburg, "Single-hole tunneling through a two-dimensional hole gas in intrinsic silicon", Appl. Phys. Lett. **102**, 192105 (2013).

[3] H. Irie, K. Kita, K. Kyuno and A. Toriumi, "In-Plane Mobility Anisotropy and Universality Under Uni-axial Strains in n- and p-MOS Inversion Layers on (100), (110), and (111)Si", *Electron Devices Meeting, 2004. IEDM Technical Digest. IEEE International,* (2004).

[4] K. Miyaji, M. Saitoh, and T. Hiramoto, "Voltage gain dependence of the negative differential conductance width in silicon single-hole transistors", Appl. Phys. Lett., **88** (14), 143505 (2006).

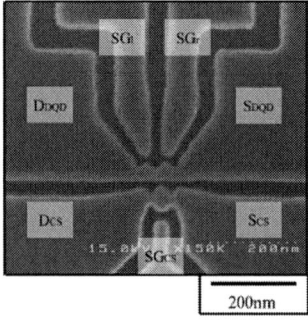

Fig.1. SEM image of p-channel DQD device. The QDs and side gates are fabricated with electron beam lithography (EBL) and dry etching. Dark region is buried oxide (BOX), and bright region is silicon on insulator (SOI). The thickness of SOI and BOX layer is 45 nm and 145 nm, respectively.

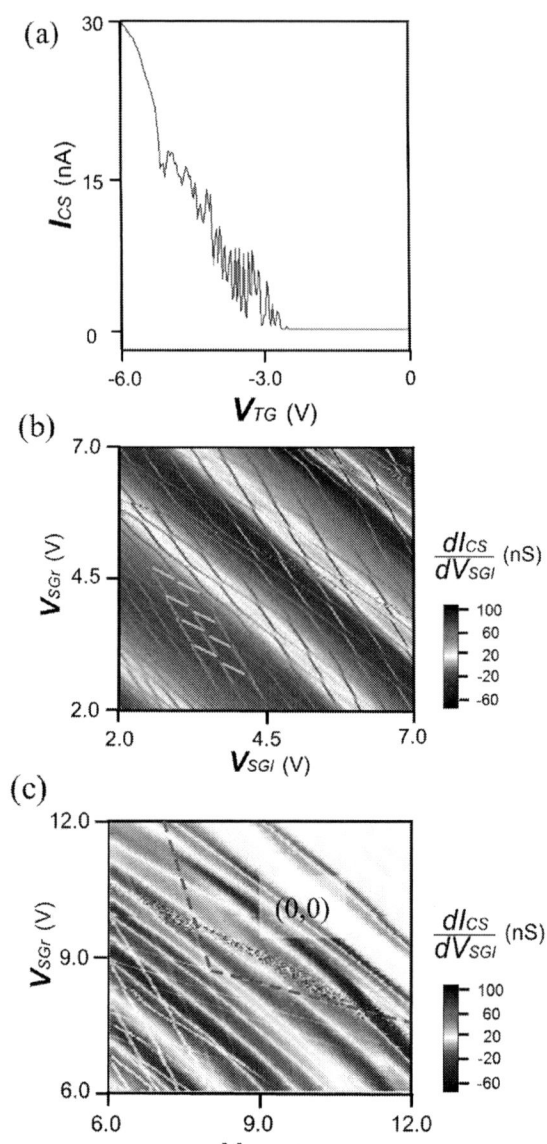

Fig.2: (a) I_{CS}-V_{TG} characteristic of the CS. Coulomb peaks are observed below V_{TG}= -3.8V. (b), (c) Charge stability diagram of the DQDs obtained by the CS when V_{TG}= -3.8V. (b) Honeycomb structure is observed at whole region, which is typical for DQDs. (c) The diagram of more positive side gate voltage region, obtained by sweeping V_{SGl} and V_{SGr} from 6.0 to 12.0 V. Pinch-off in the DQD, indicated by (0,0), is clearly observed at the upper right region.

Fluctuations and Relaxation in Graphene

D. K. Ferry, B. Liu, and R. Akis

School of Electrical, Computer, and Energy Engineering, Arizona State University
Tempe, AZ 85287-5706 USA
Email: ferry@asu.edu

Abstract — **As semiconductor devices continue to get smaller, it is clear that quantum effects will begin to play a larger role in their behavior. Phase interference and the energy relaxation via electron-electron interactions are key quantum effects which can occur in small devices. These are studied best at low temperature where other processes are quenched. Here, we discuss these effects in graphene.**

I. INTRODUCTION

The continued evolution of Moore's Law means that the size of semiconductor devices is decreasing toward the atomic spacing in the crystal. Quantum effects have already been shown to be important in the performance of these devices [1,2]. Continued reduction in device size is sure to accentuate the role played by such quantum effects. Two important areas are disorder induced fluctuations in conductance and carrier-carrier interactions. Both of these effects can be studied effectively at low temperatures, where other processes are quenched. Here, we study these two effects in graphene, although the effects can be seen in most semiconductors. Graphene has been suggested for a variety of semiconductor devices due to its unique zero-gap band structure and the expected high mobility of the carriers.

II. FLUCTUATIONS

Conductance fluctuations (CF) due to a disorder potential arising from the impurities was first discovered in small Si MOSFETs. These reproducible CF are thought to arise from phase interference effects as one varies either the density or an applied magnetic field. It was earlier thought that these effects were *universal* (ubiquitous and appearing in every device) and *ergodic* (appeared the same whether gate voltage or magnetic field was varied). However, it has been shown that neither of these effects holds, as the degree of disorder is very important [3]. We study these fluctuations via an atomic basis tight-binding formulation, in which the random disorder potential arises from impurities remote from the graphene sheet itself. The computed CF are compared with those reported for monolayer and bilayer graphene [4]. Results are shown in Figs. 1-3 below. It is clear that there is no universality, nor ergodicity, in either the experimental results or in the simulations. The CF are, however, very specific to the details of the device.

III. ENERGY RELAXATION

The manner in which the energy imparted to the carriers in the device is relaxed to the lattice is very important. This behavior can indicate the primary interaction by which the energy is relaxed. In graphene, there has been some disagreement over the process by which this energy is relaxed, and many experiments do not agree with the expected behavior for phonon scattering. We find that the energy is relaxed via an electron-plasmon interaction. Here, the calculated loss rate per electrons and the energy relaxation time are both found to agree with recent experimental results [5], with no adjustable parameters other than material properties. Here, we find that there is a large reduction in the energy relaxation time as the Dirac point is approached from either the electron or the hole side.

REFERENCES

[1] D. Vasileska, D. Schroder, and D. K. Ferry "Scaled Silicon MOSFETs: Degradation of the total gate capacitance," *IEEE Trans. Electron Dev.*, vol. 44, no. 4, pp. 584-587, April 1997.

[2] D. K. Ferry, R. Akis, and D. Vasileska, "Quantum effects in MOSFETs: Use of an effective potential in 3D Monte Carlo Simulation of ultra-short channel devices," *IEDM Proc.* (IEEE Press, New York, 2000).

[3] Bobo Liu, R. Akis, and D. K. Ferry, "Conductance fluctuations in semiconductor nanostructures," *J. Phys. Condensed Matt.*, vol. 25, 395802, 2013.

[4] G. Bohra, R. Somphonsane, N. Aoki, Y. Ochiai, R. Akis, D. K. Ferry, and J. Bird, "Nonergodicity and microscopic symmetry breaking of the conductance fluctuations in disordered mesoscopic graphene," *Phys. Rev. B*, vol. 86, 161405, 2012.

[5] R. Somphonsane, H. Ramamoorthy, G. Bohra, G. He, D. K. Ferry, Y. Ochiai, N. Aoki, and J. P. Bird, "Fast energy relaxation of hot carriers near the Dirac point of graphene," *Nano Letters*, vol. 13, pp. 4305-4310, August 2013.

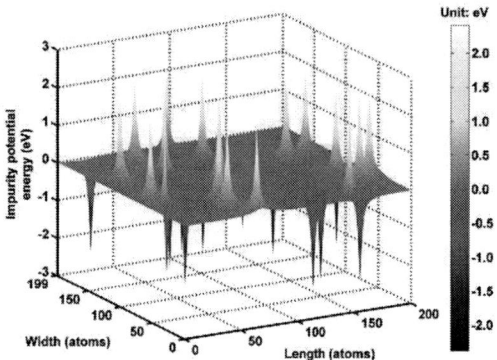

Fig. 1 Typical random potential for an array of positively and negatively charged impurities sited below a monolayer of graphene.

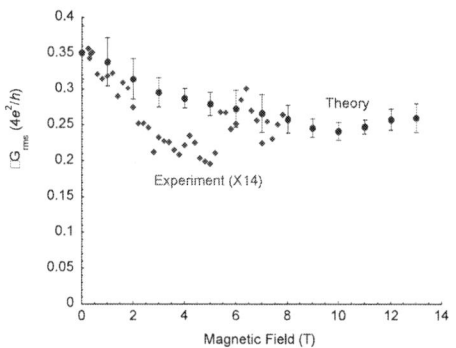

Fig. 2 Conductance fluctuations as a function of an applied magnetic field normal to the sheet of graphene. Each data point represents a sweep of the Fermi energy over the range 50-250 meV for many different samples. The experimental data is from ref. [4].

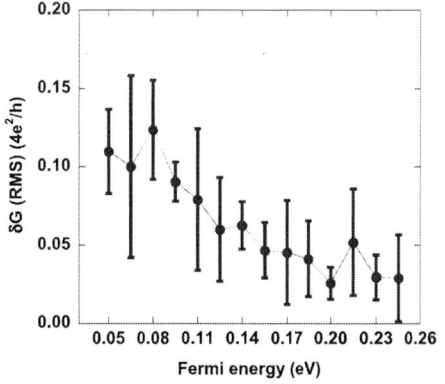

Fig. 3 Conductance fluctuations for a sweep of the magnetic field for a range of densities. Clearly, the disorder is changed as the density is increased due to screening.

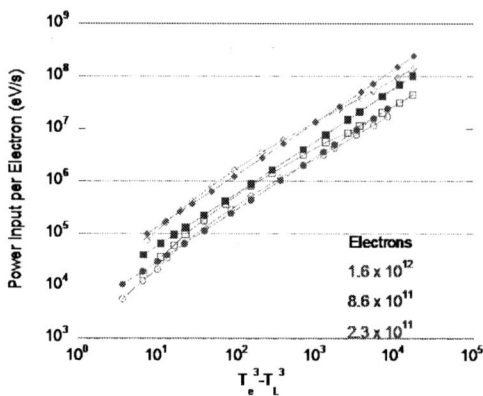

Fig. 4 The energy loss rate per electron at a lattice temperature of 1.8 K. The solid symbols are the theory and the open symbols are experimental data from ref. [5].

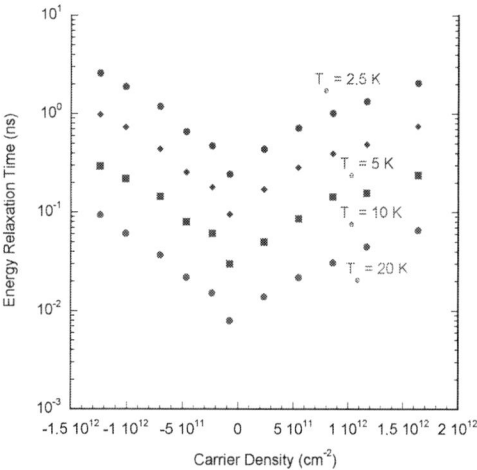

Fig. 5 The energy relaxation time for electrons and holes in graphene as a function of the carrier density. The electron temperature is given as a parameter. The Dirac point occurs at zero density.

Low pull-in voltage graphene nanoelectromechanical switches

M. Manoharan, T. Chikuba, N. Kanetake, J. Sun, H. Mizuta

[1]School of Material Science, Japan Advanced Institute of Science and Technology, Japan
[2]Nanoelectronics and Nanotechnologies Research Group, University of Southampton, U. K.
Email: mano@jaist.ac.jp

Abstract — In this work, we report double-clamped beam and cantilever graphene nanoelectromechanical (GNEM) switches with local top actuation electrode. The low pull-in voltage of below 5 V was realized in both of the switches, which is compatible with the conventional complementary metal-oxide-semiconductor circuit requirements. Additionally, the naturally formed chromium oxide was introduced at the contact interface to prevent the formation of chemical bonds between graphene and metal electrode. As consequence, the reversible switching operation was achieved.

I. INTRODUCTION

Graphene has an ultra-high Young's modulus of 1 TPa, making it a promising candidate for future nanoelectromechanical (NEM) applications. The GNEM switches have potential to realize minimized electrical leakage, sharp switching response, low actuation voltage and high on/off ratio [1]. Fig. 1 shows the proposed GNEMS three-terminal switching transistor. However, only few reports show multiple cycles of switching operation. The common failure is that graphene is stuck on electrode due to stiction and chemical bonds [2]. Moreover, the heavily doped silicon substrate has been commonly utilized as the actuation electrode. In this case, GNEM switches suffer from the relatively large pull-in voltage [3]. Thus, the local actuation electrode is necessary to reduce pull-in voltage. In this work, we report a simple method to fabricate graphene switches with local actuation electrode and enhanced reliability. The results presented in this paper are stepping-stones towards the final goal of realizing abrupt switching transistor with S << 60 mV/dec.

II. FABRICATION

Figure 2 shows the fabrication process: (a) mechanically exfoliate graphene, then, define the bottom electrode with Cr/ Au stack via conventional fabrication techniques; (b) spin coat hydrogen-silsesquioxane (HSQ) resist layer, then, pattern it with electron-beam lithography (EBL) into the desired shape as etching mask; (c) etch graphene by oxygen plasma; (d) cover graphene/HSQ with SiO_2 sacrificial layer; (e) define top contact (local actuation electrode) with Cr/Au; (f) release graphene in buffered hydrofluoric, and dry the sample in a critical point drier. Note, in fig 2, we illustrate a switch with the cantilever-type graphene element. In the double-clamped beam, the bottom electrodes are on both ends of graphene beam (Fig. 3(a)).

III. RESULTS AND DISCUSSIONS

First, we show the characterization of a double-clamped beam switch (Fig. 2(a)) in the rough vacuum (~0.1 Pa). The two-terminal configuration is employed, where top contact works as both current drain and actuation electrode (inset of Fig. 4). A low current of pico-ampere level is measured as leakage before pull-in (switch-on). At ~3.3 V, the physical pull-in of monolayer graphene beam is noticed as an abruptly current increase, namely, the "switch-on" status (Fig. 4). In the reverse operation, the pull-out effect is observed at ~2.7 V, which is ascribed to the oxidized chromium at the contact interface. It effectively stops the formation of chemical bonds and reduces the interfacial stiction. Fig. 5 shows a failed switch with graphene sticking to the top electrode, in which the chromium oxide layer is not introduced. It highlights the function of the oxide. A cantilever-type switch of four layers graphene is also characterized (Fig. 3(b)). Due to the lower spring constant in the cantilever, the reduced pull-in voltage of ~2.6 V is measured (Fig. 6). The pull-out shows up at ~1.5 V, indicating the reversible operation.

In summary, we demonstrated the graphene switches with low pull-in voltage less than 5 V. The reversible operation is realized by introducing oxide layer at the contact interface.

ACKNOWLEDGEMENTS: This work was supported by the Grant-in-Aid for Scientific Research No. 25220904 from Japan Society for the Promotion of Science (JSPS) and the Center of Innovation Program from Japan Science and Technology Agency (JST).

REFERENCES

[1] O. Y. Loh and H. D. Espinosa, "Nano electro mechanical contact switches," Nature Nanotech., vol. 7, no. 5, pp. 283–295, May 2012.

[2] J. Sun *et al*, "Low pull-in voltage graphene

electromechanical switch fabricated with a polymer sacrificial spacer," Appl. Phys. Lett., vol. 105, pp. 033103, 2014

[3] P. Li, Z. You, and T. Cui, "Raman spectrum method for characterization of pull-in voltages of graphene capacitive shunt switches," Appl. Phys. Lett., vol. 101, pp. 263103, 2012.

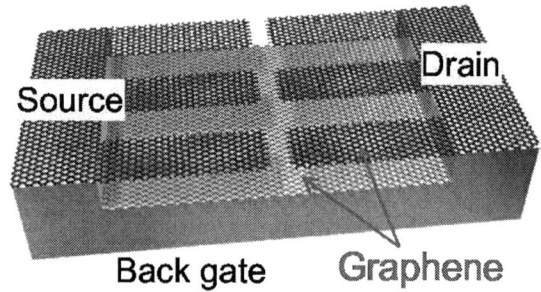

Fig. 1: Schematic diagram of GNEMS three terminal switching transistor.

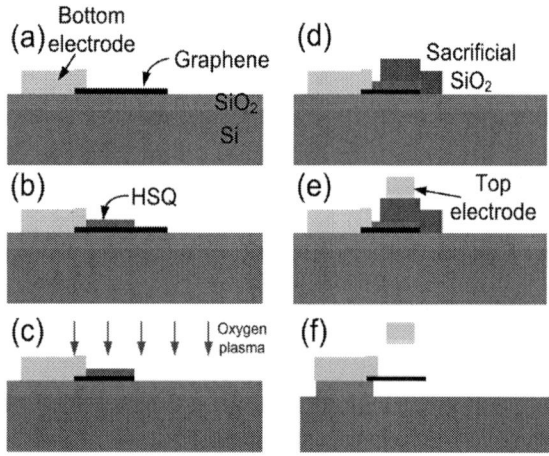

Fig. 2: Schematics of fabrication procedure of graphene switch with a cantilever-type moving element with local top actuation electrode.

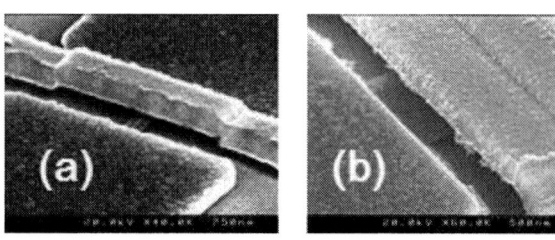

Fig. 3: SEM image of graphene (a) monolayer double-clamped beam (b) four layers cantilever type switches with local top actuation electrode.

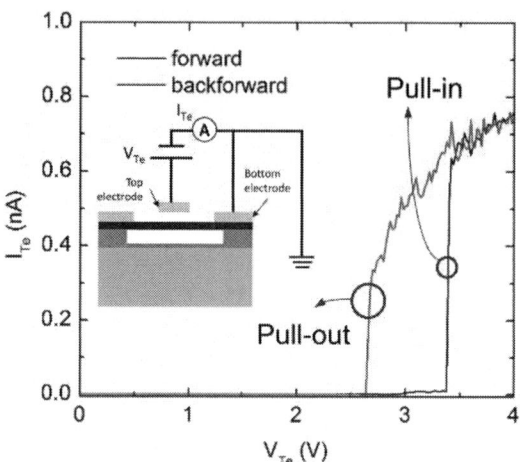

Fig. 4: Switching performance of the double-clamped monolayer graphene nanoribbon (Fig. 3(a)). Inset: two-terminal configuration.

Fig. 5: SEM image of a failed switch with graphene collapsing on the local top actuation electrode.

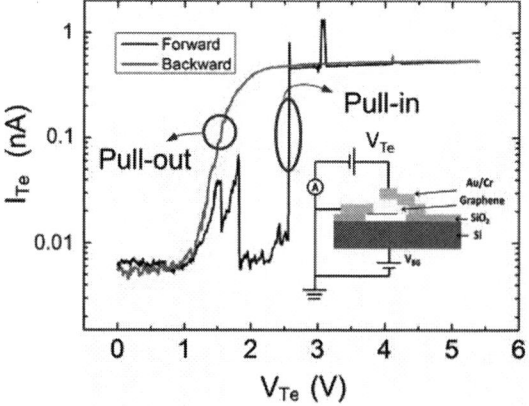

Fig. 6: Switching performance of the four layers graphene cantilever as shown in Fig. 3(b). Inset: two-terminal configuration.

Challenges of 3D VLSI-CoolCube[TM] process
with p-Ge-OI and n-InGaAs-OI for Ultimate CMOS Nodes

F. Nemouchi[1], L. Hutin[1], H. Boutry[1], P. Rodriguez[1], E. Ghegin[1,2],
J. Borrel[1,2], Y. Morand[2], S. Kerdiles[1], P. Batude[1], M. Vinet[1]

[1] CEA, Leti, Minatec Campus, 17 rue des Martyrs, 38054 Grenoble Cedex 9 - FRANCE
[2] STMicroelectronics, 850 rue Jean Monnet, 38926 Crolles – FRANCE
Email: fabrice.nemouchi@cea.fr

Abstract — **In this paper, we evaluate the various technological solutions and roadblocks for co-integrating p-Ge and n-InGaAs MOSFETs in a 3-D monolithic CoolCube[TM] technology. In particular, the process sequence (Ge-p-MOS-1st or III-V-n-MOS-1st) is examined in the light of thermal budget limitations arising from junctions definition.**

I. INTRODUCTION

High mobility channels for advanced CMOS nodes have sparked significant interest over the past decade, since a larger drive current compared to Si-channel devices at a fixed gate overdrive allows maintaining the performance-power dissipation trade-off while increasing circuit density. Based on their bulk electron and hole mobility [1], III-V materials (*eg* InGaAs) and Ge channels are generally considered among the most serious contenders, respectively for nFETs [2] and pFETs [3]. Whereas planar co-integration might add some significant process complexity, a 3D sequential integration with separately optimized nFET and pFET stages essentially boils down to solving issues of thermal budget compatibility between stages. Furthermore, the reduction of interconnects length and associated parasitics granted by the CoolCube[TM] technology [4] enables slightly relaxing the transistor dimensions, which is a considerable asset when processing more exotic materials. Recent demonstrations of 3D stacked InGaAs-OI nFETs on top of (Si)GeOI pFETs were carried out [4-5] using metallic Ni-based Source and Drain (S/D) electrodes. In the following, we will review some of the thermal budget constraints related to forming junctions on InGaAs and Ge (doping, epitaxy, intermetallic compound formation), and evaluate the corresponding impact on either a Ge-p-MOS-1st or III-V-nMOS-1st integration schemes (**Fig. 1**).

II. COOLCUBE™ CO-INTEGRATION OF GE AND INGAAS

Ge-p-MOS-1st integration (**Fig. 1 a**): Junction formation in InGaAs is usually achieved via substitutional incorporation of either Si or Ge atoms. Two strategies can be considered: i) ion implantation followed by activation annealing ii/ In Situ Doped Epitaxial Growth of the Raised S/D regions (ISDEG-RSD). In the former case, the typical range of temperatures for junction activation is above 600°C [6]. A VPD-ICPMS contamination analysis has shown As and P diffuses even through a thick oxide layer at 600°C (**Fig. 2**). Unless a cold junction activation process is developed, this is a significant roadblock for the Ge-p-MOS-1st integration scheme which would be certainly killed by As & P contamination during III-V doping junction step. The ISDEG RSD alternative offers to this day marginally broader process window. Indeed, in order to achieve selectivity of the epitaxial growth versus spacers, hard mask and isolation, temperatures greater than 500°C are typically needed [7]. Even if the As out-diffusion might be mitigated at this temperature, it is incompatible with a germanidation of the access regions in the lower stage. The NiGe integrity is indeed compromised above a thermal budget of 500°C 30s (**Fig. 3**). Thus, the Ge-p-MOS-1st meets some process limitations.

InGaAs-p-MOS-1st integration (**Fig. 1 b**): The thermal stability of InGaAs intermetallic compound is subjected to even tighter restrictions (**Fig. 4**). This implies that the Ge pFET process should be performed entirely below 450°C: spacers, doping and RSD epitaxy. Currently, Solid Phase Epitaxial Regrowth for Ge junctions doping was demonstrated at 380°C [8] In addition, both low-k spacers and Ge epitaxy are close to the upper temperature limit (~500°C) but still need improvement. Although, this approach offers less limitations further works aiming at reducing the MOSFET process temperature are still needed.

III. DOPING-FREE JUNCTIONS FORMATION

In the meantime, germanidation/III-V-idation step to form Schottky junctions can be a solution to get rid of the doping [4-5]. However surface preparation and metallic diffusion control are challenging. **Fig. 5** provides XPS analysis of various InGaAs surface treatments which exhibits major oxide reduction using wet cleaning and innovative He plasma. Finally to control the metallic diffusion into InGaAs channel, insertion of an ultra-thin dielectric layer between metal and S/D terminals is an attractive option. Its purpose is to reduce the Schottky Barrier Height via Fermi-level depinning without inducing a significant tunnel resistance. Experimental demonstration over a regular metallic contact on lightly doped III-V showed a 10^3 current increase [8].

IV. CONCLUSIONS

3D monolithic appears as a promising architecture pGe/nInGaAs co-integration. However, junction process thermal budget remains one of the main challenges whatever the integration scheme. Unless low temperature junction process, doping-free and Schottky contact could be the response toward co-integration using preferably InGaAs-p-MOS-1st.

ACKNOWLEDGEMENT

J. L. Labar from is MFA is grandly acknowledge for TEM analysis. This work is partly funded by the French public authorities through NANO 2017 program, EQUIPEX FDSOI11.

REFERENCES

[1] J. del Alamo, Nature, 479, 317-323 (2011) [2] M. Radosavljevic *et al.*, IEDM 2011 [3] R. Pillarisetty *et al.*, IEDM 2010 [5] P. Batude *et al.*, VLSI Tech. Symp. 2015 [4] T. Irisawa *et al.*, VLSI Tech. Symp. 2013 [5] T. Irisawa *et al.*, VLSI Tech. Symp. 2014 [6] A. G. Lind *et al.*, J. Vac. Sci. Tech. B, 33, 021206 (2015) [7] U. Singisetti *et al.*, IEEE Elec. Dev. Lett., 30 (2009) [8] J.H. Park *et al.*, IEDM 2008 [9] J. Hu *et al.*, J. App. Phys., 107 (2010)

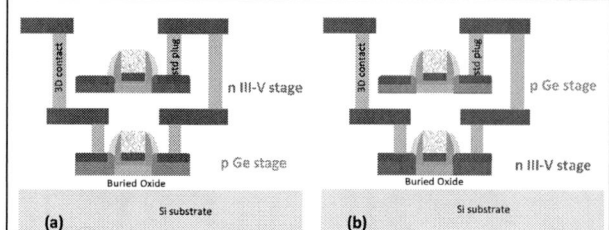

Figure 1: Schematic view of 3D-monolithically co-integrated (a) Ge-p-MOS-1st and (b) or III-V-nMOS-1st

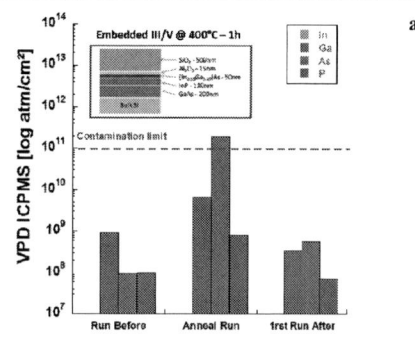

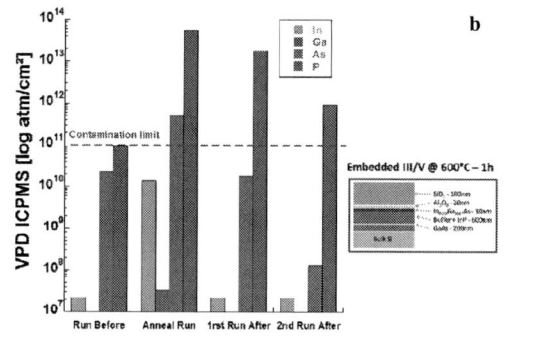

Figure 2: Histogram of Vapor Phase Decomposition Inductively Coupled Plasma Mass Spectrometry (VPD-ICPMS) analyses of embedded III/V with oxide layer after 1h furnace annealing at (a) 400°C and (b) 600°C..

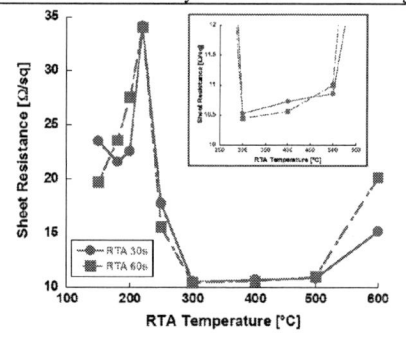

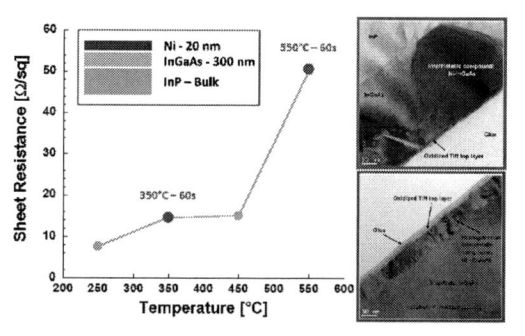

Figure 3 : Sheet resistance of Ni 12nm on Ge after RTP 30s and 60s with zoom inset of NiGe phase region

Figure 4 : Sheet resistance of Ni 20nm on InGaAs layer after RTP 60s and the corresponding TEM picture after 350°C and 550°C annealing

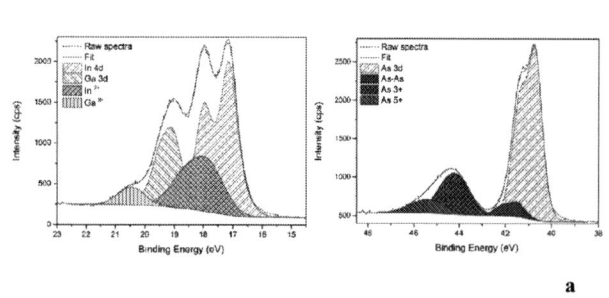

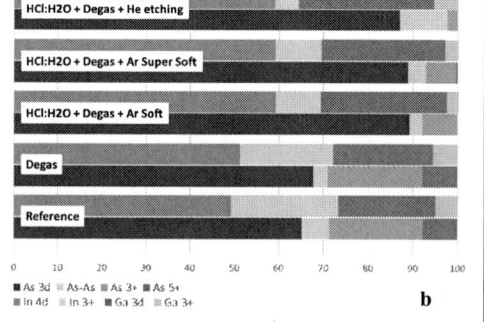

Figure 5 : (a) As 3d and Ga3d/In 4d X-ray Photoemission Spectroscopy (XPS) spectrum of untreated InGaAs (reference) and (b) the corresponding histograms after various pre-deposition surface treatment

CMOS Roadmap Analysis from the Perspective of III-V technology using MASTAR

Gaspard Hiblot[1,2], Quentin Rafhay[2], Gabriel Mugny[1], Gérard Ghibaudo[2], Frédéric Boeuf[1]

[1]STMicroelectronics, 850 rue Jean Monnet, 38920 Crolles; [2]IMEP-LAHC, 3 Parvis Louis Néel, 38016 Grenoble, France
Email: gaspard.hiblot@st.com, frederic.boeuf@st.com

Introduction

III-V materials are an attractive option for next generation MOSFET devices, essentially thanks to their excellent transport properties. The aim of this work is to benchmark the performance of III-V MOSFET technology (considering $In_{0.53}Ga_{0.47}As$ as the channel material), using the MASTAR [1] platform which includes tunneling effects, mobility physical models, ballistic transport, band-structure modification, short-channel effects, series resistance, parasitic capacitances and accurate current compact model (Fig. 1) [2]. Based on this benchmark, the inclusion of III-V MOSFET in the roadmap will be discussed. Two architectures are considered, Double-Gate (DG) and Nanowire (NW).

Mastar model for III-V MOSFET

The MASTAR model for MOSFETs includes short-channel effects (SCEs) with the Voltage-Doping Transform [3], and different tunneling mechanisms: source-drain tunneling (SDT) and band-to-band tunneling (BTBT) [4]. The mobility μ compact model (Fig. 2) includes polar optical phonons (POP) and interface traps Coulomb scattering. In short-gate lengths, μ is limited by ballistic transport [5]. At high fields, an effective velocity is introduced to make the transition between saturation velocity, overshoot and ballistic injection velocity (Fig. 3). The band-structure modification is taken into account by an empirical modification of the effective mass m_{eff} (Fig. 4 and 5), which reduces the mobility ($\mu \propto 1/m_{eff}^2$) and the Dark Space (DS) shown in Fig. 6. Two architectures are considered: Double-Gate (DG) and Nanowire (NW).

Optimal dimensions and targeted performances

The equivalent oxide thickness (EOT) is set to 0.5 nm, which is the targeted value for III-V material [6], and the supply voltage is set to $V_{dd} = 0.5V$. For each gate length L_g, there is a corresponding optimal value of the channel thickness t_{ch} which maximizes the effective current I_{eff} (Fig. 7 and 8), because a thin channel damages μ while a too thick channel leads to intolerable leakage (I_{off} is set to 10 nA/μm). Although BTBT is known to increase the minimum achievable OFF current, it is suppressed below a certain value of t_{ch}, which is always larger than the optimal t_{ch}. The optimal channel thickness will therefore be systematically used for each gate length. The parasitic capacitances are evaluated for each node, using simple design rules for the dimensions and model from [7]. The interface trap density is set to $D_{it} = 5 \cdot 10^{11} cm^{-2}$. The inverter delay can then obtained as a function of L_g (Fig. 9), assuming that the mobility of pMOSFET can match the one from III-V nMOSFET by using for example a strained Ge channel [8]. The NW architecture is always more efficient than the DG, although the advantage in effective current observed for NW in Fig. 2 is slightly mitigated by the increased parasitic capacitance of this architecture (C_{tot}/C_{gc} is approximately 17% larger in stacked NW than in DG). When this delay is compared with Silicon (Fig. 10), III-V devices have a clear advantage (+16.4% (DG) and 38.5% (NW) in frequency) if the trap density is completely suppressed ($D_{it} = 10^{10} cm^{-2}$). But this advantage completely vanishes if $D_{it} = 4 \cdot 10^{12} cm^{-2}$, due to the combined degradation of mobility subthreshold swing.

Variability of III-V MOSFETs

In undoped fully-depleted channel architectures, the main sources of variability are the workfunction (ϕ_m) fluctuations (assumed here to be given by $\sigma(\phi_m) = 0.65$ mV·μm/$\sqrt{WL_g}$, W being the electrical width), and the line edge roughness (LER). LER has a strong impact on the access resistance R_{sd} of very thin devices which hence becomes a considerable source of variability. MASTAR resistance model is calibrated on the data reported in [9] (Fig. 11), and the impact on $I_d(V_g)$ curves obtained with Monte-Carlo simulations (assuming a normal distribution of t_{ch}) is clearly visible when compared to the case where a constant value of R_{sd} is entered (Fig. 12 and 13). When all the parameters vary (assuming normal distributions given by the ITRS Lithography tables), the degradation of the SNM/σ(SNM) ratio when L_g decreases can be observed in Fig. 14. Due to its high aspect ratio, the DG has a very large W, which makes it more resilient to variability for long L_g. However, in short channels the NW outperforms the DG due to its immunity to SCEs-induced variability.

III-V MOSFET in the Roadmap

Assuming that D_{it} may be maintained at $5 \cdot 10^{11} cm^{-2}$, a new set of simulations is performed in order to determine how III-V materials could be inserted in the ITRS roadmap. The new delay is obtained using the ITRS parameters (Tab. 1) from the HP table ($I_{off} = 100$ nA/μm), with the exception of t_{ch} which is still set at its optimum value (Fig. 15). In order to maintain the general trend of 5% reduction in device delay, III-V DG should be introduced in 2017 (corresponding to the "7 nm" node), while III-V NW would be required by 2019 ("5nm" node). At the same time, the introduction of III-V is particularly detrimental to variability, due to the higher dependence of V_{th} on t_{ch} induced by quantum confinement (Fig. 16 and 17). Special design techniques or a relaxation of the WL_g product would hence be needed at the "7 nm" node, before the introduction of III-V NW ("5 nm" node), which alleviates this issue by suppression of the variability related to SCEs. Finally, it can be observed in Tab. 1 that, at the end of the roadmap, the ratio of total over intrinsic gate capacitance C_{tot}/C_{gc} will considerably increase due to the cumulative impact of NW large parasitic capacitance and high Dark Space (DS) of III-V materials.

Conclusion

Compared with Si technology, the frequency of optimally designed III-V RO oscillators was found to be 16.4% higher in DG architecture, and 38.5% higher in NW architecture at V_{dd}=0.5V , this last one featuring acceptable variability in SRAM cells. These gains are however subjected to the condition that the interface traps issue be solved. In this case, III-V DG devices could be introduced at the "7 nm" node and III-V NW at the "5 nm" node to pursue CMOS performance scaling.

References

[1] http://www.itrs.net/models.html; [2] J. Lacord et al. SSE 2014 p 137-146; [3] T. Skotnicki et al. EDL 1988 p 109-112; [4] Q. Rafhay et al. SSDM 2008 p 36-37; [5] M. Shur et al. EDL 2002 p 511-513; [6] D. Zadeh et al. IEDM 2013 2.4.1-2.4.4; [7] J. Lacord et al. TED p 1332-1344 ; [8] R. Zhang et al. IEDM 2013 p 26.1.1-26.1.4; [9] F. Monsieur et al. ESSDERC 2014 p 254 257; [10] A. Alian et al. IEDM 2013 p 317-323; [11] J. Del Alamo Nature 2011 p 317-323; [12] Y. M. Niquet et al. PRB 2006 p 165319.1-13

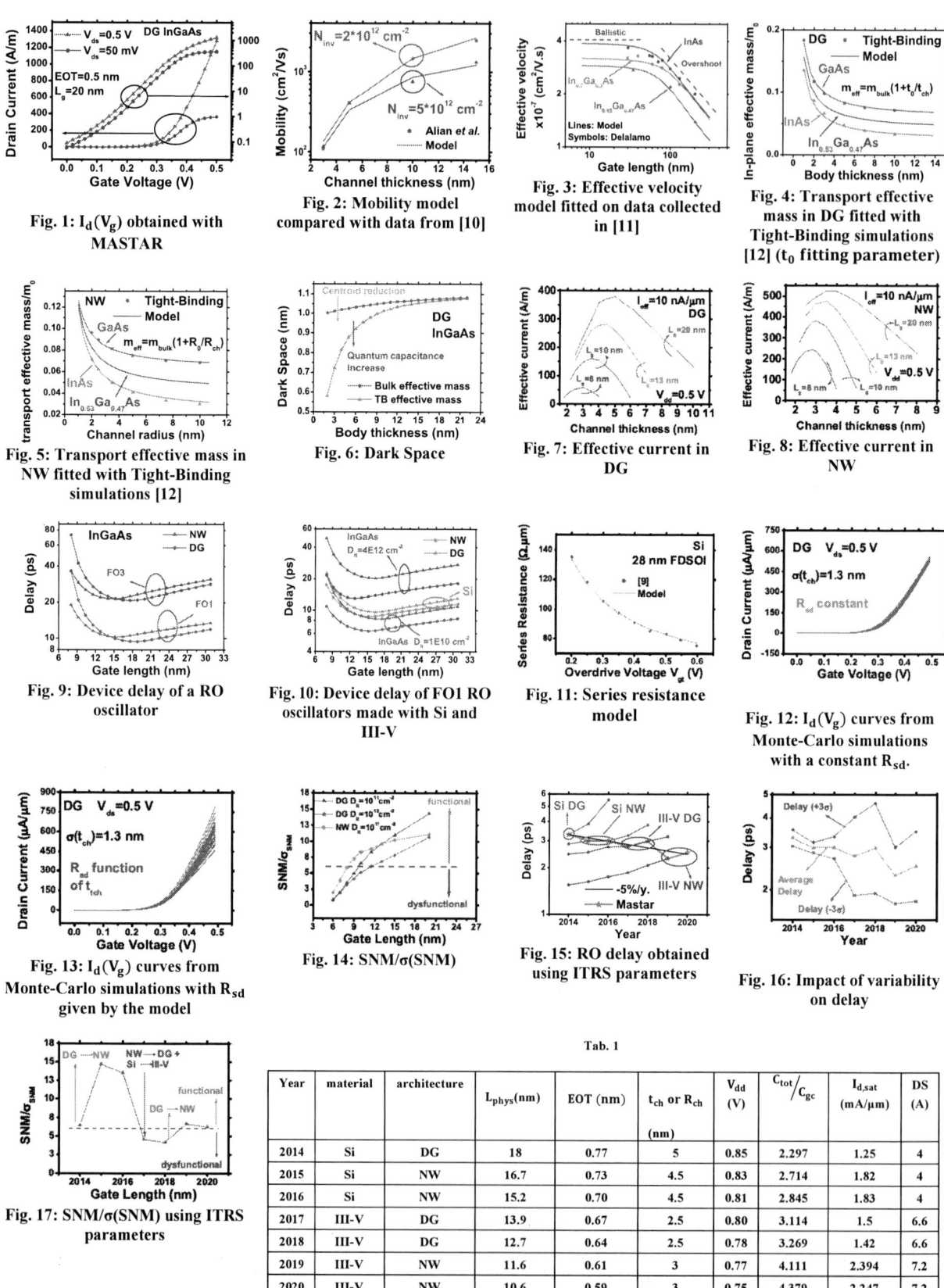

Fig. 1: $I_d(V_g)$ obtained with MASTAR

Fig. 2: Mobility model compared with data from [10]

Fig. 3: Effective velocity model fitted on data collected in [11]

Fig. 4: Transport effective mass in DG fitted with Tight-Binding simulations [12] (t_0 fitting parameter)

Fig. 5: Transport effective mass in NW fitted with Tight-Binding simulations [12]

Fig. 6: Dark Space

Fig. 7: Effective current in DG

Fig. 8: Effective current in NW

Fig. 9: Device delay of a RO oscillator

Fig. 10: Device delay of FO1 RO oscillators made with Si and III-V

Fig. 11: Series resistance model

Fig. 12: $I_d(V_g)$ curves from Monte-Carlo simulations with a constant R_{sd}.

Fig. 13: $I_d(V_g)$ curves from Monte-Carlo simulations with R_{sd} given by the model

Fig. 14: SNM/σ(SNM)

Fig. 15: RO delay obtained using ITRS parameters

Fig. 16: Impact of variability on delay

Fig. 17: SNM/σ(SNM) using ITRS parameters

Tab. 1

Year	material	architecture	L_{phys}(nm)	EOT (nm)	t_{ch} or R_{ch} (nm)	V_{dd} (V)	C_{tot}/C_{gc}	$I_{d,sat}$ (mA/µm)	DS (A)
2014	Si	DG	18	0.77	5	0.85	2.297	1.25	4
2015	Si	NW	16.7	0.73	4.5	0.83	2.714	1.82	4
2016	Si	NW	15.2	0.70	4.5	0.81	2.845	1.83	4
2017	III-V	DG	13.9	0.67	2.5	0.80	3.114	1.5	6.6
2018	III-V	DG	12.7	0.64	2.5	0.78	3.269	1.42	6.6
2019	III-V	NW	11.6	0.61	3	0.77	4.111	2.394	7.2
2020	III-V	NW	10.6	0.59	3	0.75	4.379	2.247	7.2

Acknowledgements: This work was supported by the research project COMPOSE3 number 619325, and the research project NOODLES ANR-13-NANO-0009.

Effect of Free Carrier Accumulation or Depletion on Zone-center Vibrational Mode in Ge

Shoichi Kabuyanagi, Tomonori Nishimura, Takeaki Yajima and Akira Toriumi

[1]Department of Materials Engineering, The University of Tokyo

7-3-1 Hongo, Bunkyo-ku, Tokyo 113-8656, Japan

Phone: +81-3-5841-1907, E-mail: kabuyanagi@adam.t.u-tokyo.ac.jp

1. Introduction

Heavy impurity doping is known to soften the lattice system in Si and Ge, experimentally [1, 2]. Although it is theoretically explained by the Fano-type electron-phonon interaction [3], it is practically important as well whether this effect may come from heavy impurity doping or high carrier density in semiconductors. Moreover, it may become more important in nano-devices, because the whole channel will be either in volume inversion or volume accumulation. It is, however, difficult to experimentally distinguish free carrier effects from dopant atom ones in heavily doped semiconductors. Then, the objective of this work is to clarify which plays a more significant role, by studying electron-phonon interactions in lightly doped (carrier accumulation) as well as in heavily doped Ge (carrier depletion), experimentally.

2. Experiment

Both lightly ($\sim 1 \times 10^{15}$/cm^3) and heavily doped ($\sim 1 \times 10^{19}$/cm^3) p-type GeOI were investigated. The device structure is schematically shown in **Fig. 1**. This is basically the back-gated GeOI MOSFET, in which the carrier density can be controlled by the back gate. The top surface of Ge was covered with Y_2O_3 to minimize the interface defects [4]. Electron-phonon interaction was characterized by the microscopic Raman measurement with Ar laser (λ=488 nm, 0.3 mW), in which gate bias can be applied while source and drain were grounded.

2. Results and Discussion

Raman spectroscopy measurement is quite powerful for analyzing the zone-centered phonon energy. We set the electrical probing stage under the microscope for the Raman measurement. **Fig. 2** shows the back gate voltage dependence of the Raman peak position in lightly doped p-Ge. The significant red-shift is clearly observed by applying the negative V_g (on state of the FET). It demonstrates for the first time that the zone-centered phonon energy decreases with the hole accumulation at a fixed lightly doped Ge. Furthermore, we also inspected the depletion effect of carriers in heavily doped Ge. **Fig. 3** shows I_d-V_g characteristics in a FET fabricated on 13-nm-thick p$^+$-GeOI, indicating that free carriers in p$^+$-Ge are completely depleted by applying positive V_g. This FET was inspected by Raman measurement, in which a clear blue-shift is observed by depleting holes in **Fig. 4**. To our knowledge, these results are the first and direct evidence that the softening of zone-center vibrational mode is mainly caused by free carriers rather than dopants.

How can we intuitively interpret the free carrier effect on the phonon softening? Here, we propose a model from the viewpoint of a modification of the covalent bonding by free carriers, as shown in **Fig. 5(a)**. Suppose the simple covalent bonding of two atoms with sp3 hybridized orbital. When an additional carrier is added in this system, there is still an energy gain in the covalent bonding. It should be, however, decreased due to a reduction of the hybridization gain. In fact, since free carriers are $10^{-4 \sim -6}$ of Ge atoms, it may be more appropriate to describe the high carrier density region as the weakened covalent bonding system. This view can interestingly explain the band gap narrowing in heavily doped semiconductors as well [5]. Furthermore, the reduction of the energy gain is more realistically described by the Lennard-Jones type potential for a molecular formation. **Fig. 5(b)** shows the interaction energy of two atom systems with both w/ and w/o free carriers. Since the curvature k near the energy minimum described by kx^2 corresponds to the force constant [6], it is reasonably expected that k becomes smaller under free carriers. This is considered to be the origin for the phonon softening observed in Raman measurement.

Furthermore, this effect should be taken into consideration in discussing the phonon scattering of carriers, because the phonon deformation potential is not a constant but dependent on free carrier density. This has been totally ignored in the carrier transport analysis so far. Note that it will be more important in nano-scale devices, because the atomic nature will become more dominant.

3. Conclusion

The free carrier effect on the phonon softening has been experimentally distinguished from the dopant atom effect for the first time. This effect is explainable by considering the model that the covalency is weakened by free carriers. This view can reasonably explain the band gap narrowing as well. Furthermore, it is pointed out that the present consideration should be taken into for understanding the phonon scattering of free carriers.

References [1] C. Haas, Phys. Rev. **125**, 1965 (1962). [2] F. Cerdeira. et al., Phys. Rev. B **5**, 1440 (1972). [3] F. Cerdeira. et al., Phys. Rev. B **8**, 4734 (1973). [4] S. Kabuyanagi et al., Appl. Phys. Exp., to be published. [5] J. G. Fossum et al., IEEE Trans. ED **30**, 626 (1983). [6] N. W. Ashcroft and N. D. Mermin, "Solid State Physics", (Thomson Learning, 1976).

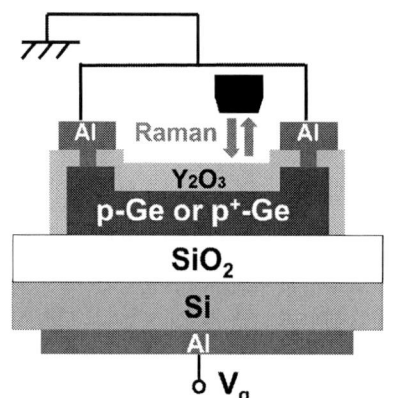

Fig. 1 The schematic image of Raman measurement with back-gate bias. The free carrier density can be controlled while the dopant atoms are fixed.

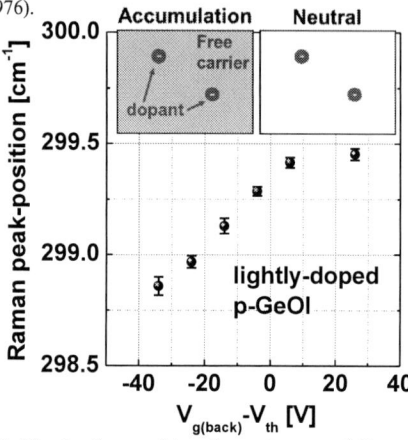

Fig. 2 The back-gate bias dependences of Raman peak position in 20-nm-thick lightly-doped p-GeOI. The phonon softening by hole accumulation was observed.

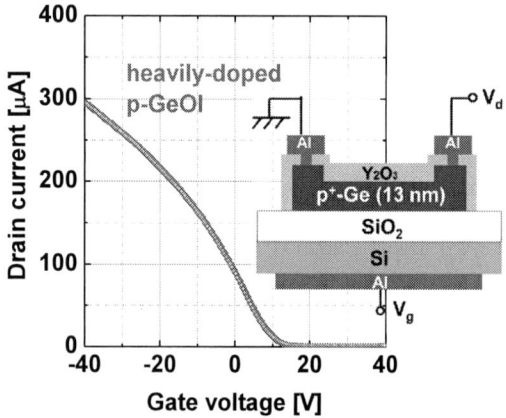

Fig. 3 The transfer characteristics of 13-nm-thick p$^+$-GeOI FET. The cut-off characteristics indicates that holes in Ge was completely depleted by applying the positive gate bias.

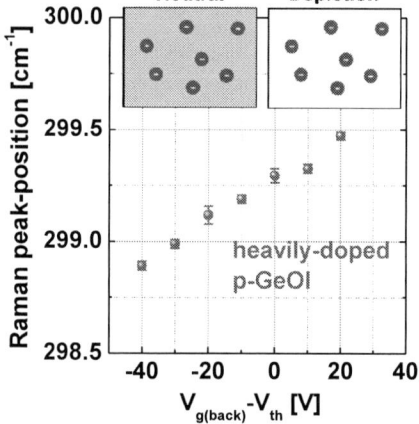

Fig. 4 The back-gate bias dependences of Raman peak position in 13-nm-thick heavily-doped p-GeOI. The phonon hardening by hole depletion was observed.

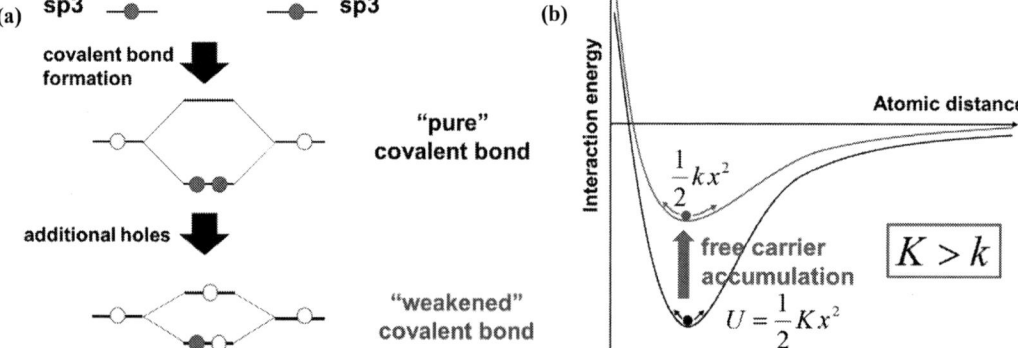

Fig. 5(a) The schematic image of the energy level before and after the covalent bond formation in Ge, with low and high hole concentration. The energy gain by covalent bond formation is considered to be reduced by increasing the hole density. **(b)** The Lennard-Jones-type potential in Ge, with low and high hole concentration. The reduction of the energy gain in bond formation may result in the decrease of the phonon frequency.

N$^+$/P Shallow Junction with High Dopant Activation and Low Contact Resistivity Fabricated by Solid Phase Epitaxy Method for Ge Technology

Pengqiang Liu, Ming Li, Xia An, Meng Lin, Yang Zhao, Bingxin Zhang, Xuyuan Xia, Ru Huang

Institute of Microelectronics, Peking University, Beijing 100871, CHINA

Email: liming.ime@pku.edu.cn, anxia@ime.pku.edu.cn, ruhuang@pku.edu.cn

Abstract — **In this paper, solid phase epitaxy (SPE) process is proposed to improve phosphorous activation and hence reduce the contact resistivity of n$^+$/p junction for Ge n-MOS technology. Over 1×10^{20} cm^{-3} electrical concentration and about 1.75×10^{-6} ohm·cm^2 contact resistivity have been achieved at P$^+$ implantation of 10keV and 5×10^{14}cm^{-2} and annealing condition of 600°C, 10seconds. The fabricated N$^+$/P diode shows 2 times higher forward current and well controlled leakage.**

I. INTRODUCTION

Germanium is an attractive channel material due to its high and symmetric mobility [1]. Substantial success has made in fabricating Ge MOSFETs [2], [3]. However, shallow n$^+$/p junction with low resistance is still difficult because of the fast diffusion and low activation rate of n-type dopants in Ge, caused by the negatively charged vacancies [4]. Many doping and annealing methods are used to enhance dopant activation [5, 6] but there still needs more efforts to achieve the requirements at low thermal budget and low cost. In this paper, the conventional SPE method is proposed to enhance the phosphorous activation concentration with small diffusion penalty.

II. EXPERIMENT

N$^+$/P diodes were fabricated with the substrate of p-type Ge (100) with a resistivity of 1.65~2Ω·cm. After cleaning, the active area patterning was defined with PECVD SiO$_2$ as isolation layer. The SPE sample firstly received the Ge$^+$ pre-amorphous implantation (PAI) at energy and dose of 15keV, 5×10^{13}cm^{-2} and then the P$^+$ implantation at 10keV, 5×10^{14}cm^{-2}. For the control sample, only P$^+$ implantation (10keV, 5×10^{14}cm^{-2}) was received. The rapid thermal annealing was then performed at 600℃ for 10 seconds. Ti/Al sputtering and lift-off processes were carried out to form the front contact. Al was also deposited for back contact.

II. RESULTS

As evident from Raman spectroscopy of SPE sample (Fig. 1), the as-implanted sample shows a broad distribution with peak position of ~270cm^{-1}, indicating that the surface crystal has become amorphous state after PAI. After annealing, the Ge-Ge optical phonon mode at about 300cm^{-1} has re-emerged and shows the exact consistence to pristine Ge sample, which means the perfect recrystallization.

The SIMS profiles in Fig. 2 show the diffusion of both samples is slow due to low thermal budget and the SPE sample shows less diffusion than control sample possibly due to better defect recovery during the epitaxy regrowth.

Fig. 3 shows experimental sheet resistance (R$_{sh}$) of the samples by four-point-probe measurement. The R$_{sh}$ of SPE sample are 65Ω/□, about 2 times lower than the control sample. The resistivity is then simply estimated to be 3×10^{-4} Ω*cm and the corresponding electrical doping concentration is over 1×10^{20} cm^{-3} as comparison to the plotting of ρ~doping concentration in Ref. [7].

The contact resistivity of Al/Ti/n$^+$-Ge was also evaluated using circular transfer length method (CTLM) as shown in Fig. 5. The SPE sample shows about 1.5 times lower contact resistivity than control sample thanks to higher active doping concentration as shown in Fig. 6.

The n$^+$/p diode forward current of SPE sample is about 2 times of the control sample as shown in Fig. 7 (b). The resistance components of the diodes are also calculated and compared in Fig.8. It's found that the major resistance is attributed to junction resistance (R$_j$) and the R$_j$ for SPE sample is about 2 times lower than control sample due to higher electrical doping concentration.

The concern about Ge$^+$ PAI may be the junction leakage due to PAI residual damages. Fortunately, with well controlled initial amorphous/crystal interface position, the residual defects can be excluded away from junction depletion region as shown in Fig. 9. The leakage of SPE sample is well controlled as shown in Fig.7 (a). The same activation energy (Ea) of two samples, closed to the band-gap (0.67 eV), shows low defect density and excellent recrystallization quality by SPE process.

III. CONCLUSION

The SPE technique in germanium is comprehensively studied and shows the capability to form high conductivity shallow n$^+$/p junction with low thermal budget and very simple process. It can be very potential for the future high-performance Ge CMOS technology.

ACKNOWLEDGEMENT

This work was supported by NSFC (61421005, 60806033 and 61474004), SRFDP (20130001110025) and National S&T Major Project 02(2013ZX02303).

REFERENCES

[1] J. Oh, et al., IEEE Electron Device Letters, vol. 28, pp. 1044, Nov. 2007.
[2] R. Zhang, et al., Proc. IEDM Tech., 28.3.1, 2011.
[3] L. Witters, et al., Proc. IEDM Tech., 15.2.1, 2013.
[4] P. Tsouroutas, et al., Journal of Applied Physics, 105, 094910, 2009.
[5] P. Bhatt, et al., Transaction on Electron Device, vol. 62, pp. 69, Jan. 2015.
[6] J. Kim, et al., Applied Physics Letters, 101, 112107, 2012.
[7] E. Liu, et al., The physics of Semiconductors, 7th Ed., Publishing House of Electronics Industry, China, pp. 109.

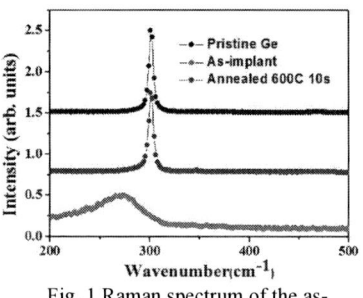

Fig. 1 Raman spectrum of the as-implanted and annealed SPE samples

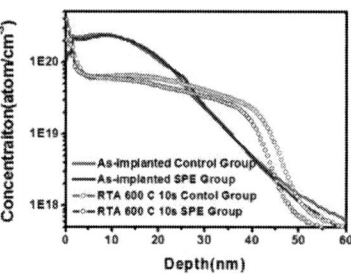

Fig. 2. SIMS profiles of the samples before and after annealing

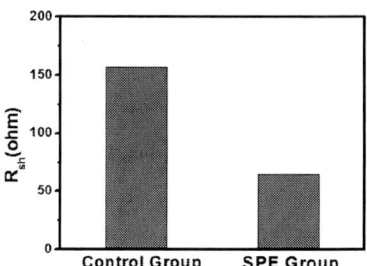

Fig. 3 Sheet resistance (R_{sh}) of the samples through four-point-probe measurement.

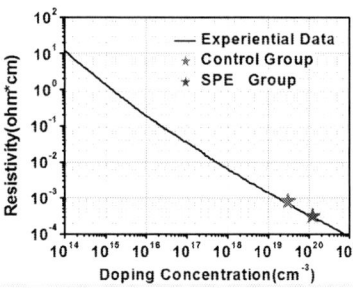

Fig. 4 The estimation of active doping concentration of the samples

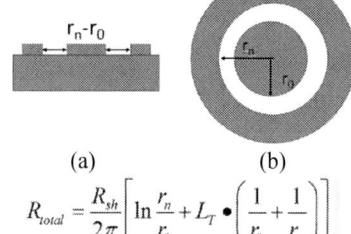

Fig. 5 CTML schematic diagram. (a) Top View (b) Cross-section View (c) Computational formula of contact resistivity

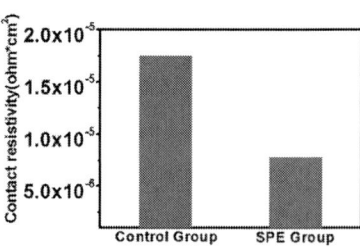

Fig. 6 The contact resistivity of metal contact to n-Ge evaluated through CTLM

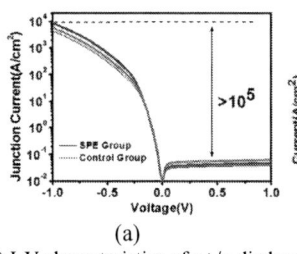

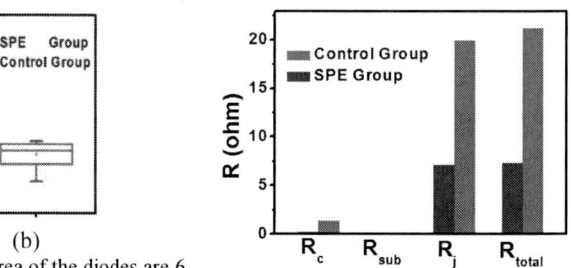

Fig. 7. (a) I-V characteristics of n+/p diodes at 300K. The active area of the diodes are 6 $\times$ 60 um^2. (b) The statistically distribution of forward current at forward bias of -1V.

Fig. 8 Comparison of R_c, R_{sub}, R_j and R_{total} of SPE diodes and control diodes

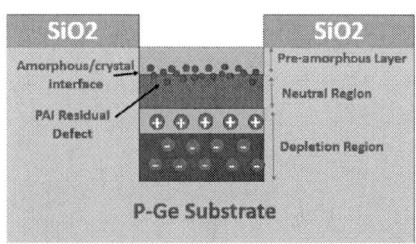

Fig. 9 Schematic diagram of the diode after PAI

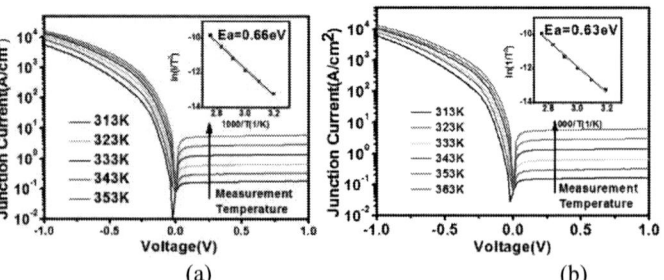

Fig. 10 Temperature-dependent n+/p diode I-V characteristics of (a) SPE sample (b) control sample. (Inset) Arrhenius plots for diodes reverse current at 1V

IEEE
445 Hoes Lane
Piscataway, NJ 08854-4141

ISBN 978-1-4673-7604-4